Advances in

ECOLOGICAL RESEARCH

VOLUME 26

Advances in
ECOLOGICAL RESEARCH

Edited by

M. BEGON

*Department of Environmental and Evolutionary Biology,
University of Liverpool, UK*

A.H. FITTER

Department of Biology, University of York, UK

VOLUME 26

ACADEMIC PRESS

Harcourt Brace & Company, Publishers
London
San Diego New York Boston
Sydney Tokyo Toronto

ACADEMIC PRESS LIMITED
24–28 Oval Road
London NW1 7DX

U.S. Edition Published by
ACADEMIC PRESS INC.
San Diego, CA 92101

This book is printed on acid free paper

A catalogue record for this book is available from the British Library

ISBN 0–12–013926–X

Typeset by Fakenham Photosetting Ltd
Fakenham, Norfolk
Printed in Great Britain by Hartnolls Limited, Bodmin, Cornwall

Contributors to Volume 26

D.D. BALDOCCHI, *NOAA/ERL, Atmospheric Turbulence & Diffusion Division, 456 South Illinois Avenue, P.O. Box 2456, Oak Ridge, TN 37831–2456, USA.*

D.J. BOOTH, *Australian Institute of Marine Science, PMB No. 3, Townsville MC, Queensland, Australia 4810.*

D.M. BROSNAN, *Sustainable Ecosystems Institute, P.O. Box 524, Lake Oswego, Oregon 97034, USA.*

A.E. DOUGLAS, *Department of Biology, University of York, York YO1 5DD, UK.*

P.G. JARVIS, *University of Edinburgh, IERM, Darwin Building, The King's Buildings, Mayfield Road, Edinburgh EH9 3JU, Scotland.*

A. ROSSITER, *Department of Zoology, College of Biological Sciences, University of Guelph, Guelph, Ontario N1G 2W1, Canada. Present address: Department of Biology, Faculty of Science, Ehime University, Bunkyo-cho 2–5, Matsuyama, Ehime 790, Japan.*

A. RUIMY, *Laboratoire d'Ecologie Végétale, Université de Paris-Sud, Bâtiment 362, 91405 Orsay Cedex, France.*

B. SAUGIER, *Laboratoire d'Ecologie Végétale, Université de Paris-Sud, Bâtiment 362, 91405 Orsay Cedex, France.*

T. TREGENZA, *Department of Genetics, Leeds University, Leeds LS2 9JT, UK.*

D.A. WARDLE, *AgResearch, Ruakura Agricultural Research Centre, Private Bag 3123, Hamilton, New Zealand.*

Preface

This volume contains six articles which cover the full range of the subject and address a number of highly topical issues. Population and community ecology is central to four of the articles, but physiological, evolutionary and systems ecologists should also find their interests well catered for. Systems studied include marine, fresh water and terrestrial, and the organisms range from micro-organisms to trees and to vertebrates.

Ruimy *et al.* address a central relationship in studies of the global carbon cycle, namely that between canopy carbon assimilation rate and the incident photon flux density. In questions such as that of the detection of the "missing sink" of atmospheric carbon dioxide, it is essential to be able to generalize that relationship. Doing so involves both technical problems of measurement of the carbon flux and interpretational problems, because of the wide range of conditions and vegetation types that must be considered. The relationship that emerges will be of great interest to all those involved in modelling the global carbon cycle.

Although symbioses involving micro-organisms are increasingly recognized as widespread and important, little ecological work has taken them into account explicitly. As Douglas shows, this is partly for technical reasons, since the microbes are usually unculturable, and partly for conceptual reasons: too many investigators have become bound up with semantic debates over the question of benefit. By cutting this Gordian knot, she gives a route that will permit others to account more easily for such symbioses, which are both ecologically significant and globally threatened, for example in coral reefs.

Food-web structure and stability is a live area that has been visited before in this series. Wardle uses data from both conventionally tilled and untilled agro-ecosystems to assess the impact of disturbance on the structure and stability of the detritus food-web in soil. The major impacts appear to be at the top of the food-web, but a mechanistic analysis is hampered by lack of knowledge of the basic biology of many of the organisms in these systems. Nevertheless, below-ground food-webs are surprisingly resistant to major disturbance, even though many of their characteristics would normally be held to promote instability.

Fish assemblages in the African Great Lakes are renowned for their species-richness and degree of endemism. Rossiter examines how this diversity has arisen and is maintained. No general rules emerge that are common to all assemblages, and, as with the detritus food-web it is often lack of knowledge of the natural history of the species involved that hampers understanding. These systems are remarkable natural laboratories, however, especially in the light of current debates over the functional role of diversity, yet they are under severe threat from human activity.

Tregenza reviews the status of the ideal free distribution and the impact that recent empirical advances have made upon its theoretical foundations. However, as so often in ecology, the match between theory and experiment is blurred by both the failure of experimenters to address model predictions with sufficient precision and modellers to take into account the complexity of the natural environment. The development of these models to reflect the reaction of individual animals to heterogeneity, as outlined here, will have wide applicability.

Finally, Booth and Brosnan address the problem of open populations, in which dispersal is a major component of the demographic equation, using examples from rocky shores and coral reefs. The principles, however, apply to many other types of system and organism. By combining population recruitment with community process models, they show that well-founded predictions can be developed and suggest that there is an urgent need for these to be tested, especially, as with the Great Lakes ecosystems, in the light of human pressures.

M. Begon
A.H. Fitter

Contents

Contributors to Volume 26   v
Preface   vii

CO_2 Fluxes over Plant Canopies and Solar Radiation: A Review

A. RUIMY, P.G. JARVIS, D.D. BALDOCCHI and B. SAUGIER

I.	Summary	2
II.	Symbols and Abbreviations	3
III.	Introduction	4
IV.	Stand CO_2 Balance	6
V.	Storage Flux	9
VI.	Respiration Flux	11
	A. Soil Respiration	13
	B. Dark Respiration	13
VII.	Canopy Photosynthesis	14
VIII.	Radiation	15
IX.	Radiation Use Efficiency	17
X.	Techniques for Measuring CO_2 Flux	18
	A. Energy Balance or Bowen Ratio Method	18
	B. Aerodynamic or Profile Method	19
	C. Eddy Correlation or Eddy Covariance	20
	D. Enclosures	20
	E. Models	21
XI.	Data Sets, Data Analysis	21
	A. Data Set Compilation	21
	B. Vegetation Classification	22
	C. Statistical Analysis of Instantaneous Data Sets	22
	D. Statistical Analysis of Grouped Data Sets	23
	E. Boundary Value Analysis	26
XII.	Effects of Various Factors on the Response Function of CO_2 Flux to PPFD	27
	A. The Micrometeorological and Enclosure Techniques	27
	B. Respiration	29
	C. Vegetation Class	32
	D. Other Variables	34
XIII.	Mean Statistics, Residuals and Boundary Value Analysis for the Complete Data Set	37

XIV. Perspectives for CO_2 Flux Measurements   40
      A.   Scaling Up in Space: Plane Mounted Eddy Flux Sensors   40
      B.   Scaling Up in Time: Instantaneous versus Daily Integrated
           Measurements   41
      C. CO_2 Enrichment   42
 XV. Conclusion   43
      A.   Specific Conclusions   43
      B.   General Conclusions   44
Acknowledgements   46
References   46
Appendix 1   52
Appendix 2   63

The Ecology of Symbiotic Micro-organisms

A.E. DOUGLAS

   I. Introduction   69
      A.   Why Study Symbiotic Micro-organisms?   69
      B.   The Scope of the Article   71
  II. The Phyletic Diversity of Endosymbiotic Micro-organisms   74
      A.   A Comparison of the Phyletic Diversity of Hosts and
           Endosymbionts   74
      B.   Taxonomic Practice and the Phyletic Diversity of Microbial
           Endosymbiosis   75
      C.   Origins and Radiations of the Partners in Endosymbioses . . .   76
 III. The Distribution of Micro-organisms in Endosymbiosis   78
      A.   Host-related Determinants of Endosymbiont Distribution . . .   78
      B.   Determinants of Endosymbiont Distribution External to the
           Symbiosis   87
 IV. Endosymbiotic Micro-organisms in the Environment   88
      A.   On the Rarity of Symbiotic Micro-organisms in the
           Environment   88
      B.   Symbiosis as a Source of Symbionts in the Environment   90
  V. The Host as a Habitat   93
 VI. Conclusions   94
Acknowledgements   95
References   95

Impacts of Disturbance on Detritus Food Webs in Agro-Ecosystems of Contrasting Tillage and Weed Managment Practices

D.A. WARDLE

Summary 105
II. Introduction 106
III. Tillage as a Disturbance Regime 109
    A. Relative Responses of Different Trophic Levels 109
    B. Overall Food Web Responses 127
    C. Response of Species Assemblages to Tillage 132
IV. Alternatives to Tillage 136
    A. Consequences of Herbicide Use 137
    B. Use of Mulches 143
    C. Alternative Agricultural Systems 147
V. Influence of Weeds on Detritus Food Webs 148
VI. Agricultural Disturbance and Food Web Theory 150
VII. Conclusions 156
Acknowledgements 158
References 158
Appendices 182

The Cichlid Fish Assemblages of Lake Tanganyika: Ecology, Behaviour and Evolution of its Species Flocks

A. ROSSITER

I. Summary 187
II. Introduction 189
    A. Biodiversity in Lake Tanganyika 189
    B. Fishes in the Lake 189
    C. The Cichlid Species Flocks of Lake Tanganyika 190
III. Speciation in Lake Tanganyika 192
    A. The Lake Environment 192
    B. Evolution of the Cichlid Fishes 199
IV. Breeding Habits 211
    A. Behaviour, Strategies and Tactics 211
    B. Substrate Spawners 212
    C. Mouthbrooders 215
    D. Commensalism and Parasitism 218
    E. Effects of Breeding Mode on Speciation 219
V. Species Diversity and Coexistence within the Rocky Shore Communities 220
    A. Mechanisms Promoting High Species Diversity and Coexistence 220

    B. Foods and Feeding 223
    C. Partitioning of Spatial Resources 229
    D. Partitioning within Communities and Effects of Relatedness . . 231
  VI. Community Composition and Species Diversity 234
    A. Importance of Size Differences for Species Diversity and
        Speciation 234
    B. Species Packing and Species Invasions 235
    C. Stability and Persistence 237
    D. Is Community Composition Predictable? 237
  VII. Concluding Remarks, Caveats and Perspectives 240
References 243

Building on the Ideal Free Distribution

T. TREGENZA

    I. Summary 253
   II. Introduction 254
  III. The Ideal Free Distribution 255
  IV. The Simplest Ideal Free Model 256
   V. Incorporating Interference 257
  VI. Incorporating Resource Wastage into Continuous Input Models . . 260
  VII. Incorporating Despotism 260
 VIII. Incorporating Unequal Competitors 262
    A. Continuous Phenotype Unequal Competitor Models 262
    B. The Isoleg Theory 265
    C. Unequal Competitor Distribution Determined by the Form of
        Competition 268
  IX. Incorporating Kleptoparasitism 269
    A. Modifying the Simple IFD Model to include Kleptoparasitism 269
    B. A Kleptoparasitic IFD Model based on a Functional Response 270
   X. Incorporating Resource Dynamics 273
    A. Continuous Input Models and Standing Crops 273
    B. Decision-making Prey 274
  XI. Incorporating Survival as a Measure of Fitness 275
  XII. Incorporating Perceptual Constraints 277
 XIII. Incorporating Patch Assessment 278
    A. Patch Assessment Models 278
    B. Experimental Investigations of Patch Assessment 282
  XIV. A Summary of IFD Models 284
  XV. Assessing the Success of Ideal Free Distribution Models 289
    A. Testing the Ideal Free Distribution 289
    B. Does the Theory Pass the Test? 289

XVI. Conclusions 300
Acknowledgements 301
References 302

The Role of Recruitment Dynamics in Rocky Shore and Coral Reef Fish Communities

D.J. BOOTH and D.M. BROSNAN

       Summary 309
  II. Introduction and Historical Perspective 310
 III. Settlement and Recruitment of Marine Organisms 314
  IV. Are Marine Systems Open? 316
   V. Recruitment: Patterns and Processes 329
       A. The Extent of Recruitment Variability in Marine Systems . . . 329
       B. Factors Affecting Recruitment Strength 332
  VI. The Influences of Recruitment on Patterns of Population and
       Community Structure 336
       A. Recruitment and Population Structure 337
       B. Recruitment Effects on Community Structure and Dynamics . 342
       C. The Role of Recruitment Variability 351
       D. Summary of Models, Predictions and Tests 358
 VII. Management Implications, Conservation Issues and Future
       Directions 360
       A. Management and Conservation 360
       B. Future Directions 363
VIII. Concluding Remarks 364
Acknowledgements 365
References 365

Index 387
Cumulative List of Titles 401

CO_2 Fluxes over Plant Canopies and Solar Radiation: A Review

A. RUIMY, P.G. JARVIS, D.D. BALDOCCHI, B. SAUGIER

I.	Summary	2
II.	Symbols and Abbreviations	3
III.	Introduction	4
IV.	Stand CO_2 Balance	6
V.	Storage Flux	9
VI.	Respiration Flux	11
	A. Soil Respiration	13
	B. Dark Respiration	13
VII.	Canopy Photosynthesis	14
VIII.	Radiation	15
IX.	Radiation Use Efficiency	17
X.	Techniques for Measuring CO_2 Flux	18
	A. Energy Balance or Bowen Ratio Method	18
	B. Aerodynamic or Profile Method	19
	C. Eddy Correlation or Eddy Covariance	20
	D. Enclosures	20
	E. Models	21
XI.	Data Sets, Data Analysis	21
	A. Data Set Compilation	21
	B. Vegetation Classification	22
	C. Statistical Analysis of Instantaneous Data Sets	22
	D. Statistical Analysis of Grouped Data Sets	23
	E. Boundary Value Analysis	26
XII.	Effects of Various Factors on the Response Function of CO_2 Flux to PPFD	27
	A. The Micrometeorological and Enclosure Techniques	27
	B. Respiration	29
	C. Vegetation Class	32
	D. Other Variables	34
XIII.	Mean Statistics, Residuals and Boundary Value Analysis for the Complete Data Set	37
XIV.	Perspectives for CO_2 Flux Measurements	40
	A. Scaling Up in Space: Plane Mounted Eddy Flux Sensors	40

ADVANCES IN ECOLOGICAL RESEARCH VOL. 26
ISBN 0–12–013926–X

    B.   Scaling Up in Time: Instantaneous versus Daily Integrated
        Measurements 41
    C. CO_2 Enrichment 42
XV. Conclusions 43
    A.   Specific Conclusions 43
    B.   General Conclusions 44
References 46
Appendix 1 52
Appendix 2 63

I. SUMMARY

In this paper we try to assess the parameters determining the shape of the response of CO_2 flux over closed plant canopies (F) to photosynthetic photon flux density PPFD (Q). Over one hundred data sets relating CO_2 flux above canopies to radiation or PPFD have been compiled, digitized, put in standard units and statistically analysed. There is a lack of data for some vegetation classes, in particular coniferous forests, tropical grasslands and mixed vegetation. Linear regressions and rectangular hyperbolic functions have been fitted through the data sets. The parameters of importance that have been extracted are the slope at the origin (defined as the apparent quantum yield, α), the CO_2 flux at maximum irradiance of $1800\,\mu mol\,m^{-2}s^{-1}$ or $80\,mol\,m^{-2}d^{-1}$ (defined as the photosynthetic capacity, F_m), the intercept on the y axis (defined as dark respiration rate, R), and the departure from linearity, calculated as the difference between the r^2 of the rectangular hyperbolic fit and the r^2 of the linear fit. The sensitivity of the F/Q relationship to various factors has been tested, using statistics on particular data sets that were obtained in similar conditions, and statistics applied to data sets for closed canopies grouped by classes, with respect to technique and vegetation class. Micrometeorological methods result in F/Q relationships closer to linearity than enclosure methods. Analysing data sets obtained with micrometeorological methods only has allowed us to distinguish between the F/Q relationships of crops and forests: the CO_2 fluxes of crops have a linear relationship with PPFD, while the CO_2 fluxes of forests have a curvilinear relationship with PPFD. The relationship between CO_2 flux corrected for dark respiration or soil respiration and PPFD is not clearly improved compared with the relationship between net ecosystem flux and PPFD. Although seasonality, nutrient availability, water availability, water vapour pressure deficit, CO_2 concentration and temperature certainly affect the F/Q relationship, we have not been able to illustrate quantitatively their role because of lack of suitable measured data. The best fit through all the data sets is a rectangular hyperbola. Scaling up in space (using aircraft mounted flux sensors) and time (daily integrated fluxes) seems to linearize the F/Q relationship.

II. SYMBOLS AND ABBREVIATIONS

CO$_2$ flux measurements (amount of CO$_2$ per unit ground area and time):
- F: net ecosystem flux, with or without addition of some respiration terms (p. 13)
- F_d: net ecosystem flux in the day
- F_n: net ecosystem flux in the night
- F_c: canopy or above-ground flux
- F_s: soil flux

Terms of the carbon budget of the stand (amount of CO$_2$ or dry mass per unit area and time):
- A_n: net photosynthetic rate or gross primary productivity
- A: gross photosynthetic rate
- R_l: leaf dark respiration rate
- R_w: dark respiration rate of above-ground woody plant parts
- R_r: root dark respiration rate
- R_a: autotrophic respiration rate
- R_h: heterotrophic respiration rate
- P_n: net primary productivity
- P_e: net ecosystem productivity
- ΔS_c: variation of CO$_2$ storage in the canopy

Radiation (amount of photons or energy per unit area and time):
- Q: photosynthetic photon flux density
- S_p: photosynthetically active radiation
- S_g: solar (global) radiation
- S_n: net radiation
- *subscripts for radiation:* 0, incident; int: intercepted; abs: absorbed; t: transmitted.
- no subscript: incident, intercepted or absorbed radiation over closed canopies (see p. 16).

Parameters of statistical regressions:
- n: number of data points
- F_∞: CO$_2$ flux at saturating photosynthetic photon flux density
- A_∞: net assimilation rate of a leaf at saturating photosynthetic photon flux density
- α: apparent quantum yield, i.e. slope of the relationship at $Q = 0$
- R: value of F at $Q = 0$
- r^2: non-linear coefficient of determination
- D: departure of the relationship from linearity, i.e. $r^2(H) - r^2(L)$
- F_m: photosynthetic capacity, i.e. F at $Q = 1800\,\mu\mathrm{mol\,m^{-2}\,s^{-1}}$ for instantaneous data or F at $Q = 80\,\mathrm{mol\,m^{-2}\,d^{-1}}$ for daily data sets

Other symbols:
  e: conversion efficiency of absorbed radiation into dry matter
  e' : photosynthetic efficiency
  f : radiation absorption efficiency
  R: radiation extinction coefficient
  C: atmospheric CO_2 concentration

Abbreviations for statistical treatments:
  L: linear best fit
  H: rectangular hyperbola best fit
  N: no statistical relationship

Other abbreviations:
  GPP: gross primary production
  NPP: net primary production
  NEP: net ecosystem production
  NEE: net ecosystem exchange
  PAR: photosynthetically active radiation
  PPFD: photosynthetic photon flux density
  LAI: leaf area index
  VPD: vapour pressure deficit
  CBL: convective boundary layer

III. INTRODUCTION

The first measurements of CO_2 fluxes over plant canopies are those reported by Thomas and Hill (1949) who used large plastic enclosures over
fields of alfalfa, sugar beet and wheat. At the time they used a chemical
method for measuring CO_2 concentration. However, this work remained
isolated and it was only in the 1960s that new measurements were obtained
with infra-red gas analysers. Beside canopy enclosures (e.g. Musgrave and
Moss, 1961; Eckardt, 1966), micrometeorological methods were also developed, using vertical profiles of CO_2 concentration and associated
measurements in the aerodynamic method and in the energy balance
method (e.g. Lemon, 1960; Monteith and Szeicz, 1960; Monteith, 1962;
Inoue, 1965; Saugier, 1970). Eddy correlation methods were first developed for measuring heat and water vapour fluxes, and extended to CO_2
fluxes when reliable fast response CO_2 sensors became available in the
1980s (e.g. Anderson *et al.*, 1984; Verma *et al.*, 1986). CO_2 flux measurements are useful in various fields:

(i) *Annual net carbon balance of ecosystems.* In the context of rising
    atmospheric CO_2 concentration as a result of the burning of fossil
    fuels and land use changes, the question of closing the carbon budget

of the global biosphere has become critical. CO_2 flux between an ecosystem and the atmosphere is a direct indication of whether that terrestrial ecosystem is a source or a sink of CO_2 over a certain period. Based on measurements of net ecosystem flux, for example, Wofsy *et al.* (1993) measured carbon storage rates by a broadleaf, deciduous forest that are somewhat larger than those currently assumed in global carbon studies.

(ii) *Canopy and stand physiology.* CO_2 flux measurements provide insight into canopy and stand physiology, i.e. the response of gas exchange at the ecosystem scale to environmental variables. Similar to analyses at leaf scale, relationships between canopy photosynthetic rate or net ecosystem flux and absorbed photosynthetic photon flux density (PPFD), or respiration rate of the canopy or the ecosystem and temperature, have been established and the effect of environmental variables, such as atmospheric water vapour pressure deficit (VPD), on stand or canopy CO_2 flux have been investigated.

(iii) *Testing "bottom-up" plant production models.* Various models were developed in the 1960s to understand patterns of radiation penetration and of photosynthesis in homogeneous plant canopies (e.g. Saeki, 1960; Monteith, 1965; Duncan *et al.*, 1967) and were later extended to heterogeneous canopies (see Sinoquet, 1993, for a recent review). Computed PPFD reaching a given leaf level in the canopy, along with data on leaf photosynthetic rate, have been used to model canopy photosynthetic rate. Similarly, models of autotrophic respiration were developed (e.g. McCree, 1974). Before the availability of CO_2 flux measurements, the only data useful for testing these models were biomass increments of plants and vegetation. Canopy CO_2 flux data are intermediate in spatial and temporal scales between ecophysiological measurements of leaf photosynthesis over short periods, and measurements of plant production made over a whole growing season, and are suitable for the purpose. Flux measurements put a "lid" on the system and, therefore, provide a test of our understanding of processes within the system, and of "bottom-up" *process models* (e.g. Amthor *et al.*, 1994)

(iv) *Parameterizing "top-down" plant production models using remote sensing.* For many crops not short of water, the production of plant biomass is proportional to accumulated absorbed photosynthetically active radiation or PAR (Monteith, 1972; Monteith, 1977; Varlet-Grancher, 1982). The ratio between these two quantities has been called the conversion efficiency of absorbed radiation into dry matter, and is used in many simple models of crop growth, i.e. by-passing the complex processes of photosynthesis and of respiration known to depend on many environmental variables. A linear relationship between plant biomass production and absorbed PAR implies a linear relationship between canopy CO_2 flux and absorbed PPFD. In addition, the efficiency of radiation absorption by vegetation can be derived from remotely sensed vegetation indices. These coefficients are used in "top-down" *parametric models* (e.g. Kumar and Monteith, 1981).

The response of the CO_2 flux between a *leaf* and the atmosphere to PPFD can be described by a rectangular hyperbola. There is a respiratory flux in the dark; at low, limiting irradiance the CO_2 flux is proportional to PPFD, the slope being defined as the apparent quantum yield of photosynthesis; at high, saturating irradiance the response curve levels off, the asymptotic value being defined as photosynthetic capacity. The response of the CO_2 flux of a closed canopy is qualitatively similar. However, when the upper leaves are PPFD saturated, the lower leaves may still be PPFD limited, and the response of the CO_2 flux between a *canopy* and the atmosphere to PPFD saturates less rapidly than in the case of a single leaf, even being linear over the whole range of PPFD in some cases.

In this paper we aim to assess the possibility of obtaining general relationships between the CO_2 flux over canopies and absorbed PPFD. In particular, we aim to determine whether the relationship for a closed canopy is linear, consistent with the Monteith model, or curvilinear, consistent with the results of many mechanistic canopy models. Both general relationships and relationships for different vegetation classes are sought. Some methodology-related differences are also investigated, including comparison between micrometeorological and enclosure methods, the effect of taking into account respiration rate, and the effect of integrating measurements in time. For this we use, whenever possible, statistics on *similar data sets* (i.e., except for the variable under study, all environmental conditions were similar, and the data were obtained on the same site, with the same measuring technique, by the same scientific team), and statistics on *grouped data sets* (i.e. all the data sets on closed canopies satisfying the conditions described in the data analysis section are grouped by vegetation class or method of measurement or representation). Some effects of environmental variables are also investigated, but not quantitatively analysed because of lack of information in the original papers.

IV. STAND CO_2 BALANCE

The flux of CO_2 measured across a plane above a stand is the net result of all the CO_2 fluxes occurring within the system. In this respect the stand is analogous to a leaf. Within the stand there are photosynthetic organs and respiratory sources just as there are photosynthetic organelles and non-photosynthetic, respiring tissues in a leaf. The net flux of CO_2 across the plane bounding the stand system is analogous to the net flux of CO_2 across the surface of a leaf through the stomata. The normal convention in micrometeorology is that fluxes downwards are negative and fluxes upwards positive. The convention we have followed in this paper is the convention used in ecophysiology, i.e. both photosynthetic and respiratory fluxes at the organ scale are treated as positive. At the stand scale, downward fluxes into the stand are treated as positive and upward fluxes as negative. The change in the amount of CO_2 stored in the column of air between the system boundaries, ΔS_c, is treated as either positive if S_c increases, or negative if S_c decreases.

Because the flux across the system is the algebraic sum of the fluxes within the system, if we consider instantaneous fluxes within the forest depicted in Figure 1, and assume a one-dimensional, horizontally uniform

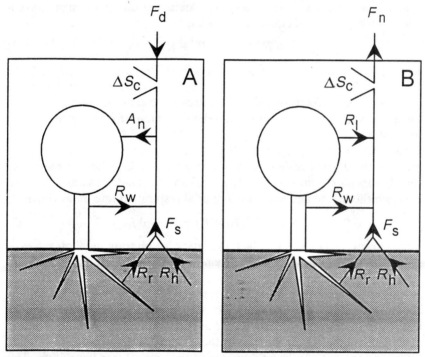

Fig. 1. CO_2 budget of a stand (A) during the day, and (B) during the night. F_d is the net ecosystem flux during daylight, F_n is the net ecosystem flux during darkness (night-time respiration), F_s is the soil CO_2 flux measured below the canopy assuming there is no photosynthesizing understorey vegetation, ΔS_c is the change of CO_2 stored in the canopy. The components of the carbon budget of the stand are: leaf net photosynthesis (A_n), leaf dark respiration (R_l), wood respiration (R_w), root respiration (R_r), and heterotrophic microbial decomposition (R_h).

system, then we can write

for the day: $F_d = A_n - (R_w + R_r + R_h) + \Delta S_c$

for the night: $F_n = - (R_l + R_w + R_r + R_h) + \Delta S_c$ (1)

where F_d is net CO_2 flux above the canopy during the day, F_n is net CO_2 flux above the canopy during the night, A_n is rate of net photosynthesis, R_l is respiratory flux from leaves, R_w from wood, R_r from roots and R_h from microorganisms in the soil and soil fauna.

From the point of view of the carbon balance of the system as a whole, the net flux across the upper system boundary, integrated over time, defines the gain or loss of carbon by the system, and thus determines whether the forest system is a sink or a source of carbon. The net flux, when integrated over time, gives the *Net Ecosystem Production*, NEP.

The physiological capability of the forest canopy to assimilate CO_2 in daytime in relation to weather and other environmental variables is determined by net photosynthetic rate, A_n. Assuming the above sign convention, rearrangement of equation (1) gives:

$$A_n = F_d + R_w + R_r + R_h - \Delta S_c \qquad (2)$$

This flux, integrated over time, is the *Gross Primary Production*, GPP. Although A_n depends on all the fluxes in the system and will only be approximated by F_d if all the other fluxes in respiration and storage are small by comparison, attempts to relate F_d alone to PPFD have been successful.

Subtracting the average CO_2 flux in the night from instantaneous CO_2 fluxes in the day (or adding the absolute values of these fluxes) gives a so-called *gross photosynthetic rate A*, which can be regarded as the sum of net photosynthetic rate and leaf mitochondrial respiration rate in the daytime:

$$A = A_n + R_l = F_d - F_n - \Delta S_c \qquad (3)$$

This, however, ignores the fact that dark respiration rate of leaves is significantly lower in the daytime than in the nighttime (e.g. Villar *et al.*, 1994). True gross photosynthesis is not measurable by flux techniques.

Another commonly reported flux is the difference between the CO_2 flux above and below the canopy, measured simultaneously. This flux is the above-ground flux, or *canopy flux*, F_c. The CO_2 flux measured below the canopy or *soil flux*, F_s is the sum of the root and heterotrophic respiration rates, and the net photosynthetic rate of the understorey vegetation. This latter term is usually neglected.

$$F_s = - (R_h + R_r) \qquad (4)$$

$$F_c = A_n - R_w = F_d - F_s - \Delta S_c \qquad (5)$$

F_c has also been successfully related to radiation, and has been regarded as *canopy net photosynthetic rate*, even though some wood respiration is included.

If one could easily separate roots and microorganisms in the soil respiration term, we could rearrange equation (1):

for the day: $F_d + R_h - \Delta S_c = A_n - (R_w + R_r)$
for the night: $F_n + R_h - \Delta S_c = - (R_l + R_w + R_r)$ (6)

This flux, integrated over time, is analogous to *Net Primary Productivity*, NPP.

Table 1 summarizes the correspondence between the CO_2 flux measurements terms and the components of the CO_2 budget of a stand, omitting the storage component.

Table 1

Equivalence between CO_2 flux measurements and terms of the carbon budget of a stand

Process	Flux measurement	Carbon budget
Leaf or canopy gross photosynthesis	$F_d - F_n$	A
Leaf respiration		R_l
Leaf or canopy net photosynthesis		$A_n = A - R_l$
Respiration of above-ground woody plant parts		R_w
Root respiration		R_r
Autotrophic respiration		$R_a = R_l + R_w + R_r$
Heterotrophic respiration		R_h
Net ecosystem flux in darkness or total respiration of the system	F_n	$R_a + R_h$
Soil flux, or root + microorganism respiration	F_s	$R_h + R_r$
Net primary production		$P_n = A - R_a$
Above-ground flux or canopy flux	$F_c = F_d - F_s$	$A_n - R_w$
Net ecosystem flux or net ecosystem production	F_d	$P_e = P_n - R_h$

V. STORAGE FLUX

A so-called *Net Ecosystem Exchange* NEE, defined as the flux of the biota, is usually calculated as:

for the day: $F_d - \Delta S_c = A_n - (R_w + R_r + R_h)$
for the night: $F_n - \Delta S_c = - (R_l + R_w + R_r + R_h)$ (7)

The difference from equation (1) is that the atmospheric storage term within the system, ΔS_c, is now on the left hand side, together with the net gain or loss of CO_2 by the system, F_d or F_n, and all the terms describing immediate biological activity are on the right hand side.

An increase in storage of CO_2 in the air column can be regarded as a measure of ecosystem activity if the concentration at the reference plane remains constant. This approach was used by Woodwell and Dykeman (1966) to estimate ecosystem respiration rate at night, under stable conditions generated by temperature inversion. The temperature inversion turns what was essentially an open system into a closed system and allows a straightforward treatment that cannot be easily applied in more general circumstances, when the concentration of CO_2 within the stand may change for several reasons, which are related to both the physiology of the canopy (photosynthetic and respiration rate of the different compartments), and the meteorology (growth of the convective boundary layer, CBL, and entrainment of air).

The flux in and out of storage can be particularly important at times. For example, after an overnight temperature inversion, when the CO_2 concentration within the stand may have reached 500 or 600 μmol mol^{-1}, the rapid reduction in CO_2 concentration at dawn is partly a result of the assimilation of CO_2 stored within the system. As the system warms in the early morning and an upward heat flux develops, the overnight inversion breaks down, warm air is convected upwards, the CBL grows, and the concentration of CO_2 within the stand falls as a result.

Some recent publications present results of NEE where ΔS_c was estimated by measuring the variations of CO_2 concentration with time at a reference height in or above the canopy (e.g. Wofsy et al., 1993; Hollinger et al., 1994). Both assimilation of CO_2 by the canopy and entrainment of air of lower CO_2 concentration contribute to the change in CO_2 storage within the canopy.

During the day, the CO_2 concentration within the mixed layer varies as the result of assimilation of CO_2 by vegetation at *regional* scale. This concentration may easily change by 100 μmol mol^{-1} from dawn to mid-afternoon. The column of air that arrives at a measurement point has usually travelled a substantial distance. Whilst the CO_2 concentration within the surface layer close to the vegetation surface is determined by a comparatively local process, the concentration of CO_2 in the mixed layer reflects both entrainment and physiological activity of vegetation on a much larger scale. This change in concentration of CO_2 within the CBL may be used to estimate CO_2 assimilation integrated over the day at the regional scale (McNaughton, 1988; Raupach et al., 1992). The CO_2 concentration at the upper system boundary of a rough surface such as forest is closely coupled to the CO_2 concentration in the mixed layer and consequently reflects regional scale processes rather than local processes.

Figure 2A shows the idealized development of CO_2 concentration profiles through a canopy during the course of a day, in the absence of regional scale phenomena. The bulge to the left becomes more pronounced during the morning and retracts during the afternoon. At the same time, the CO$_2$ concentration profile as a whole moves to the left as a result of regional CO$_2$ assimilation modified by entrainment, as shown in Figure 2B, and moves back again in the evening. The locally attributable change in storage is represented only by the shaded area in Figure 2A and not by the entire shaded area in Figure 2B. Thus, assessment of the change in storage must be based on measurements of CO$_2$ concentrations down through the canopy. It is a basic premise of the flux measuring schemes in use that a steady state exists over the period of integration, as pointed out by Tanner (1968). It is also a basic assumption that transport is one-dimensional and that the area of vegetation is extensive and homogeneous: if advection occurs on a substantial scale then additional terms need to be added as shown by Tanner for water transport.

Conclusions regarding storage:
  (i) Change in storage is a component of the mass balance. Storage of CO$_2$ is important and relevant to a discussion on forest CO$_2$ fluxes. Unfortunately, most past studies have ignored this process. However, the fluxes in and out of storage are generally much smaller than the other fluxes, except in the early morning period after an inversion, and 24-hour storage in the airspace is generally zero or negligible.
  (ii) Even in a simple situation with a constant ambient CO$_2$ concentration, the fluxes in and out of storage are a complex product of physiology and meteorology.
  (iii) The apparent flux out of storage at a particular point in a region, based on a shift in the vertical profile of CO$_2$ concentration through the vegetation, may give a misleading impression of the physiological activity of the vegetation at that point.
  (iv) When the atmospheric CO$_2$ concentration is changing on a regional scale, the flux out of storage throughout the whole of the mixed layer and the surface layer represents the net CO$_2$ exchange of the vegetation on a regional scale.

Fluxes measured over the canopy may or may not have been corrected for change in storage. Appendix 1 mentions if change in storage has been taken into account in the day or in the night, in each data set. What are referred to as CO$_2$ fluxes in the rest of this paper may represent somewhat different quantities for this reason.

VI. RESPIRATION FLUX

If we do ignore CO$_2$ storage within the air column, the flux over a plant canopy, F_d, is a balance between the photosynthetic and respiration rates

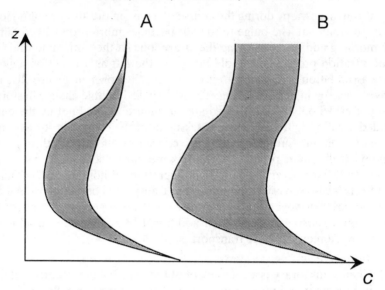

Fig. 2. Idealized development of CO_2 concentration profiles through a canopy, during the course of a day (A) in the absence of regional scale CO_2 depletion, (B) with regional scale CO_2 depletion. z is the height from the ground, C is the CO_2 concentration. The concentration profile moves from right to left within the shaded area during the morning, and moves back during the afternoon.

of the system. Total respiration rate may represent the sum of two or more of the individual respiration fluxes, depending on the definition of the system. Heterotrophic respiration depends on soil temperature and moisture content. Autotrophic respiration can be partitioned, following McCree (1974), into maintenance and growth respiration. Maintenance respiration rate depends on temperature and the protein content of organs; growth respiration rate depends on the fraction of assimilates allocated to growth and on the chemical composition of growing tissues. In considering the relationship between F_d and Q, we should bear in mind the following:

(i) In the dark, maintenance and heterotrophic respiration rates vary with air and soil temperature. This affects the dark respiration rate, i.e. CO_2 flux at $Q = 0$. For a given temperature, maintenance respiration rate depends principally on the biomass of the canopy.

(ii) In the light, maintenance and heterotrophic respiration rates follow the diurnal course of temperature. The time lag between the diurnal course of solar radiation and of air and soil temperature generates hysteresis in the response of F_d to PPFD.

(iii) Growth respiration rates vary during the day, therefore the time step and interval of measurements affect the value of F_d.

A. Soil Respiration

In the case of micrometeorological measurements, the system comprises vegetation and soil, and therefore measured F_d always includes soil CO_2 flux. In forest systems, it is possible to measure fluxes above ground level and below the canopy by placing eddy covariance sensors below the canopy (e.g. Baldocchi *et al.*, 1987). In this case the CO_2 efflux from the forest floor also includes the photosynthetic flux of the understorey vegetation.

In the case of enclosures, definition of the system depends on which parts are enclosed. In herbaceous vegetation, it is usually not easy physically to separate vegetation from soil, and therefore measured F_d includes soil respiration rate. Soil respiration rate can be measured separately with enclosures over areas where the vegetation has been removed (e.g. Puckridge and Ratkowski, 1971). In the rare cases of measurements with enclosures over areas of forests (Wong and Dunin, 1987; Mordacq *et al.*, 1991), the soil was not included in the enclosure.

B. Dark Respiration

Night time CO_2 flux, F_n, represents the rate of all the components of the system. Two ways of taking dark respiration rate into account in the calculation of CO_2 flux are to be found in the literature:

(i) The mean value of F_n has been subtracted directly from daytime CO_2 fluxes F_d, so that the PPFD response curve passes through the origin. In this case, the differences in respiration rate during the night and day, resulting from differences in temperature in particular, have been neglected.

(ii) F_n has been recorded at night with the corresponding air temperature. A respiration rate/temperature relationship has been established, used to calculate dark respiration rate at the appropriate temperature during the day, and then subtracted from daytime CO_2 flux. In this case the reported CO_2 flux is regarded as gross photosynthetic rate, A.

Appendix 1, describing the data sets, indicates whether soil flux or dark respiration rate was measured by the authors, and whether the daytime fluxes reported have had some respiration terms subtracted. In this study, there has been no attempt to standardize all the CO_2 fluxes compiled, in order to retrieve, for instance, the flux of CO_2 assimilated by the canopy, A_n. Therefore, what is referred to as CO_2 flux, "F", in the rest of this paper may represent very different quantities: net ecosystem flux alone, or net ecosystem flux plus the absolute value of some respiration terms.

VII. CANOPY PHOTOSYNTHESIS

Canopy net photosynthetic rate is determined by leaf photosynthesis, respiration of twigs and branches, and by radiative transfer through the canopy. The response of leaf photosynthetic rate to PPFD is asymptotic, the parameters depending on leaf photosynthetic properties and the environment. The result of radiative transfer is to linearize the response curves of CO_2 flux to PPFD at the canopy scale. This is achieved in canopies in three ways (Jarvis and Leverenz, 1983):

  (i) Grouping of the foliage into shoots, branches and crowns results in efficient transmission of direct radiation through the canopy and effective scattering where it is intercepted.

 (ii) Spatial distribution of the leaves, together with the distribution of their inclination angles, results in exposure of the majority of leaves to intermediate PPFD so that the leaves are not PPFD saturated.

(iii) Compensation in physiological and anatomical properties for the gradient of PPFD through the canopy as a result of "shade" acclimation: photosynthesis of shade leaves is more efficient at low PPFD, largely because of lower dark respiration rate (Osmond et al., 1980).

Saugier (1986), Wong and Dunin (1987) and Jarvis and Leverenz (1983) show comparisons between the PPFD response curves of leaf and canopy photosynthetic rate for, respectively, a grass crop of *Dactylis*, a broadleaf forest of *Eucalyptus maculata*, and a needleleaf forest of Sitka spruce (*Picea sitchensis*). In the case of the Sitka spruce, a curve for an intermediate photosynthetic element, the shoot, was also presented. Figure 3 shows the response of photosynthetic rate to PPFD for a leaf (expressed per unit projected leaf area), a shoot (expressed per unit shoot silhouette area) and a canopy (expressed per unit ground area) of Sitka spruce. Visual analysis of the regression curves through the data points presented in these papers leads to the following conclusions:

  (i) Dark respiration rate (CO_2 flux at $Q = 0$) is less for a leaf than for a shoot, and for a shoot than for the canopy. This results probably from the difference in the relative respiring biomass of these elements.

 (ii) Maximum photosynthetic rate at high PPFD is lower for a leaf than for a shoot, and for a shoot than for the canopy.

(iii) The response curve of photosynthetic rate to PPFD departs further from linearity for a leaf than for a shoot, and for a shoot than for the canopy. Points (ii) and (iii) are the result of more efficient utilization of photons by groups of photosynthetic elements than by a single leaf.

(iv) The slope of the PPFD response curve at $F_d = 0$ shows no general trend: in the case of the grass crop, the initial slope was steeper for the

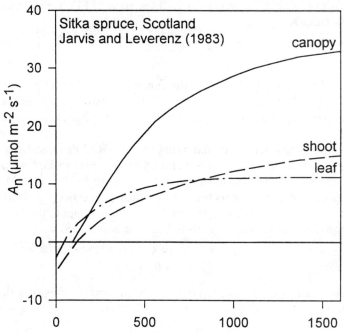

Fig. 3. The relationship between rate of net photosynthesis (A_n) and incident quantum flux density (Q_0) in Sitka spruce, for "sun" needles, a "sun" shoot and a forest canopy. Replotted from Jarvis and Leverenz (1983).

canopy than for the leaf, but the slope was less steep for the canopy than for the leaf in the case of the broadleaf forest, and all three slopes were almost identical for the needleleaf forest (Fig. 3). In all three cases, the slope is close to the quantum yield for a C3 leaf at normal CO$_2$ concentration, i.e. $0.05 \, \mu$mol (CO$_2$) μmol^{-1} (photons) (Ehleringer and Pearcy, 1983).

VIII. RADIATION

In the literature, CO$_2$ fluxes have originally been reported in relation to photosynthetically active radiation (S_p), photosynthetic photon flux density (Q), solar radiation (S_g), or net radiation (S_n). Radiation is either incident (subscript: 0), intercepted by the canopy (subscript: int) or absorbed by the canopy (subscript: abs). For a review and a definition of these terms, see the articles by Varlet-Grancher et al. (1989) or Goward and Huemmrich (1992). Intercepted and absorbed PPFD, for instance, are defined as follows:

$$Q_{int} = Q_0 - Q_t$$
$$Q_{abs} = Q_{int} - Q_r + Q_{rs} \tag{8}$$

with Q_t, PPFD transmitted through the canopy, Q_r, PPFD reflected by the canopy and Q_{rs}, PPFD reflected by the soil or understorey, and reabsorbed by the canopy. Q_{rs} is small compared with the other fluxes, and is usually neglected.

The relationships between solar radiation, PAR, PPFD and net radiation are variable, depending on several environmental variables, such as the proportions of diffuse and direct radiation (see for instance McCree 1972) and on the canopy structure and optical properties. To transform radiation ($W\,m^{-2}$) into PPFD ($\mu mol\,m^{-2}\,s^{-1}$) in this paper, the following assumptions have been made (Varlet-Grancher et al., 1981):

$$S_p = 0.48\, S_g$$
$$Q = 4.6\, S_p \tag{9}$$

One single, statistical relationship has also been used to transform the few data expressed in net radiation into solar radiation:

$$S_n = 0.7\,(S_g - 30) \tag{10}$$

We have focused our review on *closed canopies*, for which we consider that all the incident photon flux is intercepted, and we have neglected the difference between absorbed and intercepted photon flux. Leaf area index (LAI) is the primary factor determining the absorption of photon flux by a canopy. However, variations in the extinction coefficient can be important at times: e.g. Breda (1994) reports extinction coefficients of solar radiation varying from 0.46 to 0.32 for thinned oak stands. Another issue is that absorption or interception of radiation by photosynthetic organs is different for different expressions of radiation: for example, photosynthetically active radiation is more atttenuated through the canopy than solar radiation: Hutchison and Baldocchi (1989) report extinction coefficients for an oak-hickory forest of 0.65 for PAR and 0.51 for solar radiation. The absorbed PPFD depends on the albedo of the canopy, and albedo varies with vegetation class and structure: Sellers (1965) reported albedos for solar radiation of 5–15% for coniferous forests, 10–20% for deciduous forests, and 15–25% for crops. Albedos for PPFD are less than 5%.

In this study, these discrepancies have been neglected, and there has been no attempt to standardize the radiation flux compiled, in order to retrieve, for instance, PPFD absorbed by the canopy. Therefore, what is referred to as photosynthetic photon flux density, "Q", in the rest of this paper may represent incident, absorbed or intercepted PPFD in closed canopies. Appendix 1, describing the data sets, indicates the expression of radiation that was originally reported by the authors.

IX. RADIATION USE EFFICIENCY

Part of this study is intended to define whether, and on what time and space scales, a linear relationship between CO_2 fluxes over closed plant canopies and absorbed PPFD exists. In these cases, we can calculate a *photosynthetic efficiency*, as the slope of the linear regression between F and Q.

Usefulness of a linear relationship. A linear relationship between absorbed PPFD and canopy photosynthetic rate would be very useful in doing "biology from space". Kumar and Monteith (1981) derived a "top-down" plant production model for remote sensing of crop growth. This model has been used to estimate NPP over various surfaces, ranging from the field scale to the whole biosphere:

$$P_n = e f S_p, o \tag{11}$$

where P_n is net primary productivity, f is the efficiency of absorption of incident PAR, and e is the efficiency of conversion of absorbed PAR into dry matter.

The coefficient f can be related to combinations of reflectances obtained by remote sensing, called vegetation indices. The most commonly used are the Normalized Difference Vegetation Index, NDVI, and the Simple Ratio, SR, both of them being combinations of reflectances in the red and near infra-red channels of the NOAA-AVHRR radiometer. Monteith (1977) found that the shape of the relationship of annual NPP versus absorbed PAR integrated over a year is linear, and that the slope, e, is constant over a wide range of crops and climatic conditions in England. This constant slope of 3 g of dry matter produced per MJ of PAR absorbed has been used by Heimann and Keeling (1989) to estimate seasonal variations of global NPP.

Ruimy *et al.* (1994) showed that e is much more variable between different natural vegetation types, and used a conversion efficiency varying with vegetation class for global NPP estimates. One of the factors with which e varies is the proportion of assimilates lost by autotrophic respiration. Monteith (1972) expressed NPP as:

$$P_n = \epsilon_r \epsilon_q \epsilon_d f S_{p,0} = \epsilon_r A_n \tag{12}$$

where ϵ_q is quantum efficiency, ϵ_d efficiency of diffusion of CO_2 molecules from the atmosphere to the sites of photosynthesis in the leaves, and ϵ_r fraction of assimilates not used for respiration.

A linear relationship between annual NPP and integrated absorbed PAR requires that there is a linear relationship between GPP and absorbed PPFD, at least on an annual time-scale. In this case, we can define a *photosynthetic efficiency e'*, so that

$$A_n = e' f Q_0 \tag{13}$$

Linear relationships imply that, on a seasonal time-scale, i.e. weekly or monthly, mean solar radiation can be multiplied by mean absorption efficiency and mean conversion efficiency (respectively photosynthetic efficiency) to retrieve NPP (respectively GPP), whereas a curvilinear relationship implies that NPP (respectively GPP) has to be computed instantaneously or hourly, and this dramatically increases computing time.

Are linear relationships artefacts? Three arguments may lead to the suggestion that linear relationships between F and Q are artefacts.

(i) Model computations of canopy photosynthesis, ranging from instantaneous to daily scale, when plotted versus radiation always show curvilinear relationships, however far from photon saturation (e.g. Baldocchi, 1994; Sellers, 1985; Oker-Blom, 1989; Wang et al., 1992).

(ii) The mean initial slopes of regressions through data sets for which the best fits are curvilinear are often higher than the slopes of linear regressions through data sets for which the best fits are linear (see Appendix 2).

(iii) Some linear best fits are poorly defined or the result of environmental stresses: a large amount of scatter in the data characterized by low r^2 below 0·40 (e.g. Allen and Lemon, 1976); a small number of data points, fewer than 10 (e.g. Desjardins et al., 1985); relatively low maximum PPFD, below $1000 \, \mu$mol photons m^{-2} s^{-1} (e.g. Valentini et al., 1991); water-stressed vegetation (e.g. Kim and Verma, 1990).

X. TECHNIQUES FOR MEASURING CO$_2$ FLUX

A. Energy Balance or Bowen Ratio Method

This method was first used to measure water vapour fluxes over a surface (Bowen, 1926), and has then been applied to the measurement of CO$_2$ fluxes (see Jarvis et al., 1976 for examples). The method is based on the energy balance at a surface:

$$S_n = G + H + LE \tag{14}$$

where G is sensible heat flux into the ground, H, sensible heat flux into the air, and LE, latent heat flux into the air (L, latent heat of vaporization of water; E, rate of evaporation at the surface). S_n and G are measured quantities. To partition between the two other quantities, the Bowen ratio, β, can be measured as:

$$\beta = \frac{H}{LE} = \frac{c_p \, \Delta T}{L \, \Delta q} \tag{15}$$

where c_p is specific heat of air at constant pressure, ΔT, difference of air temperature between two reference heights and Δq, difference of specific humidity of the air over the same height interval above the vegetation. Then:

$$E = \frac{S_n - G}{(\beta + 1) \, L} \tag{16}$$

Knowing the evaporation rate E, we can calculate CO_2 flux F from:

$$\frac{F}{E} = \frac{\Delta C}{\Delta q} \tag{17}$$

where ΔC is the difference of specific CO_2 concentration over the same height interval.

B. Aerodynamic or Profile Method

The theory of this method and some applications were given by Monteith and Szeicz (1960). The method is based on the assumption of similarity of the turbulent transfer of mass and momentum, i.e. equality of the turbulent transfer coefficients for CO_2, K_c, and for momentum, K_M. The CO_2 flux can be written with our sign convention as:

$$F = \rho_c K_c \frac{\Delta C}{\Delta z} \tag{18}$$

where ρ_c is the density of CO_2 and K_c the turbulent transfer coefficient for CO_2.

K_M is calculated from the logarithmic profiles of wind speed through the canopy as follows:

$$K_M = \frac{k^2 z \Delta u}{\ln\left(\dfrac{z_2}{z_1}\right)^2} \tag{19}$$

where k is von Karman's constant, Δu the difference in horizontal wind speed between the two measurement heights, z_1 and z_2. If z is measured from the ground, it is necessary to subtract from z the displacement height of the wind-speed profile, d. Substituting the expression for K_M in equation

(19) for K_c in equation (18), incorporating d and integrating yields the final equation:

$$F = \frac{\rho_c k^2 \; (u_2 - u_1)(C_1 - C_2)}{\ln \left(\dfrac{z_2 - d}{z_1 - d}\right)^2} \qquad (20)$$

Equation (20) assumes that the atmosphere is in neutral equilibrium. Various corrections have been proposed for unstable or stable conditions (e.g. Monin and Yaglom, 1971).

C. Eddy Correlation or Eddy Covariance

This method was first used to measure water vapour fluxes, and has been extended to CO_2 fluxes only recently (Anderson et al., 1984; Verma et al., 1986). The time average of the vertical flux of CO_2 at a fixed point above a surface can be expressed as:

$$F = \overline{\rho_c w' C'} \qquad (21)$$

where $C' = C - \overline{C}$ is the instantaneous deviation from the mean CO_2 concentration measured at a reference height and w' is the instantaneous deviation of vertical wind speed from the mean wind speed.

CO_2 concentration is measured with a fast-response CO_2 gas analyser and wind speed with a sonic anemometer. Because this method requires observations at only one reference height, instruments can be placed on an aircraft as well as on a tower.

D. Enclosures

Details on this method were given by Musgrave and Moss (1961). A portion of the vegetation is enclosed from the outside environment in a large chamber. The chamber may or may not include the soil surface. In a *closed system*, the change of CO_2 concentration in the chamber with time is used to calculate CO_2 flux as:

$$F = V\frac{\Delta C}{\Delta t} \qquad (22)$$

where V is chamber volume, and $\Delta C/\Delta t$ the rate of decrease of CO_2 concentration.

In the case of *open systems*, there is a constant flow rate of air (d) through the enclosure, and the difference in concentration of the air entering and leaving the system ($\Delta C''$) is monitored and used to calculate CO_2 flux:

$$F = d\,\Delta C''\tag{23}$$

E. Models

The shape of the response of canopy GPP to intercepted PPFD can be modelled knowing the PPFD response curve of a leaf and the attenuation of PPFD through the canopy (e.g. Monteith, 1965). There are several models that combine leaf photosynthesis and radiative transfer through a canopy (e.g. de Wit *et al.*, 1978; Wang and Jarvis, 1990; Wang *et al.*, 1992). PPFD response curves of leaves can be modelled with a rectangular hyperbola:

$$A_n = \frac{\alpha Q_0 A_\infty}{\alpha Q_0 + A_\infty} - R\tag{24}$$

where A_∞ is leaf assimilation rate at saturating PPFD, α is apparent quantum yield and R is dark respiration rate.

In closed canopies with randomly distributed leaves, the curve of PPFD attenuation versus LAI is asymptotic, and PPFD interception efficiency can be simply modelled with a Beer-Lambert law:

$$f = 1 - \exp^{-KL}\tag{25}$$

where K is radiation extinction coefficient of the canopy and L is leaf area index.

Models can be effectively used to investigate, for example, the effect of stand structure on radiation interception (e.g. Oker-Blom, 1989), of soil warming on soil respiration rate (e.g. Baldocchi, 1994), of leaf photosynthetic properties on canopy photosynthesis (e.g. Jarvis and Leverenz, 1983), and of all variables affecting the relationship between canopy photosynthetic rate and absorbed PPFD.

XI. DATA SETS, DATA ANALYSIS

A. Data Set Compilation

This study is intended first to be an exhaustive collection of data sets on CO_2 fluxes over plant canopies in relation to the radiation environments. However, because of the relatively large number of data sets on crops, not all data on crops have been reviewed. As eddy correlation measurements develop and become more widely used, especially in large field campaigns such as BOREAS (BOReal Ecosystem Atmosphere Study), many more data sets will become available and the conclusions of this study may be altered. We compiled only original data; the following data sets were not compiled in this study:

(i) data sets on photosynthetic conversion efficiency already calculated as the ratio of CO_2 flux integrated over a certain period to integrated radiation (e.g. Rauner, 1976);

(ii) data sets where canopy photosynthetic rate was modelled and not measured (e.g. Oker-Blom, 1989);

(iii) curves given in original papers as fitted curves with no data points (e.g. Uchijima, 1976).

Appendix 1 lists the data sets collected and digitized, along with relevant information for each data set that was available in the original paper. Data sets are identified by the authors and year of publication, plus a digit identifying the data sets obtained in different treatments or conditions within the same article: for instance, "Hollinger *et al.*, 1994 −1 to −5" are data sets published by Hollinger *et al.* (1994) obtained at five different times of year.

B. Vegetation Classification

The following simple functional classification of vegetation classes has been applied to the data sets: broadleaf forests, coniferous forests, C3 grasslands, C4 grasslands, C3 crops, C4 crops and mixed vegetation. Because of the disproportionate number of data sets for some vegetation classes, for some statistical procedures the data sets are also grouped into *broad vegetation classes*: forests (broadleaf and coniferous), grasslands (C3 and C4), crops (C3 and C4) and mixed vegetation. Some vegetation classes have received more attention than others: the most numerous studies are on crops, followed by forests, grasslands, and mixed vegetation. Among the crops studied, there are more C3 than C4 types; among the grasslands there are about the same amount of each; and among the forests there are more broadleaves than conifers. Overall, there is a lack of data for some vegetation classes, in particular coniferous forests, tropical grasslands and mixed vegetation.

C. Statistical Analysis of Instantaneous Data Sets

All the radiation response curves were digitized, using a digitizing table (ALTEK), and converted into the following standard units: CO_2 flux density in $\mu mol\, m^{-2} s^{-1}$ and photon flux density in $\mu mol\, m^{-2} s^{-1}$. Statistical regression analysis has been applied to all the resulting files, using the SIGMA-PLOT ® curve fitter. In all cases, the following two models have been fitted:

$$F = \alpha Q - R \qquad \text{(model L)} \quad (26)$$

$$F = \frac{\alpha Q F_\infty}{\alpha Q + F_\infty} - R \qquad \text{(model H)} \quad (27)$$

where F_∞ is F at saturating Q, α is apparent quantum yield (i.e. dF/dQ at Q = 0), and R is dark respiration rate (i.e. F at Q = 0).

We have applied the following constraints to these models: $0 < \alpha < 1$, $F_\infty > 0$, $R > 0$. If a value of R is given in the original paper, this value is used to constrain R in the statistical model. For all the regressions, the non-linear coefficient of determination, r^2, was calculated using the SIGMA-PLOT iterative procedure. The *best fit* has been defined as follows:

(i)   if the r^2 of the two models (L and H) are different at 1%, the relation-ship having the highest r^2 is the best fit;
(ii) if the two r^2 are identical at 1%, the best fit is the simplest relation-ship, i.e. L;
(iii) if F_∞ is "unrealistically" high (i.e. $F_\infty > 100 \, \mu\text{mol m}^{-2}\text{s}^{-1}$ for ambient atmospheric CO_2 concentration), L is regarded as the best fit.

To facilitate comparison among the data sets, some derived parameters have also been calculated. The *departure of the relationship from linearity* has been expressed as:

$$D = r^2 \, (\text{H}) - r^2 \, (\text{L}) \tag{28}$$

The *photosynthetic capacity* (F_m) of the vegetation has been defined as the calculated value of F at $Q = 1800 \, \mu\text{mol m}^{-2}\text{s}^{-1}$.

The results of the statistical regressions for all individual data sets are shown in Appendix 2.

D. Statistical Analysis of Grouped Data Sets

The statistical analysis described above has also been applied to *grouped data sets*, i.e. to all the data sets obtained in given conditions (for example, all the data sets obtained with micrometeorological methods; all the data sets obtained with enclosures), that satisfy the following criteria.

(i) The canopy is *closed*: forests are considered as closed canopies, unless otherwise stated; grasslands, both natural and cultivated, are con-sidered as having closed canopies if the paper explicitly mentions a leaf area index $L > 3$. This arbitrary convention may not be valid for

Table 2
Statistics on the relationship between CO_2 flux (F) and PPFD (Q).
2.1. Instantaneous data sets, F_∞, R and F_m are in μmol m^{-2} s^{-1}.

Description	n	F_∞	α	R	r^2 (best fit)	D	F_m	best fit
All data sets (Fig. 10)	1362	43·35	0·044	4·29	0·57	0·06	23·72	H
					mean			
effect of vegetation class								
Broadleaf forest	593	39·35	0·037	3·34	0·59	0·05	21·39	H
Conifer forest	125	32·37	0·024	0·00	0·34	0·10	18·50	H
C3 grasslands	68	/	0·017	0·00	0·59	0·00	30·60	L
C4 grasslands	280	84·13	0·028	8·11	0·88	0·02	23·41	H
C3 crops	476	42·85	0·062	4·03	0·65	0·14	26·93	H
C4 crops	83	88·84	0·036	4·45	0·77	0·01	33·02	H
All forests (Fig. 8A)	718	35·25	0·040	3·46	0·57	0·08	20·20	H
All grasslands (Fig. 8C)	348	82·92	0·025	5·39	0·80	0·02	23·78	H
All crops (Fig. 8B)	560	46·87	0·056	3·94	0·66	0·10	28·05	H
effect of respiration								
Net ecosystem flux (F_d) (Fig. 7A)	978	56·79	0·034	5·17	0·70	0·03	24·29	H
Canopy flux ($F_c = F_d - F_s$) (Fig. 7B)	614	34·60	0·066	4·24	0·50	0·15	22·55	H
effect of technique								
Micrometeorological (Fig. 5A)	1120	58·27	0·030	4·48	0·70	0·03	23·79	H
Enclosure (Fig. 5B)	478	44·44	0·060	2·42	0·60	0·20	29·08	H

Table 2—*cont.*

Description	n	F_∞	α	R	r^2 (best fit)	D	F_m	best fit
		effect of technique and vegetation class						
Micrometeorological, forests (Fig. 9A)	654	39·13	0·035	3·90	0·61	0·06	20·23	H
Micrometeorological, crops (Fig. 9B)	225	/	0·023	3·21	0·84	0·00	37·2 6	L
		envelope (10% upper values)						
All data sets (Fig. 10)	159	55·78	0·081	1·62	0·58	0·27	38·72	H
		effect of vegetation class						
Forests	70	47·36	0·071	0·00	0·89	0·43	34·55	H
Grasslands	35	70·09	0·044	2·90	0·85	0·07	34·28	H
Crops	56	57·30	0·084	0·00	0·68	0·44	41·55	H
		effect of respiration						
Net ecosystem flux (F_d)	94	71·13	0·046	0·00	0·90	0·14	38·26	H
Canopy flux ($F_c = F_d - F_s$)	59	51·48	0·100	0·89	0·65	0·50	39·14	H

2.2. Daily data sets, F_∞, R and F_m are in mol m^{-2} d^{-1}

Description	n	F_∞	α	R	r^2 (best fit)	D_∞	F_m	best fit
		mean						
All data sets (Fig. 13)	66	/	0·020	0·02	0·54	0·00	1·58	L

Data sets are grouped by the variable described in column 1. Statistics were applied on full data sets ("mean") or on the 10% upper boundary values ("envelope"). n is number of data points. A rectangular hyperbola (H) and a linear model (L) are tested and best fit is defined by the relationship having the highest coefficient of determination (r^2). The parameters of the statistical models are initial slope (α), intercept on the y axis (R), and for the hyperbolic model value of F at saturating Q (F_∞). Derived parameters are departure from linearity (D), defined as r^2(H)–r^2(L), and photosynthetic capacity (F_m), defined as F at maximum Q (1800 μmol m^{-2} s^{-1} for instantaneous data sets, 80 mol m^{-2} d^{-1} for daily data sets). $R = 0.00$ indicates that the intercept would be positive (therefore the value of R negative), but is constrained by the model.

some erectophile crops such as corn, where PPFD is not fully intercepted even at $L = 3$, or for vegetation with highly clumped foliage
such as some forests.

(ii) *Instantaneous* data sets: the data are half or one hour average values.
Data sets where F and Q are expressed as daily means are analysed
separately.

(iii) *Local* data sets: the data apply to a stand with a scale of metres to
hundreds of metres. Data sets obtained with aircraft-mounted eddy
correlation instruments over a large region have been analysed separately.

(iv) The following data sets have been arbitrarily discarded: data sets
showing no statistical relationship (best fit with a coefficient of determination $r^2 < 0.4$); data sets obtained in non-ambient CO_2 concentrations; data sets obtained on water-stressed vegetation; data sets
obtained in artificial environments; and data sets where F is negative
in high PPFD (i.e. $Q > 1000\,\mu\mathrm{mol\,m^{-2}\,s^{-1}}$) (Hollinger *et al.*, 1994;
Price and Black, 1990).

When instantaneous data sets are integrated over time (usually a day), they
have been analysed independently, using the same procedure, in standard
units for F, R, F_∞, Q of $\mathrm{mol\,m^{-2}\,d^{-1}}$. Photosynthetic capacity F_m is then
defined as F at $Q = 80\,\mathrm{mol\,m^{-2}\,d^{-1}}$.

Mean statistics on grouped instantaneous data sets are summarized in
Table 2.1. Mean statistics on grouped daily data sets are summarized in
Table 2.2.

One must note that grouped data sets have nothing in common except
for the variable under study. They are not *similar*, in the sense that,
generally, environmental conditions were different, and the data were
obtained on different sites, with different measuring technique, by different scientific teams.

E. Boundary Value Analysis

For grouped data sets, the *upper boundary values* of the data have been
determined as follows: a histogram has been made of the distribution of
data in eight equal size classes of PPFD: 0–250, 250–500, 500–750, 750–
1000, 1000–1250, 1250–1500, 1500–1750, 1750–2000 $\mu\mathrm{mol\,m^{-2}\,s^{-1}}$ (points
above $2000\,\mu\mathrm{mol\,m^{-2}\,s^{-1}}$ have been discarded), and in each class the
highest 10% of the CO_2 flux values have been retained. The resulting data
sets have been analysed using the statistical procedures described above,
the resulting equation being the *upper envelope* of each data set. Statistics
on boundary values of grouped data sets are summarized in Table 2.1.

XII. EFFECTS OF VARIOUS FACTORS ON THE RESPONSE FUNCTION OF CO_2 FLUX TO PPFD

A. The Micrometeorological and Enclosure Techniques

Denmead (1991) has noted that the relationship between F and Q departs further from linearity when using enclosure methods than when using micrometeorological methods. The hypotheses to explain this are:

(i) The enclosure methods increase the amount of diffuse radiation in the canopy, so that maximum assimilation is obtained at lower PPFD (Denmead, 1991). This is supported by the finding that micrometeorological methods result in a lower initial slope, whereas photosynthetic capacity determined with the two methods is similar.

(ii) Enclosures usually isolate a plant, a few plants or a mini-ecosystem, allowing more PPFD onto the sides of the canopy than in natural conditions, so that more leaves become PPFD saturated in the enclosure than in nature, especially at low solar elevation.

(iii) Micrometeorological methods integrate CO_2 fluxes over larger areas than enclosures, and spatial integration could lead to more linear PPFD response curves (see XIV.A).

Example 1. Eucalyptus forest in Australia (Denmead, 1991). Figure 4 shows that the linear model gives the best fit to the micrometeorological data, while the hyperbolic model fits the enclosure measurements best, with $r^2 = 0.96$ and 0.94, respectively. Denmead suggests that enclosures

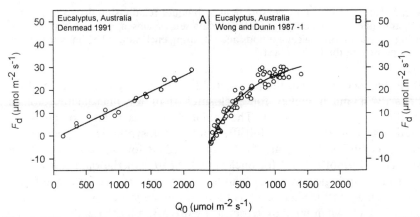

Fig. 4. Effect of CO_2 flux measurement technique: relationship between net CO_2 flux (F_d) over a Eucalyptus forest in Australia and incident quantum flux density (Q_0) measured with (A) a micrometeorological method and (B) an enclosure. The lines through the data points are the lines of best fit. Replotted from Denmead (1991) and Wong and Dunin (1987).

diffuse radiation, so that 90% of the assimilation rate is obtained at about half of the saturating PPFD. This is supported by the finding that at $Q = 1800\,\mu\text{mol}\,\text{m}^{-2}\,\text{s}^{-1}$, the CO_2 flux is similar in the two cases ($F_m = 24\cdot1$ for the micrometeorological method and $22.4\,\mu\text{mol}\ CO_2\,\text{m}^{-2}\,\text{s}^{-1}$ for the enclosure). This is the only linear PPFD response function that we found for forests that is unlikely to be an artefact (i.e. r^2 too low, or maximum Q not saturating).

Statistics on grouped data sets. We plotted all data sets on closed canopies, for all vegetation classes, distinguishing between data sets obtained using micrometeorological methods and sets obtained using enclosures (Fig. 5). The micrometeorological regression curve is closer to linearity

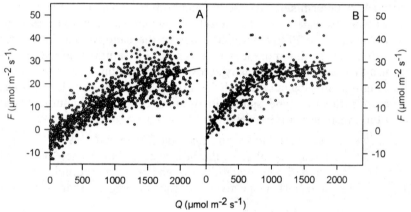

Fig. 5. Effect of CO_2 flux measurement technique: relationship between CO_2 flux (F) and quantum flux density (Q) for all data sets on closed canopies obtained (A) using micrometeorological methods and (B) using enclosures. The line through the data points are the lines of best fit.

than the enclosure one: the parameter D representing the departure from linearity is much higher for enclosures than for micrometeorological methods ($0\cdot20$ versus $0\cdot03$). The initial slope α is higher for enclosure methods ($0\cdot060$ versus $0\cdot030$), but photosynthetic capacities at $1800\,\mu\text{mol}\,\text{m}^{-2}\,\text{s}^{-1}$, F_m, are fairly similar ($23\cdot8$ for micrometeorological versus $29\cdot1\,\mu\text{mol}\,\text{m}^{-2}\,\text{s}^{-1}$ for enclosures). Scatter of the data is lower for micrometeorological methods than for enclosures ($r^2 = 0\cdot70$ versus $0\cdot60$), but this may not be significant as there are many more points for the first case. In fact, when we look closer at Figure 5B, a large number of points seem to be concentrated further from linearity than the line of best fit, saturating at PPFD approximately $1000\,\mu\text{mol}\,\text{m}^{-2}\,\text{s}^{-1}$ at a value of CO_2 flux around $25\,\mu\text{mol}\,\text{m}^{-2}\,\text{s}^{-1}$. However, some points are way off this tendency, and this increases the variability and lowers D.

In conclusion, both Example 1 and the statistics on grouped data sets show that micrometeorological methods lead to a relationship between CO_2 flux and PPFD closer to linearity than enclosures.

Artificial environments. When the environment is controlled, high PPFD can be obtained artificially, independently of other variables which are usually correlated in the field, e.g. high PPFD is usually associated with high temperature and high VPD, both of which tend to reduce CO_2 flux by either reducing photosynthetic rate or increasing respiration rate. The response of CO_2 flux to PPFD by these canopies is an indication of what the *potential* PPFD response of canopy photosynthetic rate would be.

Among the compiled data sets, several come from crops cultivated in artificial environments (Sheehy, 1977; Jones *et al.*, 1984; Baker *et al.*, 1990; Warren-Wilson *et al.*, 1992). All the curves obtained on forage grass species from Sheehy (1977) are curvilinear, but there is no indication of LAI in the paper. The relationship for a rice crop in an early stage (Baker *et al.*, 1990 −1) is also curvilinear. For the other data sets, LAI was above three and relationships are linear. Slopes of these linear relationships are fairly high ($\alpha = 0·048 \, \text{mol mol}^{-1}$ for Baker *et al.*, 1990, −2, 0·041 for Warren-Wilson *et al.*, 1992), i.e. of the same order of magnitude as the quantum efficiencies determined from rectangular hyperbolic fits, and similar to the effective quantum efficiency for a C3 leaf ($\alpha = 0·05$). In natural environments, linear relationships are usually found with lower slopes (α around 0·02). These findings suggest that firstly the response of F to PPFD for closed canopies would be close to linearity with an initial slope similar to the quantum efficiency for a leaf, if it were not for other limitations that occur at high PPFD. In natural conditions, distance from linearity could result partly from the fact that high temperature and high VPD usually occur together, so that F is further limited by both high soil respiration rate and stomatal closure. Secondly linear response functions with a slope much smaller than the leaf quantum efficiencies may be artefacts (for instance because of large data scatter).

Because data sets obtained in artificial environments are obtained in conditions that may be far from the conditions encountered in the field, they are not comparable to the field data, and have not been included in the grouped data sets.

B. Respiration

As shown above, soil respiration rate or dark respiration rate can be measured and subtracted from daily CO_2 flux to give an estimate of canopy net or gross photosynthetic flux. The expected effects on the F/Q relationship are:

(i) an upward shift of the F/Q relationship;

(ii) an increase in r^2, because hysteresis between photosynthetic and respiration rates generates scatter in the data;

(iii) a relationship further from linearity if temperature, and therefore respiration rate, increased with increasing PPFD; but a relationship closer to linearity, if hysteresis between photosynthetic and respiration rates increased the data scatter on each side of the regression curve.

Example 2. Tundra in Alaska (Coyne and Kelley, 1975) (Figure 6).

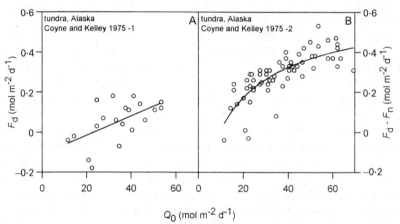

Fig. 6. Effect of dark respiration: relationship between daily CO_2 flux over tundra in Alaska, measured with a micrometeorological method and daily incident quantum flux density (Q_0). (A) net daytime ecosystem flux (F_d), (B) gross canopy photosynthesis, $A = F_d - F_n$ for daytime hours. The lines shown are the lines of best fit. Replotted from Coyne and Kelley (1975).

Daily net ecosystem flux, F_d, was measured, and dark respiration rate, F_n, was calculated from the CO_2 flux in the dark and an exponential relationship between respiration rate and temperature. Analysis of the resulting statistics indicates that subtracting F_n from F_d to get the "gross canopy photosynthesis" improves the statistical relationship ($r^2 = 0.61$ versus 0.34), transforms a linear best fit into a rectangular hyperbola, and increases the CO_2 flux ($R = 0.11$ versus 0.32, $F_m = 0.44$ versus $0.24 \, mol \, m^{-2} \, d^{-1}$).

Statistics on grouped data sets (Figure 7). We plotted all data sets for closed canopies, for all vegetation classes, distinguishing between data sets in which F_d, the ecosystem CO_2 flux, and F_c, the canopy CO_2 flux, were reported. Analysis of the statistical parameters leads to the following conclusions.

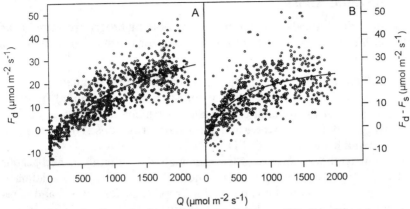

Fig. 7. Effect of soil respiration: relationship between CO$_2$ flux (F) and quantum flux density (Q) for all data sets on closed canopies. (A) net ecosystem flux (F_d), (B) above-ground flux or canopy flux, $F_c = F_d - F_s$ for daytime hours. The lines shown are the lines of best fit.

(i) The relationship between F_d and PPFD is less scattered than for F_c (r^2 = 0·70 against 0·50). This contradicts the idea that F_d, being not wholly dependent on PPFD, would create errors in PPFD response curves, as seen in Example 2.

(ii) CO$_2$ flux at $1800\,\mu\mathrm{mol\,m^{-2}\,s^{-1}}$, as well as CO$_2$ flux at $0\,\mu\mathrm{mol\,m^{-2}\,s^{-1}}$, is not significantly different in the two cases (for F_d, $F_m = 24\cdot3$ and $R = 5\cdot2$; for F_c, $F_m = 22\cdot6$, and $R = 4\cdot2\,\mu\mathrm{mol\,m^{-2}\,s^{-1}}$).

(iii) The relationship between F_d and Q is closer to linearity than the relationship between F_c and Q ($D = 0\cdot03$ versus $0\cdot15$), in agreement with Example 2.

In conclusion, subtracting respiration rate from the net ecosystem CO$_2$ flux has effects on the F/Q relationship that are difficult to analyse. In both Example 2 and the grouped data sets, subtracting respiration rates generates data sets further from linearity. In Example 2, the linearity of the response of net ecosystem flux to PPFD seems to be an artefact caused by large data scatter. For grouped data sets, however, there is more scatter in the F_c than in the F_d data. In Example 2, the effect of respiration is to lower both apparent dark respiration rate and photosynthetic capacity, while the differences are not significant for the grouped data sets. The reasons for these seemingly inconsistent findings could be that CO$_2$ flux measurements at night time and below the canopy are technically problematic and therefore not reliable, or that, for grouped data sets, made up of data sets that are not similar, comparisons are biased.

C. Vegetation Class

Looking through the data sets compiled, we note that there are more linear relationships between F and Q for crops than for forests. This may be for the following reasons.

(i) PPFD is the overwhelming factor controlling CO_2 assimilation in well-watered crops, the CO_2 flux of forests being also limited by other factors such as stomatal conductance, or much larger losses of CO_2 from the boles of the trees and particularly from soils at high temperature and hence at high PPFD.
(ii) Photosynthetic capacity of leaves is higher in general for herbaceous crops than for forest species (Larcher, 1975; Saugier, 1983), leading to higher photosynthetic capacity for crops than for forests and a relationship with PPFD closer to linearity.
(iii) The structure of grassland canopies is more erectophile, whereas canopies for most broadleaf forest species are more planophile: erectophile leaves enable the transmission of photons to the lower layers of the canopy, distributing them more evenly through the canopy, so that the canopy behaves more like a single leaf and saturates at lower PPFD.

Statistics on data sets grouped by vegetation class (Fig. 8). Because of the small number of data sets for certain vegetation classes (especially coniferous forests, C3 grasslands, C4 crops, see Table 2) the resulting statistics for these vegetation classes are less reliable than for the other vegetation classes. The data sets have, therefore, been grouped in broader vegetation classes of forests, grasslands and crops. Even so, the statistics on grasslands are less robust than for forests and crops, as there are fewer points in the data set despite the grouping of C3 and C4 vegetation classes ($n = 348$ for grassland versus 718 for forests and 560 for crops). Analysis of the regression parameters (Table 2.1) indicates that:

(i) Values of F_m decrease in the following order: crops 28·0, grasslands 23·8, forests 20·2 μmol m^{-2} s^{-1}. This agrees with our *a priori* idea that forests have higher respiration rates than stands of herbaceous vegetation because of higher respiring biomass, and that herbaceous crops have higher photosynthetic rates than forests and natural grasslands. However, this idea is not supported by the values of R, which are similar for forests and crops, and higher for natural grasslands.
(ii) Forests have a more curvilinear relationship between CO_2 flux and PPFD than grasslands (departure of relationship from linearity, $D = 0·08$ and 0·02, respectively), which is in accordance with our *a priori* knowledge on the differences between forests and grasslands, but the relationship for crops is as curvilinear as for forests ($D = 0·10$), which is surprising.

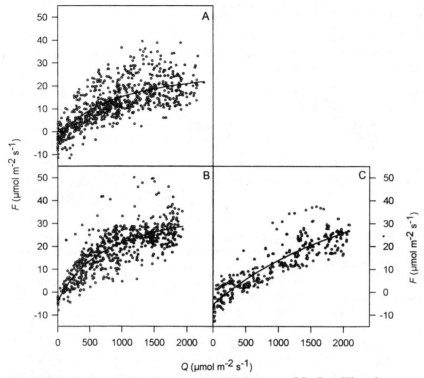

Fig. 8. Effect of vegetation class: relationship between CO_2 flux (F) and quantum flux density (Q) for all data sets on closed canopies of (A) forests, (B) crops, and (C) grasslands. The lines shown are the lines of best fit.

(iii) Grasslands have a relatively low initial slope: $\alpha = 0 \cdot 025$, while forests and crops have similar apparent quantum yields, closer to $0 \cdot 05$ mol-mol^{-1}.

In conclusion, forests and grasslands have similar photosynthetic capacities, but the relationship between CO_2 flux and PPFD is further from linearity for forests than for grasslands. Crops have a higher photosynthetic capacity than both natural vegetation classes, but do not seem to have a more linear relationship with PPFD than forests, in contradiction with the individual examples.

Statistics on data sets grouped by vegetation class and technique used to measure CO_2 flux (Fig. 9). Because the technique used to measure CO_2 flux has an effect on the shape of the F/Q relationship (see Section A), and because the CO_2 fluxes of forests have mostly been studied using micrometeorological methods, while the fluxes of crops and grasslands have more often been studied using enclosures, the statistics on data sets

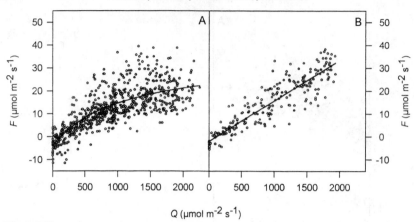

Fig. 9. Effect of vegetation type and CO_2 flux measurement technique: relationship between CO_2 flux (F) and quantum flux density (Q) for all data sets on closed canopies obtained with micrometeorological methods for (A) forests, (B) crops. The lines shown are the lines of best fit.

grouped by vegetation class could be biased. Hence, to investigate F/Q relationship independently of method of measurement, we have included only data sets obtained with micrometeorological methods and compared forests and crops. The resulting relationships are very different, with a curvilinear best fit for forests $(D = 0.06)$ and a linear best fit for crops. The apparent quantum yield is higher for forests than for crops (0.035 for forests, 0.023 mol mol^{-1} for crops). The intercepts on the y axis are quite similar (3.9 for forests, 3.2 μmol m^{-2} s^{-1} for crops): but despite forests having a higher biomass and more soil organic matter dark respiration rates are similar. However, an effect of nutrient fertilization on autotrophic and heterotrophic respiration in crops could compensate for the effect of higher carbon content of the forests. The r^2 is much higher for crops (0.84) than for forests (0.61), indicating that the forests may have been more limited than the crops by other variables besides PPFD.

In conclusion, analysing data sets obtained with micrometeorological methods only has allowed us to distinguish between the F/Q relationships of crops and forests: the CO_2 fluxes of crops have a linear relationship with PPFD, while the CO_2 fluxes of forests have a curvilinear relationship with PPFD. There are too few data sets on grasslands to analyse their behaviour in this way.

D. Other Variables

In this section we address other identified sources of variation influencing net ecosystem flux in natural conditions. Since F is largely the algebraic sum of photosynthetic and respiratory fluxes (equation 1), it is affected by all the variables affecting the individual processes. These variables have not been studied in detail in this review: whenever possible an example is given but no statistical analyses on grouped data sets have been made.

1. Seasonality

There are two classes of seasonal effects on F. Changes in leaf area index affect the amount of photons absorbed by the canopy, and the phenology of growth affects canopy CO$_2$ flux independently of LAI. Leaf area index determines firstly the area of leaves available to absorb photons, and secondly the canopy structure (i.e. arrangement of the canopy leaves), both of which determine radiation reaching the soil, multiple scattering within the canopy and the PPFD absorbed.

(a) Effect of LAI on absorption of PPFD. As LAI increases during a growing season, the fraction of incident PPFD absorbed by the canopy increases in a hyperbolic manner. The relationship between canopy fluxes (measured throughout a season with changing LAI) and absorbed PPFD is closer to linearity than the relationship between F and incident PPFD.

Example 3. Salt marsh in New York state, USA (Bartlett *et al.*, 1990). F was measured at intervals of three weeks over a whole growing season, together with PPFD transmitted by the canopy (Q_t) and incident PPFD (Q_0). The relationship between F and intercepted PPFD was linear, whereas the relationship between F and incident PPFD was curvilinear: e' defined as the ratio F/Q_0 decreased linearly with Q_0.

(b) Effect of canopy structure on CO$_2$ flux. Example 4. Baldocchi (1994) compared CO$_2$ fluxes measured over a closed canopy C3 (wheat) crop and an open canopy C4 (*corn*) crop, in Oregon, USA, on the same site and with similar techniques. Plotting F versus Q_{abs} led to apparent quantum yields of 0·026 for wheat and 0·020 for corn, in contradiction with other studies in which photosynthetic efficiency was significantly higher in C4 crops than in C3 crops. Baldocchi hypothesized that there was a bias because of the effect of LAI on radiative transfer and photosynthetic rate, which could be normalized by plotting F/L against Q_{abs}. The resulting slope of the linear relationship between CO$_2$ flux per unit leaf area and PPFD was 17% higher for the corn than for the wheat, as expected.

LAI also affects the ratio between canopy assimilation and respiration rates, because canopy closure leads to lower soil temperatures, and therefore to lower soil and root respiration rates. This effect can be taken into account by modelling respiration as a function of temperature.

(c) Phenological effects other than variations in LAI. Example 5. Wheat in

Australia (Puckridge and Ratkowski, 1971 -1, -3, -5, -7 and -10). F and Q were measured at different times throughout the season. As LAI increased steadily (0·8, 0·9, 1·2, 1·4, 1·7), photosynthetic capacity and departure of the relationship from linearity changed in an unpredictable way ($F_m = 9·6$, 16·0, 13·4, 14·4, 21·2 μmol m^{-2}s^{-1}, respectively, and $D = 0·09$, 0·40, 0·00, 0·22, 0·48, respectively).

In conclusion, to compare CO_2 fluxes measured at different times of year, it is better to relate F to absorbed PPFD rather than to incident PPFD. In order to compare CO_2 fluxes across different vegetation classes having different LAI, or CO_2 fluxes over the same vegetation class in different phenological stages having different LAI, it may be helpful to normalize F by dividing it by the LAI, i.e. express F on a leaf area basis. For simplicity this study has concentrated on closed canopies.

2. Nutrient Availability

Carbon assimilation rates of leaf and canopy depend very closely on nutrient availability. Rubisco accounts for a significant fraction of the nitrogen content of a leaf, and thus the Rubisco content of leaves is very closely related to leaf nitrogen content. Schulze et al. (1994) have related canopy CO_2 assimilation rate to leaf nitrogen content in an indirect way: they established linear relationships between maximum stomatal conductance and leaf nitrogen concentration, between canopy conductance and stomatal conductance, and between canopy photosynthetic rate and canopy stomatal conductance, all relationships having relatively high linear determination coefficients. Although nutrient availability is almost certainly one of the most important variables determining canopy photosynthetic rate, we have not been able to illustrate its role because of lack of suitable measured data.

3. Water Availability

Example 6. Tallgrass prairie in kansas, USA (Kim and Verma, 1990). F/Q relationships were given for three ranges of VPD, and for non-limiting and limiting soil water. When soil water was non-limiting, F_m was 25.5 μmol m^{-2}s^{-1} for VPD of 0 to 1·5 kPa, and 20·9 for VPD of 1·5 to 3 kPa; when soil water was limiting, F_m was 21·1 and 7·4, respectively. For the low ranges of VPD, with limiting soil water and VPD between 3–4·5 kPa, there was strong hysteresis in the F/Q relationship, some F values in high PPFD even being negative. Similar results were found by Price and Black (1990) with Douglas-fir. Jarvis (1994) described negative values of F with Sitka spruce in high PPFD and VPD, even when soil water was non-limiting.

In conclusion, soil water shortage and high VPD can reduce canopy photosynthetic rate to almost zero, particularly when acting together. An additional effect of VPD is to increase scatter in the data, because of hysteresis generated by the daily course of VPD.

In the grouped data sets, we have eliminated data sets on canopies limited by water shortage in the soil, whenever stated by the authors. However, we have not been able to investigate a possible strong VPD effect in the variability of the F/Q relationships, as most studies have not segregated their data into VPD classes.

4. Temperature

Besides the effects of temperature on both autotrophic and heterotrophic respiration, air temperature can affect the CO_2 flux over an ecosystem through effects on canopy photosynthesis, and particularly photorespiration (e.g. Long, 1991). Not many studies report the effect of temperature on canopy photosynthesis, and when reported, the effects are not significant. For example, Mordacq *et al.* (1991) report no temperature effect on the response of canopy CO_2 flux of chestnut coppice to PPFD, with mean air temperatures ranging from 19 to 30 °C.

XIII. MEAN STATISTICS, RESIDUALS AND BOUNDARY VALUE ANALYSIS FOR THE COMPLETE DATA SET

Figure 10 shows the F/Q relationships for the *complete* data set, i.e. all data sets on closed canopies satisfying the conditions for grouping data sets (see above). The resulting best fit is curvilinear:

$$F = \frac{0{\cdot}044Q \times 43{\cdot}35}{0{\cdot}044Q + 43{\cdot}35} - 4{\cdot}29 \qquad (n = 1362,\ r^2 = 0{\cdot}57) \qquad (29)$$

The residual variability can be explained in two alternative ways: either the *mean* relationship represents the response of CO_2 fluxes over canopies to PPFD, variation around the mean being attributable to experimental errors and natural variability amongst study sites, or the *upper envelope* of the points represent the response of CO_2 flux in *optimal* conditions, all the points below this envelope being the result of reductions of net ecosystem exchange by various stresses such as shortage of water or nutrients, and high temperature and VPD.

The variability in the mean F/Q relationship is strong, but not excessive, with a r^2 of nearly 0·6. Visual analysis of the figure shows that maximum F lies between 10 and 40 μmol m^{-2} s^{-1}, whereas maximum leaf photosynthetic rate per unit leaf area in optimum temperature and low VDP conditions generally lies between 2 and 25 μmol m^{-2} s^{-1} in tree species

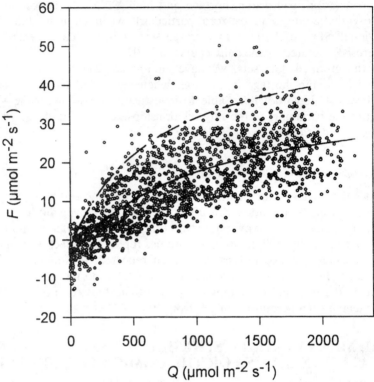

Fig. 10. Relationship between CO_2 flux (F) and quantum flux density (Q) for all data sets on closed canopies. The solid line is the line of best fit through all the data, the dotted line is the line of best fit through the 10% upper boundary values.

(Ceulemans and Saugier, 1991). We analysed the residuals of the rectangular hyperbolic fit through the complete data set. If the variability results from random error around a mean relationship, distribution of residuals would be Gaussian.

Figure 11 shows that the distribution of residuals around the mean relationship is indeed close to a Gaussian distribution. As we have shown above, taking into account factors such as the technique used to measure F or the vegetation class could reduce the variability of this relationship (i.e. increase r^2). Some variables that have not usually been measured, such as nutrient or water availability, could reduce the variability still further.

The upper envelope of the complete data set shown in Figure 10 is:

$$F = \frac{0 \cdot 081 Q \times 55.78}{0 \cdot 081 Q + 55.78} - 1 \cdot 62 \qquad (r^2 = 0 \cdot 57) \qquad (30)$$

The initial slope of this envelope curve ($0 \cdot 08 \, \mu\text{mol} \, \mu\text{mol}^{-1}$) is somewhat

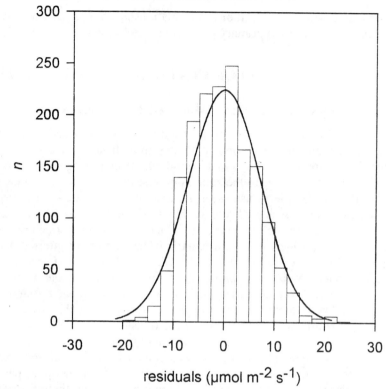

Fig. 11. The frequency of residuals of the rectangular hyperbola fitted through all data sets on closed canopies in figure 10. The line is a Gaussian relationship fitted through the histogram.

higher than typical values for the quantum yield of a C3 leaf (i.e. 0·05) and, therefore, it is not altogether realistic (an apparent quantum requirement of 12). Departure from linearity is high ($D = 0·27$) which could be the result of mixing data sets derived from micrometeorological and enclosure methods: at low F, the upper envelope is dominated by enclosure measurements, while at high F it is dominated by micrometeorological measurements.

In conclusion, the mean relationship for all data sets is an acceptable predictor of the relationship between CO$_2$ flux and PPFD, with relatively limited and randomly distributed variation around it. This single relationship could be very useful in modelling of net ecosystem productivity at regional to global scale, where information on vegetation-related parameters and reliable climate data bases are usually lacking. If we suppose that autotrophic and heterotrophic respiration affect only the value of the intercept of this relationship (which would be the case if temperature were not correlated with solar radiation), this relationship could also be adapted to the estimation of gross primary productivity at regional to global scale.

XIV. PERSPECTIVES FOR CO_2 FLUX MEASUREMENTS

A. Scaling up in Space: Plane Mounted Eddy Flux Sensors

The general relationships given above result from data obtained over supposedly homogeneous vegetation, covering areas from a scale of a few metres (enclosures) to a few tens to hundreds of metres (ground-based, eddy-flux measurements). In the context of upscaling tools such as remote sensing, it is preferable to use measurements that integrate several vegetation classes over scales of tens to hundreds of kilometres. Aircraft-mounted eddy covariance instruments, which have these characteristics, are starting to be widely used, particularly in the context of international campaigns such as FIFE, HAPEX-SAHEL and BOREAS. Large scale CO_2 fluxes can be used to do "biology from space". For example, aircraft-based CO_2 fluxes over tallgrass prairie in Kansas, USA, were found to correlate well with satellite-derived vegetation indices over the same sites, at least on a short time scale. No linear relationship was found, however, on a seasonal basis during the FIFE experiment (Cihlar et al., 1992).

We have also included two sets of measurements with aircraft-mounted instruments, which we use to investigate the effects of scaling up in space, even though there were no ground-based measurements on the same sites for comparison. The two examples are wheat fields in Canada (Desjardins, 1991), and mixed forest stands of conifers and broadleaves in Canada (Desjardins et al., 1985). In both cases, the best fit of the F/Q relationship is linear, with $r^2 = 0.72$ and 0.56, respectively. Apparent quantum yields are low ($\alpha = 0.017$ for the wheat fields, 0.014 for the mixed forest).

In conclusion, CO_2 fluxes measured over large, heterogeneous areas seem to be linearly related to PPFD. With respect to scaling up, we can alternatively regard the differences between F/Q relationships obtained with enclosures, ground-based micrometeorological measurements and aircraft-based flux measurements as a result of increase in spatial scale. We would then conclude that:

(i) The best fit becomes closer to linearity as the scale increases: it is linear for aircraft-based measurements, and closer to linearity for ground-based micrometeorological measurements ($D = 0.03$) than for enclosures ($D = 0.20$).
(ii) The apparent quantum yield decreases as the scale increases: $\alpha = 0.06$ for enclosures, 0.03 for ground-based micrometeorological measurements and around 0.015 for aircraft-based measurements.

However, scatter in the data increases at larger scales, and this could also explain points (i) and (ii).

B. Scaling up in Time: Instantaneous versus Daily Integrated Measurements

Models of canopy photosynthesis have been used to estimate daily integrated canopy photosynthetic rate and daily integrated absorbed PPFD. The resulting curves are usually closer to linearity than instantaneous curves, but not fully linear (see for example Wang *et al.*, 1992, who used the BIOMASS model to simulate a hypothetical *Pinus radiata* stand with various LAIs over an eight-year period).

Example 7. Winter wheat in Kansas, USA (instantaneous data reported by Wall and Kanemasu, 1990, daily integrals reported by Wall *et al.*, 1990) (Figure 12). The instantaneous measurements resulted in a curvilinear

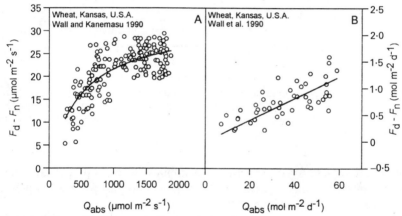

Fig. 12. Effect of scaling up in time: relationship between canopy gross photosynthesis $A = F_d - F_n$ of winter wheat in Kansas, USA measured with enclosures and absorbed quantum flux density (Q_{abs}), (A) instantaneous data, (B) data integrated over daytime hours. The lines shown are the lines of best fit. Replotted from Wall and Kanemasu (1990) and Wall *et al.* (1990).

relationship ($D = 0.11$), while the daily means resulted in a best fit to the linear model (see also Fig. 6). Even though both r^2 are low (0·59 and 0·58), we note that scaling up in time—in this case averaging instantaneous measurements over a day—linearized the PPFD response curve, and decreased the apparent quantum yield ($\alpha = 0.027 \, \text{mol mol}^{-1}$ for the daily data set, 0·065 for the instantaneous data set).

Statistics on grouped data sets. The instantaneous and daily integrated data sets have been plotted separately. The daily data sets are from CO_2

fluxes measured throughout a growing season with a changing LAI, and plotted against absorbed or intercepted radiation. These data sets are fairly rare and comprise data for herbaceous crops only: winter wheat (Wall *et al.*, 1990) and wheat (Whitfield, 1990). The resulting best fit (Fig. 13) is linear:

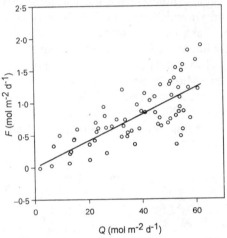

Fig. 13. Effect of scaling up in time: relationship between CO_2 flux integrated over daytime hours (F) and quantum flux density (Q) for all data sets on closed canopies (Wall *et al.*, 1990; Whitfield, 1990). The line shown is the line of best fit with $F = 0.020 \, Q$ ($n = 66$, $r^2 = 0.54$).

The quantum yield for daily data sets is about half the apparent quantum yield for instantaneous data sets ($\alpha = 0.020$ versus 0.044 for the complete, instantaneous data set, Figure 10).

In conclusion, scaling up in time tends to linearize the relationship between CO_2 flux and PPFD, and to decrease dramatically the apparent quantum yield. This is supported both by the example presented and by statistics on grouped data sets. A general, linear relationship can be derived between daily integrated F and Q, but it may not be representative, because of the small number of points and great scatter, and because all the vegetation classes are not represented.

C. CO_2 Enrichment

In the context of modelling vegetation responses to global change, it is also essential to assess the response of canopy gas exchange to increase in atmospheric CO_2. Some studies have been done using enclosures to enrich the atmosphere in CO_2 and simultaneously to measure canopy gas exchange (see Appendix 1).

Eucalyptus trees in Australia (Wong and Dunin, 1987) and herbaceous crops, e.g. soybeans in Florida, USA in controlled environment (Jones *et al.*, 1984) have been exposed to elevated CO_2 concentration. In both cases the apparent quantum yield was nearly doubled in doubled atmospheric concentration of CO_2 ($\alpha = 0.032$ in normal CO_2, 0.050 in doubled CO_2 for Eucalyptus and $\alpha = 0.049$ in normal CO_2, 0.102 in doubled CO_2 for soybean).

These experiments were done, however, on a very small scale with enclosures. It would be preferable, in the context of modelling effects of climate change, to be able to do such experiments on a large spatial scale, and over extended periods of time with fully acclimated trees, grasslands and crops.

XV. CONCLUSIONS

A. Specific Conclusions

1. Effects of Methodology

(i) Micrometeorological methods seem better adapted than enclosures to obtain relationships useful for "top down" modelling, because they integrate over a larger scale and do not disturb the environment within the canopy. Micrometeorological methods result in F/Q relationships closer to linearity than enclosure methods. Apparent quantum yields obtained with these methods were of the order of 0.02–0.03 mol mol^{-1}, and were significantly lower than the effective quantum yield of leaves of about 0.05 mol mol^{-1}.

(ii) Subtracting respiration rates from net ecosystem CO_2 fluxes to retrieve canopy CO_2 flux does not seem to improve relationships between CO_2 flux and PPFD: this could be a result of the difficulties in measuring accurately soil CO_2 flux or CO_2 flux in the dark. We would recommend that both net ecosystem flux and canopy flux be explicitly published, whenever soil or night-time CO_2 flux were measured.

(iii) The best fit through the daily data sets ($n = 66$) is linear, with an apparent quantum yield of $\alpha = 0.020$, and photosynthetic capacity of 1.58 mol m^{-2} d^{-1} at $Q = 80$ mol m^{-2} d^{-1}. The best fits through the few data sets obtained with aircraft-mounted eddy-correlation sensors are also linear. We suspect that scaling up in time and space tends to linearize the F/Q relationship, but there are too few data sets, and scatter in the data is too high, to draw a definite conclusion. As statistics on daily integrated curves would be very useful for large-scale modelling, we recommend that more data sets be presented this way, in particular in the case of long-term experiments on forests.

2. Effects of Vegetation Class

We have only been able to study the differences between broad vegetation classes (crops, grassland and forests), as there were not enough data sets to segregate them into more detailed classes. When both were measured with micrometeorological methods, crops are found to differ significantly from forests. The F/Q relationship is linear for crops and curvilinear for forests; the apparent quantum yield is 0·035 for forests and 0·023 for crops; the photosynthetic capacity is lower for forests than for crops ($F_m = 20·2$ for forest and 37·2 μmol m^{-2} s^{-1} for crops), and there is more scatter in the data for forests.

3. Effects of Environment and Season

The effects of water, nutrient availability and phenology were assessed qualitatively and show that CO_2 fluxes are very sensitive to these variables. Further statistical studies could be done using the methodology described in this paper, e.g. grouping data sets by classes of VPD, soil water content, leaf nitrogen content and phenological stage, provided that relevant information is given by authors. We would recommend that as many data as possible be explicitly published to increase understanding of canopy and stand physiology.

B. General Conclusions

The best fit through all the instantaneous data sets is a rectangular hyperbola. The apparent quantum yield ($\alpha = 0·044$) is close to the value for leaves of C3 plants of 0·05 mol mol^{-1}. Photosynthetic capacity at 1800 μmol m^{-2} s^{-1} is $F_m = 23·7 \mu$mol m^{-2} s^{-1}. The apparent dark respiration rate is 4·3 μmol m^{-2} s^{-1}, i.e. approximately 15% of photosynthetic capacity of the ecosystem, which is higher than what is found at the leaf scale (approximately 10%, e.g. Ceulemans and Saugier, 1991).

The data sets were very heterogeneous, with respect to vegetation class, environmental conditions, and methodology. This caused problems in trying to group them and analyse the grouped data sets: "CO_2 flux", F, represents either net ecosystem flux, above-ground flux or calculated canopy photosynthetic rate; "photosynthetic photon flux density", Q, is either absorbed, intercepted or incident PPFD, and was derived from PAR, solar radiation or net radiation. Besides, the techniques used to measure CO_2 flux and radiation, and the choice of presentation of data (e.g. instantaneous data, daily means or means per radiation class) vary greatly amongst authors. The general relationships obtained are likely to be biased by the data sets with the largest number of data points; this could be improved by using curve fitting weighted by the number of data points in each data set. Similarly, the comparisons of data sets grouped by factor could be biased because the different groups are formed of data sets that are not similar; this could be improved by careful sampling of data sets, so that in each group there are only strictly comparable data sets.

As a consequence, the variability within the grouped data sets is relatively important. For the complete instantaneous data set, the standard deviation of the residuals is $7 \cdot 2 \, \mu\text{mol m}^{-2} \text{s}^{-1}$, which leads to error in the estimation of F_m, calculated as the ratio of the standard deviation of the residuals to F_m, of 30%. Analysis of the residuals of the best fit of the complete instantaneous data set suggests that this variability is the result of randomly distributed error around the mean. The variability could be reduced by:

(i) reducing the effects of methodology-related errors by better standardizing data sets;
(ii) identifying data sets in optimal conditions, e.g. without constraints of water or nutrients;
(iii) characterizing the potential of each vegetation class and phenological stage;
(iv) quantifying the effects of water, nutrient and other stresses.

We recommend that CO_2 flux measurements are published whenever possible with the following supplementary information: leaf area index, radiation absorption efficiency, nitrogen content of leaves, and the occurrence of various stresses during measurements (e.g. water stress, high VPD).

The relationships obtained in this study can be used for doing "biology from space", either by applying them to the estimation of net ecosystem productivity, or by adapting them to the estimation of gross primary productivity, by making some assumptions about respiration of the total system.

The relationship through all the instantaneous data sets is far from linear and cannot reasonably be simplified to a linear relationship. Indeed, linear regression through this data set results in $\alpha = 0 \cdot 015 \, \text{mol mol}^{-1}$ ($r^2 = 0 \cdot 51$), which is unrealistically low. The relationship between daily integrated CO_2 flux and PPFD is linear, but is probably less reliable because it has been obtained from a limited data set and results in a slope of $0 \cdot 020 \, \text{mol mol}^{-1}$ ($r^2 = 0 \cdot 54$). An alternative way forward is to run the hyperbolic model with actual hourly solar radiation for some sites, and to derive a relationship between CO_2 flux and PPFD at a longer time scale, e.g. daily or monthly.

ACKNOWLEDGEMENTS

We thank all those who have assisted in this compilation by the provision of references, offprints, pre-publication and unpublished manuscripts. We sincerely hope that no authors will feel that they, or their data have been misrepresented.

REFERENCES

Allen, L.H. and Lemon, R. (1976). Carbon dioxide exchange and turbulence in a Costa Rican tropical rainforest. In: *Vegetation and the Atmosphere*. Vol. 2 Case Studies. (Ed. by J.L. Monteith), pp. 265–308. Academic Press, London, New York, San Francisco.

Amthor, J.S., Goulden, M.L., Munger, J.W. and Wofsy, S.C. (1994). Testing a mechanistic model of forest-canopy mass and energy exchange using eddy correlation: carbon dioxide and ozone uptake by a mixed oak-maple stand. *Aust. J. Plant Physiol.* **21**, 623–651.

Anderson, D.E. and Verma, S.B. (1986). Carbon dioxide, water vapor and sensible heat exchanges of a grain sorghum canopy. *Bound-Lay. Meteorol.* **34**, 317–331.

Anderson, D.E., Verma, S.B. and Rosenberg, N.J. (1984). Eddy correlation measurements of CO_2, latent heat, and sensible heat fluxes over a crop surface. *Bound-Lay. Meteorol.* **29**, 263–272.

Baker, J.T., Allen, L.H., Boote, K.J., Jones, P. and Jones, J.W. (1990). Rice photosynthesis and transpiration in subambient and superambient CO_2 concentration. *Agron. J.* **82**, 834–840.

Baldocchi, D.D. (1994). A comparative study of mass and energy exchange rates over a closed C_3 (wheat) and an open C_4 (corn) crop: II. CO_2 exchange and water use efficiency. *Agr. Forest Meteorol.* **67**, 291–321.

Baldocchi, D.D., Verma, S.B. and Rosenberg, N.J. (1981a). Environmental effects on the CO_2 flux and CO_2–water flux ratio of alfalfa. *Agr. Forest Meteorol.* **24**, 175–184.

Baldocchi, D.D., Verma, S.B. and Rosenberg, N.J. (1981b). Mass and energy exchanges of a soybean canopy under various environmental regimes. *Agron. J.* **73**, 706–710.

Baldocchi, D.D., Verma, S.B. and Anderson, D.E. (1987). Canopy photosynthesis and water-use efficiency in a deciduous forest. *J. Appl. Ecol.* **24**, 251–260.

Bartlett, D.S., Whiting, G.J. and Hartman, J.M. (1990). Use of vegetation indices to estimate intercepted solar radiation and net carbon dioxide exchange of a grass canopy. *Remote Sens. Environ.* **30**, 115–128.

Belanger, G. (1990). Incidence de la fertilisation azotée et de la saison sur la croissance, l'assimilation et la répartition du carbone dans un couvert de fétuque élevée en conditions naturelles. PhD. thesis, Université de Paris-Sud, Orsay, 168 pp.

Biscoe, P.V., Scott, R.K. and Monteith, J.L. (1975). Barley and its environment. III: carbon budget of the stand. *J. Appl. Ecol.* **12**, 269–293.

Bowen, I.S. (1926). The ratio of heat losses by conduction and by evaporation from any water surface. *Phys. Rev.* **27**, 779–787.

Breda, N. (1994). Analyse du fonctionnement hydrique des chênes sessile (*Quercus *petraea*) et pédonculé (*Quercus robur*) en conditions naturelles; effets des facteurs du milieu et de l'éclaircie. PhD. thesis, Université de Paris-Sud, Orsay, 233 pp.

Brunet, Y., Berbigier, P. and Daudet, F.A. (1992). Carbon dioxide exchanges between a temperate pine forest and the atmosphere: Landes project. Poster presented at the first IGAC scientific connference, Eilat, Israël, 18–22 April 1992.

Ceulemans, R.J. and Saugier, B. (1991). Photosynthesis. In: *Physiology of Trees*. (Ed. by A.S. Raghavendra), pp. 21–50. John Wiley and Sons, New York.

Cihlar, J., Caramori, P.H., Schjuepp, P.H., Desjardins, R.L. and MacPherson, J.I. (1992). Relationship between satellite-derived vegetation indices and aircraft-based CO$_2$ measurements. *J. Geophys. Res.* 97(D17), 18515–18521.

Coyne, P.I., and Kelley, J.J. (1975). CO$_2$ exchange over the Alaskan arctic tundra: meteorological assessment by an aerodynamic method. *J. Appl. Ecol.* **12**, 587–611.

den Hartog, G., Neumann, H.H. and King, K.M. (1987). Measurements of ozone, sulphur dioxide and carbon dioxide fluxes to a deciduous forest. *18th Conf. Agr. Forest Meteorol., AMS, W. Lafayette, Indiana 1987*, 206–209.

Denmead, O.T. (1976). Temperate cereals. In: *Vegetation and the Atmosphere*. Vol 2 Case Studies (Ed. by J.L. Monteith), pp. 1–31. Academic Press, London New York, San Francisco.

Denmead, O.T. (1991). Sources and sinks of greenhouse gases in the soil-plant environment. *Vegetatio* **91**, 73–86.

Desjardins, R.L. (1991). Review of techniques to measure CO$_2$ fluxes densities from surface and airborne sensors. *Adv. Bioclimatol.* **1**, 1–25.

Desjardins, R.L., Buckley, D.J. and St. Amour G. (1984). Eddy flux measurements of CO$_2$ above corn using a microcomputer system. *Agr. Forest Meteorol.* **32**, 257–265.

Desjardins, R.L., McPherson, J.L., Alvo, P. and Schuepp, P.H. (1985). Measurements of turbulent heat and CO$_2$ exchange over forests from aircraft. In: *The Forest-Atmosphere Interaction*. (Ed. by B.A. Hutchison and B.B. Hicks), pp. 645–658. D. Reidel Publishing Company, Dordrect, Holland.

de Wit, C.T. *et al.* (1978). *Simulation of Assimilation, Respiration and Transpiration of Crops*. Pudoc, Wageningen, 141 pp.

Drake, B.G. (1984). Light response characteristics of net CO$_2$ exchange in brackish wetland plant communities. *Oecologia* **63**, 263–270.

Duncan, W.G., Loomis, R.S., Williams, W.A. and Hanau, R. (1967). A model for simulating photosynthesis in plant communities. *Hilgardia* **38**, 181–205.

Eckardt, F.E. (1966). Le principe de la soufflerie climatisée, appliqué à l'étude des échanges gazeux de la couverture végétale. *Oecologia Plantarum* **1**, 369–400.

Ehleringer, J. and Pearcy, R.W. (1983). Variation in quantum yield for CO$_2$ uptake among C$_3$ and C$_4$ plants. *Plant Physiol.* **73**, 555–559.

Fan, S-M., Wofsy, S.C., Bakwin, P.S. and Jacob, D.J. (1990). Atmosphere-biosphere exchange of CO$_2$ and O$_3$ in the central Amazon forest. *J. Geophys. Res.* **95**(D10), 16851–16864.

Goward, S.N. and Huemmrich, K.F. (1992). Vegetation canopy PAR absorptance and the normalized difference vegetation index: an assessment using the SAIL model. *Remote Sens. Environ.* **39**, 119–140.

Heimann, M. and Keeling, C.D. (1989). A three-dimensional model of atmospheric CO$_2$ transport based on observed winds. 2. Model description and simulated tracer experiments. In: *Aspects of Climate Variability in the Pacific and the*

Western Americas, Geophysical Monographs 55. (Ed. by D.H. Peterson), pp. 237–274.

Hollinger, D.Y., Kelliher, F.M., Byers, J.N., Hunt, J.E., McSeveny, T.M. and Weir, P.L. (1994). Carbon dioxide exchange between an undisturbed old-growth temperate forest and the atmosphere. *Ecology* **75**, 134–150.

Houghton, R.A. and Woodwell, G.M. (1980). The flax pond ecosystem study: exchanges of CO_2 between a salt marsh and the atmosphere. *Ecology* **61**, 1434–1445.

Hutchison, B.A. and Baldocchi, D.D. (1989). Analysis of biogeochemical cycling processes in walker branch watershed. In: *Forest Meteorol.* Chap. 3. (Ed. by D.W. Johnson and R.I. Hook), pp. 22–95. Springer-Verlag, New York.

Inoue, E. (1965). On the CO_2-concentration profiles within plant canopies. *J. Agr. Meteorol. of Japan* **20**, 141–146.

Jarvis P.G. (1994). Capture of carbon dioxide by a coniferous forest. In: *Resource Capture by Forest Crops.* (Ed. by J.L. Monteith, R.K. Scott and M.H. Unsworth), pp. 351–374. Nottingham University Press.

Jarvis, P.G., James, G.B. and Landsberg, J.J. (1976). Coniferous forests. In: *Vegetation and the Atmosphere.* Vol 2. Case Studies. (Ed. by J.L. Monteith), pp. 171–240. Academic Press, London, New York, San Francisco.

Jarvis, P.G. and Leverenz, J.W. (1983). Productivity of temperate, deciduous and evergreen forests. In: *Encyclopaedia of Plant Physiology*, new series Vol. 12D. (Ed. by O.L. Lange, P.S. Nobel, C.B. Osmond and H. Ziegler), pp. 233–280. Springer-Verlag, Berlin.

Jones, J.W., Zur, B. and Bennett, J.M. (1986). Interactive effects of water and nitrogen stresses on carbon and water vapor exchange of corn canopies. *Agr. Forest Meteorol.* **38**, 113–126.

Jones, P., Allen, L.H., Jones, K.W., Boote, K.J. and Campbell, W.J. (1984). Soybean canopy growth, photosynthesis and transpiration responses to whole-season carbon dioxide enrichment. *Agron. J.* **76**, 633–637.

Kelliher, F.M., Hollinger, D.Y. and Whitehead, D. (1989). Water vapour and carbon dioxide exchange in a beech forest. *New Zealand Society of Plant Physiologists meeting, 1989*, University of Otago, Dunedin, New Zealand.

Kim, J. and Verma, S.B. (1990). Carbon dioxide exchange in a temperate grassland ecosystem. *Bound-Lay. Meteorol.* **52**, 135–149.

Kumar, M. and Monteith, J.L. (1981). Remote sensing of crop growth. In: *Plants and the Daylight Spectrum.* (Ed. by H. Smith), pp. 133–144. Academic Press, London.

Larcher, W. (1975). *Physiological Plant Ecology.* Springer-Verlag, Berlin, pp. 35.

Lemon, E.R. (1960). Photosynthesis under field conditions. II. *Agron. J.* **52**, 697–703.

LeRoux, X. and Mordelet, P. (1995). Leaf and canopy CO_2 assimilation in a West African homid savanna during the early growing season. *J. Trop. Ecol.* **11** (in press).

Long, S.P. (1991). Modification of the response of photosynthetic productivity to rising temperature by atmospheric CO_2 concentration: has its importance been underestimated? *Plant Cell Environ.* **14**, 657–677.

McCree, K.J. (1972). Test of current definitions of photosynthetically active radiation against leaf photosynthesis data. *Agr. Meteorol.* **10**, 442–453.

McCree, K.J. (1974). Equation for the rate of dark respiration of white clover as function of dry weight, photosynthesis rate and temperature. *Crop Sci.* **14**, 509–514.

McNaughton, K.G. (1988). Regional interactions between canopies and the atmosphere. In: *Plant Canopies: their Growth, Form and Functions.* (Ed. by G. Russel, B. Marshall and P.G. Jarvis), pp. 63–81. Cambridge University Press.

Monin, A.S. and Yaglom, A.M. (1971). *Statistical Fluid Mechanics: Mechanics of Turbulence.* M.I.T. Press. 769 pp.

Monteith, J.L. (1962). Measurement and interpretation of carbon dioxide fluxes in the field. *Neth. J. Agr. Sci.* **10**, 334–346.

Monteith, J.L. (1965). Light distribution and photosynthesis of field crops. *Ann. Bot.* **29**, 17–37.

Monteith, J.L. (1972). Solar radiation and productivity in tropical ecosystems. *J. Appl. Ecol.* **9**, 747–766.

Monteith, J.L. (1977). Climate and the efficiency of crop production in Britain. *Philos. T. Roy. Soc. Lon. B.* **281**, 277–294.

Monteith, J.L. and Szeicz, G. (1960). The carbon-dioxide flux over a field of sugar beet. *Q. J. Roy. Meteor. Soc.* **86**, 205–214.

Monteny, B.A. (1989). Primary productivity of a *Hevea* forest in the Ivory Coast. *Ann. Sci. Forest.* **46** (suppl.), 502s–505s.

Mordacq, L., Ghasghaie, J. and Saugier, B. (1991). A simple field method for measuring the gas exchange of small trees. *Funct. Ecol.* **5**, 572–576.

Morgan, J.A. and Brown, R.H. (1983). Photosynthesis and growth of Bermuda grass swards. I. Carbon dioxide exchange characteristics of swards mowed at weekly and monthly intervals. *Crop Sci.* **23**, 347–352.

Moss, D.N., Musgrave, R.B. and Lemon, E.R. (1961). Photosynthesis under field conditions. III. Some effects of light, carbon dioxide, temperature, and soil moisture on photosynthesis, respiration, and transpiration of corn. *Crop Sci.* **1**, 83–87.

Musgrave, R.B. and Moss, D.N. (1961). Photosynthesis under field conditions. I. A portable, closed system for determining net assimilation and respiration of corn. *Crop Sci.* **1**, 37–41.

Ohtaki, E. (1980). Turbulent transport of carbon dioxide over a paddy field. *Bound-Lay. Meteorol.* **19**, 315–336.

Oker-Blom, P. (1989). Relationship between radiation interception and photosynthesis in forests canopies: effect of stand structure and latitude. *Ecol. Model.* **49**, 73–87.

Osmond, C.B., Björkman, O. and Anderson, D.J. (1980). Photosynthesis. In: *Physiological Processes in Plant Ecology.* (Ed. by C.B. Osmond, O. Björkman and D.J. Anderson), pp. 291–377. Springer-Verlag, Berlin, Heidelberg, New York.

Pettigrew, W.T., Hesketh, J.D., Peters, D.B. and Wooley, J.T. (1990). A vapor pressure deficit effect on crop canopy photosynthesis. *Photosynth. Res.* **24**, 27–34.

Price, D.T. and Black, T.A. (1990). Effects of short-term variation in weather on diurnal canopy CO$_2$ flux and evapotranspiration of a juvenile Douglas-fir stand. *Agr. Forest Meteorol.* **50**, 139–158.

Puckridge, D.W. (1971). Photosynthesis of wheat under field conditions. III. Seasonal trends in carbon dioxide uptake of crop communities. *Austr. J. Agr. Res.* **22**, 1–10.

Puckridge, D.W. and Ratkowski, D.A. (1971). Photosynthesis of wheat under field conditions. IV. The influence of density and leaf area index on the response to radiation. *Austr. J. Agr. Res.* **22**, 11–20.

Rauner, J.L. (1976). Deciduous forests. In: *Vegetation and the Atmosphere.* Vol. 2

Case Studies. (Ed. by J.L. Monteith), pp. 241–264. Academic Press, London, New York, San Francisco.

Raupach, M.R., Denmead, O.T. and Dunin, F.X. (1992). Challenges in linking surface CO_2 and energy fluxes to larger scales. *Austr. J. Bot.* **40**, 697–716.

Ruimy, A., Dedieu, G. and Saugier, B. (1994). Methodology for the estimation of terrestrial net primary production from remotely sensed data. *J. Geophys. Res.* **99**(D3), 5263–5283.

Saeki, T. (1960). Interrelationships between leaf amount, light distribution and total photosynthesis in a plant community. *Bot. Mag. Tokyo* **73**, 55–63.

Saugier, B. (1970). Transports turbulents de CO_2 et de vapeur d'eau au dessus et à l'intérieur de la végétation. Méthodes de mesure micrométéorologiques. *Oecologia Plantarum* **5**, 179–223.

Saugier, B. (1983). Aspects écologiques de la photosynthèse. *B. Soc. Bot. Fr.* **130**(1), 113–128.

Saugier, B. (1986). Productivité des écosystèmes naturels. *Biomasse Actualités* **9**, 42–49.

Schulze, E.D., Kelliher, F.M., Körner, C., Lloyd, J. and Leuning, R. (1994). Relationships between maximum stomatal conductance, ecosystem surface conductance, carbon assimilation rate and plant nitrogen nutrition: a global ecology scaling exercise. *Annu. Rev. Ecol. Syst.* **25**, 629–660.

Sellers, P.J. (1985). Canopy reflectance, photosynthesis and transpiration. *Int. J. Remote Sens.* **6**, 1335–1372.

Sellers, W.D. (1965). *Physical Climatology*. The University of Chicago Press. 270 pp.

Sheehy, J.E. (1977). Microclimate, canopy structure and photosynthesis in canopies of three contrasting forage grasses. III. Canopy photosynthesis, individual leaf photosynthesis and the distribution of current assimilates. *Ann. Bot.* **41**, 593–604.

Sinoquet, H. (1993). Modelling radiative transfer in heterogeneous canopies and intercropping systems. In: *Crop structure and Light Microclimate. Characterisation and Applications*. (Ed. by C. Varlet-Grancher, R. Bonhomme and H. Sinoquet), pp. 229–252. INRA. Paris.

Tanner, C.B. (1968). Evaporation of water from plants and soil. In: *Water Deficits and Plant Growth Vol 1. Development, Control and Measurement*. (Ed. by T.T. Kozlowski), pp. 74–106. Academic Press, New York and London.

Thomas, M.D. and Hill, G.R. (1949). 2. Photosynthesis under field conditions. In: *Photosynthesis in Plants*. (Ed. by J. Franck and W.E. Loomis). A monograph of the American Society of Plant Physiologists, The Iowa State College Press, Ames, Iowa.

Uchijima, Z. (1976). Maize and rice. In: *Vegetation and the Atmosphere*. Vol. 2 Case Studies. (Ed. by J.L. Monteith), pp. 33–64. Academic Press, London, New York, San Francisco.

Valentini, R., Scarascia Mugnozza, G.E., DeAngelis, P. and Bimbi, R. (1991). An experimental test of the eddy correlation technique over a Mediterranean macchia canopy. *Plant Cell Environ.* **14**, 987–994.

Varlet-Grancher, C. (1982). Analyse du rendement de la conversion de l'énergie solaire par un couvert végétal. PhD. thesis, Université Paris-Sud, Orsay, 144 pp.

Varlet-Grancher, C., Chartier, M., Gosse, G. and Bonhomme, R. (1981). Rayonnement utile pour la photosynthèse des végétaux en conditions naturelles: caractérisation et variations. *Oecol. Plant.* **2**(16), 189–202.

Varlet-Grancher, C., Gosse, G., Chartier, M., Sinoquet, H., Bonhomme, R. and

Allirand, J.M. (1989). Mise au point: rayonnement solaire absorbé ou inter-cepté par un couvert végétal. *Agronomie* **9**, 419–439.

Verma, S.B. and Rosenberg, N.J. (1976). Carbon dioxide concentration and flux in a large agricultural region of the great plains of North America. *J. Geophys. Res.* **81**, 399–405.

Verma, S.B., Baldocchi, D.D., Anderson, D.E., Matt, D.R. and Clement, R.J. (1986). Eddy fluxes of CO$_2$ water vapour and sensible heat over a deciduous forest. *Bound-Lay. Meteorol.* **36**, 71–96.

Verma, S.B., Kim, J. and Clement, R.J. (1989). Carbon dioxide, water vapor and sensible heat fluxes over a tallgrass prairie. *Bound-Lay. Meteorol.* **46**, 53–67.

Villar, R., Held, A.A. and Merino, J. (1994). Comparison of methods to estimate dark respiration in the light in leaves of two woody species. *Plant Physiol.* **105**, 167–172.

Wall, G.W. and Kanemasu, E.T. (1990). Carbon dioxide exchange rates in wheat canopies. Part 1. Influence of canopy geometry on trends in leaf area index, light interception and instantaneous exchange rates. *Agr. Forest Meteorol.* **49**, 81–102.

Wall, G.W., Lapitan, R.L., Asrar, G. and Kanemasu, E.T. (1990). Spectral esti-mates of daily carbon dioxide exchange rates in wheat canopies. *Agron. J.* **82**(4), 819–826.

Wang, Y.P. and Jarvis, P.G. (1990). Description and validation of an array model—MAESTRO. *Agr. Forest Meteorol.* **51**, 257–280.

Wang, Y.-P., McMurtrie, R.E. and Landsberg, J.J. (1992). Modelling canopy photosynthetic productivity. In: *Crop Photosynthesis: Spatial and Temporal Determinants*. (Ed. by N.R. Baker and H. Thomas), pp. 43–67. Elsevier, Oxford.

Warren-Wilson, J., Hand, D.H. and Hannah, M.A. (1992). Light interception and photosynthetic efficiency in some glasshouse crops. *J. Exp. Bot.* **43**, 363–373.

Whitfield, D.M. (1990). Canopy conductance, carbon assimilation and water use in wheat. *Agr. Forest Meteorol.* **53**, 1–18.

Whiting, G.J., Bartlett, D.S., Fan, S-M., Bakwin, P.S., and Wofsy, S.C. (1992). Biosphere/atmosphere CO$_2$ exchange in tundra ecosystems: community charac-teristics and relationships with multispectral reflectance. *J. Geophys. Res.* **97**(D15), 16671–16680.

Wofsy, S.C., Goulden, M.L., Munger, J.W., Fan, S-M., Bakwin, P.S. and Daube, B.C. (1993). Net exchange of CO$_2$ in midlatitude forests. *Science* **260**, 1314–1317.

Wong, S.C. and Dunin, F.X. (1987). Photosynthesis and transpiration of trees in a *Eucalyptus* forest stand: CO$_2$, light and humidity responses. *Austr. J. Plant Physiol.* **14**, 619–632.

Woodwell, G.M. and Dykeman, W.R. (1966). Respiration of a forest measured by carbon dioxide accumulation during temperature inversion. *Science* **154**, 1031–1034.

APPENDIX 1: DESCRIPTION OF DATA SETS
COLLECTED AND DIGITIZED, CLASSIFIED BY
VEGETATION KIND, AND IN EACH CLASS BY
ALPHABETICAL ORDER OF AUTHORS

column 1: Kind of vegetation, characterized by English name of species (when monospecific) or ecosystem type (when composed of several species: e.g. rainforest).

column 2: Location, characterized by the country, and sometimes the region.

column 3: Indication of phenological stage or season when available: early growth (" ↑ ") / peak growth ("→") / senescence (" ↓ ") (usually characterized by the seasonal course of leaf area index), or "spring/summer/autumn/winter", or day of year, and year of measurement.

column 4: Leaf area index (L) of the period of measurement (in $m^2 m^{-2}$).

column 5: Techniques of measurement of CO_2 flux: micrometeorological methods: energy balance ("en. balance"), aerodynamic ("gradient"), eddy correlation ("eddy corr"); canopy enclosure techniques ("enclosure"). When two sets of eddy correlation instruments have been used, one above and one below the canopy, this is indicated in this column ("above-below"). For enclosure measurements, there is usually no indication in the original article of which portion of the ecosystem was enclosed. Otherwise, the part of the ecosystem enclosed is mentioned in this column ("ecosystem/above-ground").

column 6: Measure of radiation reported in the original paper: photosynthetically active radiation or photosynthetic photon flux density (Q), global radiation ("S_g"), or net radiation ("S_n"). Radiation is either incident (subscript: "0"), intercepted by the canopy (subscript: "int") or absorbed by the canopy (subscript: "abs").

column 7: Respiration measurement mode: whether soil respiration ("F_s") and/or night time respiration ("F_n") were measured, and whether these measurements have already been subtracted from the reported CO_2 flux measurements ("subt.", or "not subt."). When CO_2 storage in the canopy is accounted for, "ΔS_c" is mentioned in this column.

column 8: Time-step of the CO_2 flux measurements, and duration of the experiment. Several cases can occur: instantaneous, raw measurements are reported; instantaneous measurements are averaged over periods varying from 15 minutes to one week; data sets that do not report raw measurements, but report mean CO_2 fluxes in classes of solar radiation ("means/classes").

column 9: Treatment, conditions; this column is used to indicate the different treatment or conditions, between two or more data sets reported in the same article. ppmv $\equiv \mu$mol mol^{-1}.

column 10: Identification code of the data sets: names of authors, year, curve number.

Appendix 1—*cont.*

Vegetation class	Location	Phenology date	L	Flux measurements	Radiation measurements	Respiration measurements	Time-step, duration	Treatment, conditions	Reference
broadleaf forests									
Rainforest	Costa Rica	Nov 14 67	?	en. balance	Q_0	?	½ hour; 1 day		Allen & Lemon 1976
Mixed oak, hickory	Tennessee, US	summer 92	≈ 5	eddy corr.	Q_0	F_s (subt.)	½ hour; 1 month		Baldocchi & Harley (unpublished data)
Mixed oak, hickory	Tennessee, US	Jul 24–Aug 10 84	4.9	eddy corr. (above–below)	Q_0	F_s (subt.)	½ hour		Baldocchi et al. 1987
Deciduous forest	?	Sep 3 86	?	?	$S_{n,0}$	?	½ hour; 1 day		den Hartog et al. 1987
Eucalyptus	New S. Wales Australia	Nov 16–Mar 16	3.3	gradient	$S_{g,0}$	no	?		Denmead 1991
Rainforest	Manaus, Brazil	Apr 22–May 8 87	≈ 7	eddy corr.	$S_{g,0}$	F_s (not subt.) ΔS_c	1 hour		Fan et al. 1990
Rainforest	Rondonia, Brazil	Sep 92	?	eddy corr.	$S_{g,0}$	F_s (subt.)	?		Grace (unpublished data)
Evergreen beech	New Zealand	Jul (winter) 89	5	eddy corr.	Q_0	F_n (not subt.) ΔS_c	½ hour; several days		Hollinger et al. 1994–1
Evergreen beech	New Zealand	Sep (early spring) 89	5	eddy corr.	Q_0	F_n (not subt.) ΔS_c	½ hour; several days		Hollinger et al. 1994–2
Evergreen beech	New Zealand	Dec (late spring) 89	6.8	eddy corr.	Q_0	F_n (not subt.) ΔS_c	½ hour; several days		Hollinger et al. 1994–3
Evergreen beech	New Zealand	Jan (summer) 90	7.0	eddy corr.	Q_0	F_n (not subt.) ΔS_c	½ hour; several days		Hollinger et al. 1994–4
Evergreen beech	New Zealand	Mar (late summer) 90	6.3	eddy corr.	Q_0	F_n (not subt.) ΔS_c	½ hour; several days		Hollinger et al. 1994–5
Evergreen beech	New Zealand	Jul 89–Mar 90	6.3	eddy corr.	Q_0	F_n (not subt.) ΔS_c	½ hour; several days		Hollinger et al. 1994–6
Evergreen beech	New Zealand	Feb. 18–19 89	≈ 5	eddy corr.	Q_0	no	1 hour		Kelliher et al. 1989
Hevea (young)	Abidjan, Ivory Coast	Apr–May	?	en. balance	$S_{g,0}$	F_s (not subt.)	means/classes, several days	no water stress	Monteny 1989–1
Hevea (young)	Abidjan, Ivory Coast	Apr–May	?	en. balance	$S_{g,0}$	F_s (not subt.)	means/classes, several days	water stress	Monteny 1989–2

Hevea (old)	Abidjan, Ivory Coast	Dec–Jan	3·5–4·5	en. balance	$S_{g,0}$	F_s (not subt.)	means/classes, several days	no water stress	Monteny 1989–3
Chestnut coppice	Paris, France	Aug 21–Sep 9 84	≈ 4	enclosure	Q_0	F_s (subt.)	4 days		Mordacq et al. 1991
Macchia	Italy	Nov 11–14 89	4·5	eddy corr.	Q_0	no	1 hour 3 days		Valentini et al. 1991
Mixed oak, hickory	Tennessee, US	Aug 2–Aug 9	4·9	eddy corr.	$S_{g,0}$	no	?	VPD 0–0·6 kPa	Verma et al. 1986–1
Mixed oak, hickory	Tennessee, US	Aug 2–Aug 9	4·9	eddy corr.	$S_{g,0}$	no	?	VPD 0·6–1·2 kPa	Verma et al. 1986–2
Mixed oak, hickory	Tennessee, US	Aug 2–Aug 9	4·9	eddy corr.	$S_{g,0}$	no	?	VPD 1·2–1·8 kPa	Verma et al. 1986–3
Mixed deciduous forest	Massachussets, US	Sep 18–27 81	3·5	eddy corr.	Q_0	F_n, F_s (not subt.) ΔS_c	?	afternoon	Wofsy et al. 1993–1
Mixed deciduous forest	Massachussets, US	Sep 18–27 81	3·5	eddy corr.	Q_0	F_n, F_s (not subt.) ΔS_c	?	morning	Wofsy et al. 1993–2
Eucalyptus	New S. Wales Australia	Mar 12–Apr 4 84	3·3	enclosure	$S_{g,0}$	?	?	ambient CO_2	Wong & Dunin 1987–1
Eucalyptus	New S. Wales Australia	Mar 12–Apr 4 84	3·3	enclosure	$S_{g,0}$	?	?	double CO_2	Wong & Dunin 1987–2

conifer forests

Maritime pine	Les Landes, France	summer 91 or 92	2·5–3·5	eddy corr.	Q_0	F_n (not subt.)	½ hour ? several days ?	no water stress ?	Brunet et al. 1992
Sitka spruce	Fetteresso, Scotland	summer 70	?	en. balance	Q_0	F_s (enclosure) (not subt.)	means/classes	VPD 0–0·2 kPa	Jarvis 1994–1
Sitka spruce	Fetteresso, Scotland	summer 70	?	en. balance	Q_0	F_s (enclosure) (not subt.)	means/classes	VPD 0·2–0·6 kPa	Jarvis 1994–2
Sitka spruce	Fetteresso, Scotland	summer 70	?	en. balance	Q_0	F_s (enclosure) (not subt.)	means/classes	VPD 0·6–0·8 kPa	Jarvis 1994–3
Sitka spruce	Fetteresso, Scotland	summer 70	?	en. balance	Q_0	F_s (enclosure) (not subt.)	means/classes	VPD 0·8–1 kPa	Jarvis 1994–4
Sitka spruce	Fetteresso, Scotland	summer 70	?	en. balance	Q_0	F_s (enclosure) (not subt.)	means/classes	VPD 1–2 kPa	Jarvis 1994–5
Douglas-fir	British Columbia, Canada	Jul–Aug 84	5·0	en. balance	Q_0	no	½ hour 1 day	July 29	Price & Black 1990–1
Douglas-fir	British Columbia, Canada	Jul–Aug 84	5·0	en. balance	Q_0	no	½ hour 1 day	July 24	Price & Black 1990–2
Douglas-fir	British Columbia, Canada	Jul–Aug 84	5·0	en. balance	Q_0	no	½ hour 1 day	Aug 6 (cloudy)	Price & Black 1990–3

Appendix 1—*cont.*

Vegetation class	Location	Phenology date	L	Flux measurements	Radiation measurements	Respiration measurements	Time-step, duration	Treatment, conditions	Reference
					C3 grasslands				
Fescue	Lusignan, France	day 16 ↑	0.79	enclosure	Q_0	F_n (subt.)	¼ hour 1 day	N = 0 kg/ha	Belanger 1990–1
Fescue	Lusignan, France	day 16 ↑	2.44	enclosure	Q_0	F_n (subt.)	¼ hour 1 day	N = 80 kg/ha	Belanger 1990–2
Fescue	Lusignan, France	day 16 ↑	3.57	enclosure	Q_0	F_n (subt.)	¼ hour 1 day	N = 240 kg/ha	Belanger 1990–3
Tundra	Barrow, Alaska	growing season 71	?	gradient	$S_{g,0}$	no	1 day 3 month		Coyne & Kelley 1975–1
Tundra	Barrow, Alaska	growing season 71	?	gradient	$S_{g,0}$	F_n (subt.)	1 day 3 month		Coyne & Kelley 1975–2
Mixed-grass prairie	Matador, Canada	Jul	0-9	en. balance + gradient	$S_{g,0}$	$F_n F_s$ (not subt.)	mean/classes 5 days		Ripley & Saugier–1 (unpublished data)
Mixed-grass prairie	Matador, Canada	Jun	0-8	en. balance + gradient	$S_{g,0}$	$F_n F_s$ (not subt.)	mean/classes 5 days		Ripley & Saugier–2 (unpublished data)
Mixed-grass prairie	Matador, Canada	May	0-4	en. balance + gradient	$S_{g,0}$	$F_n F_s$ (not subt.)	mean/classes 5 days		Ripley & Saugier–3 (unpublished data)
Dactylis	Montpellier, France	?	6-0	?	Q_0	no	?		Saugier 1986
Ryegrass	Berks., UK	day 36	?	enclosure	$S_{g,0}$ (artificial)	F_n (not subt.)	1 day	reerected after storm	Sheehy 1977–1
Ryegrass	Berks., UK	day 35	?	enclosure	$S_{g,0}$ (artificial)	F_n (not subt.)	1 day	lodged after storm	Sheehy 1977–2
Tall fescue	Berks., UK	day 36	?	enclosure	$S_{g,0}$ (artificial)	F_n (not subt.)	1 day	reerected	Sheehy 1977–3
Tall fescue	Berks., UK	day 35	?	enclosure	$S_{g,0}$ (artificial)	F_n (not subt.)	1 day	lodged	Sheehy 1977–4
Lolium perenne	Berks., UK	day 36	?	enclosure	$S_{g,0}$ (artificial)	F_n (not subt.)	1 day	reerected	Sheehy 1977–5
Lolium perenne	Berks., UK	day 35	?	enclosure	$S_{g,0}$ (artificial)	F_n (not subt.)	1 day	lodged	Sheehy 1977–6
Ryegrass	Berks., UK	day 15	?	enclosure	$S_{g,0}$ (artificial)	F_n (subt.)	1 day	dawn	Sheehy 1977–7
Ryegrass	Berks., UK	day 15	?	enclosure	$S_{g,0}$ (artificial)	F_n (subt.)	1 day	noon	Sheehy 1977–8
Ryegrass	Berks., UK	day 15	?	enclosure	$S_{g,0}$ (artificial)	F_n (subt.)	1 day	dusk	Sheehy 1977–9

Ecosystem	Location	Date		Method	Flux	Correction	Time	Notes	Reference
Tundra	W. Alaska, US	Jul 6–Aug 8	?	enclosure	Q_0	no ΔS_c	?	wet site	Whiting et al. 1992–1
Tundra	W. Alaska, US	Jul 6–Aug 8	?	enclosure	Q_0	no ΔS_c	?	dry site	Whiting et al. 1992–2
Tundra	W. Alaska, US	?	?	eddy corr.	Q_0	no ΔS_c	?		Whiting et al. 1992–3
C4 grasslands									
Salt marsh	Delaware, Virginia, US	May–Oct 87	?	enclosure (ecosystem)	Q_{mt}	$F_n F_s$ (not subt.)	3 weeks growing season	3 sites, control & fert.	Bartlett et al. 1990
Salt marsh	Maryland, US	Jul 19 77	?	enclosure	Q_0	no	1 hour 1 day		Drake 1984
Salt marsh	New York, US	Jul 7–19	?	gradient	$S_{g,0}$	no	1 hour 2 days		Houghton & Woodwell 1980
Tallgrass prairie	Kansas, US	↑ 87	1·9	eddy corr.	Q_0	F_n (not subt.)	?		Kim & Verma 1990–1
Tallgrass prairie	Kansas, US	→ 87	3·0	eddy corr.	Q_0	F_n (not subt.)	?		Kim & Verma 1990–2
Tallgrass prairie	Kansas, US	87 →	2·7–1·7	eddy corr.	Q_0	F_n (not subt.)	?		Kim & Verma 1990–3
Tallgrass prairie	Kansas, US	→ 87	3·0	eddy corr.	Q_0	F_n (not subt.)	?	no water stress VPD 0–1·5 kPa	Kim & Verma 1990–4
Tallgrass prairie	Kansas, US	→ 87	3·0	eddy corr.	Q_0	F_n (not subt.)	?	no water stress VPD 1·5–3·0 kPa	Kim & Verma 1990–5
Tallgrass prairie	Kansas, US	→ (end) 87	2·7	eddy corr.	Q_0	F_n (not subt.)	?	water stress VPD 0–1·5 kPa	Kim & Verma 1990–6
Tallgrass prairie	Kansas, US	→ (end) 87	2·7	eddy corr.	Q_0	F_n (not subt.)	?	water stress VPD 1·5–3·0 kPa	Kim & Verma 1990–7
Tallgrass prairie	Kansas, US	→ (end) 87	2·7	eddy corr.	Q_0	F_n (not subt.)	?	water stress VPD 3·0–4·5 kPa	Kim & Verma 1990–8
Savanna	Lamto, Ivory Coast	early growing season (Feb)	0·80	enclosure	Q_0	F_s (subt.)	instant several days		Leroux & Mordelet–1 1995

Appendix 1—*cont.*

Vegetation class	Location	Phenology date	L	Flux measurements	Radiation measurements	Respiration measurements	Time-step, duration	Treatment, conditions	Reference
Savanna	Lamto, Ivory Coast	early growing season (Mar)	1·48	enclosure	Q_0	F_s (subt.)	instant several days		Leroux & Mordelet–2 1995
Savanna	Lamto, Ivory Coast	early growing season (Apr)	1·87	enclosure	Q_0	F_s (subt.)	instant several days		Leroux & Mordelet–3 1995
Bermudagrass	Georgia, US	Jun–Oct	?	enclosure	Q_{int} (calcul.)	no	mean/class weekly	mowed monthly	Morgan & Brown 1983–1
Bermudagrass	Georgia, US	Jun–Oct	?	enclosure	Q_{int} (calcul.)	no	mean/class weekly	mowed weekly	Morgan & Brown 1983–2
Tallgrass prairie	Kansas, US	Aug 5–6 86	2·0	eddy corr.	Q_0	no	1 hour several days	no water stress VPD 0–1 kPa	Verma et al. 1989–1
Tallgrass prairie	Kansas, US	Aug 5–6 86	2·0	eddy corr.	Q_0	no	1 hour several days	no water stress VPD 1–2 kPa	Verma et al. 1989–2
Tallgrass prairie	Kansas, US	Jul 30–Aug 4 86	2·0	eddy corr.	Q_0	no	1 hour several days	water stress VPD 0–1 kPa	Verma et al. 1989–3
Tallgrass prairie	Kansas, US	Jul 30–Aug 4 86	2·0	eddy corr.	Q_0	no	1 hour several days	water stress VPD 1–2 kPa	Verma et al. 1989–4
Tallgrass prairie	Kansas, US	Jul 30–Aug 4 86	2·0	eddy corr.	Q_0	no	1 hour several days	water stress VPD 2–3 kPa	Verma et al. 1989–5
Tallgrass prairie	Kansas, US	Jul 30–Aug 4 86	2·0	eddy corr.	Q_0	no	1 hour several days	water stress VPD 3–4 kPa	Verma et al. 1989–6
C3 crops									
Soybean	Nebraska, US	Aug 22–Sep 3	≥4	eddy corr.	Q_0	no	several days	no water stress	Anderson et al. 1984
Rice	Florida, US control chamber	↑	?	enclosure	$S_{g,0}$	F_n (subt.)	1 day	early planted day 41	Baker et al. 1990–1

Rice	Florida, US control chamber	→	11·8–14·3	enclosure	$S_{g,0}$	F_n (subt.)	1 day	early planted day 74	Baker et al. 1990–2
Wheat	Oregon, US	←	3·0	eddy corr.	Q_0	F_n (not subt.)	1 day		Baldocchi et al. 1994–1
Alfalfa	Nebraska, US	Sep 1	?	gradient	$S_{n,0}$	F_n, F_s (not subt.)	1 day		Baldocchi et al. 1981–a
Soybean	Nebraska, US	summer	4·1	gradient	Q_0	no	1 day several days	T 20–30°C irrigated	Baldocchi et al. 1981–b
Barley	Leics., UK	Jun 17–18 →	5·45	en. balance + gradient	$S_{g,0}$	F_s (enclosure) (subt.)	1 hour 2 days		Biscoe et al. 1975–1
Barley	Leics., UK	Jun 28 →	5·9	en. balance + gradient	$S_{g,0}$	F_s (enclosure) (subt.)	1 hour 1 day		Biscoe et al. 1975–2
Barley	Leics., UK	Jul 5 →	5·1	en. balance + gradient	$S_{g,0}$	F_s (enclosure) (subt.)	1 hour 1 day		Biscoe et al. 1975–3
Barley	Leics., UK	Jul 12 →	4·7	en. balance + gradient	$S_{g,0}$	F_s (enclosure) (subt.)	1 hour 1 day		Biscoe et al. 1975–4
Barley	Leics., UK	Jul 19 →	3·4	en. balance + gradient	$S_{g,0}$	F_s (enclosure) (subt.)	1 hour 1 day		Biscoe et al. 1975–5
Barley	Leics., UK	Jul 26 →	2·0	en. balance + gradient	$S_{g,0}$	F_s (enclosure) (subt.)	1 hour 1 day		Biscoe et al. 1975–6
Barley	Leics., UK	Jul 14	4·05	en. balance + gradient	$S_{g,0}$	F_s (enclosure) (subt.)	1 hour 1 day	leaf water pot. normal	Biscoe et al. 1975–7
Barley	Leics., UK	Jul 17	?	en. balance + gradient	$S_{g,0}$	F_s (enclosure) (subt.)	1 hour 1 day	leaf water pot. low	Biscoe et al. 1975–8
Wheat	Australia	?	1·6	gradient	$S_{g,0}$	F_s (not subt.) (subt.)	1 hour 5 weeks	dryland	Denmead 1976–1
Wheat	Australia	?	3·2	gradient	$S_{g,0}$	F_s (not subt.) (subt.)	1 hour 5 weeks	irrigated	Denmead 1976–2
Soybean	Florida, US control chamber	Oct 6	9·0	enclosure	Q_0	no	1 hour 1 day	double CO_2	Jones et al. 1984–2
Soybean	Florida, US control chamber	Oct 6	6·9	enclosure	Q_0	no	1 hour 1 day	ambient CO_2	Jones et al. 1984–1
Rice	Japan	Sep 2–3 (ear emergence)	?	eddy corr.	$S_{n,0}$	no	2 days		Ohtaki 1980
Soybean	Illinois, US	?	?	enclosure	Q_0	no	½ hour 1 day	fertilized, irrigated	Pettigrew et al. 1990
Wheat	Australia	Sep → 67	≈4	enclosure	$S_{g,0}$	F_s (separate enclosure)	¼ hour 1 day	dry season fertilized	Puckridge 1971–1

Appendix 1—*cont.*

Vegetation class	Location	Phenology date	L	Flux measurements	Radiation measurements	Respiration measurements	Time-step, duration	Treatment, conditions	Reference
Wheat	Australia	Sep → 68	≈ 3	enclosure	$S_{g,0}$	F_s (separate enclosure)	¼ hour 1 day	wet season fertilized	Puckridge 1971–2
Wheat (Australian var)	Australia	Sep 6	0·8	enclosure	$S_{g,0}$	F_s (separate enclosure)	¼ hour 1 day	low N	Puckridge & Ratkowski 1971–1
Wheat (Australian var)	Australia	Sep 13	4·6	enclosure	$S_{g,0}$	F_s (separate enclosure)	¼ hour 1 day	high N	Puckridge & Ratkowski 1971–2
Wheat (Australian var)	Australia	Sep 12	0·9	enclosure	$S_{g,0}$	F_s (separate enclosure)	¼ hour 1 day	low N	Puckridge & Ratkowski 1971–3
Wheat (Australian var)	Australia	Sep 19	4·9	enclosure	$S_{g,0}$	F_s (separate enclosure)	¼ hour 1 day	high N	Puckridge & Ratkowski 1971–4
Wheat (Australian var)	Australia	Sep 18	1·2	enclosure	$S_{g,0}$	F_s (separate enclosure)	¼ hour 1 day	low N	Puckridge & Ratkowski 1971–5
Wheat (Australian var)	Australia	Sep 28	4·1	enclosure	$S_{g,0}$	F_s (separate enclosure)	¼ hour 1 day	high N	Puckridge & Ratkowski 1971–6
Wheat (Australian var)	Australia	Sep 24	1·4	enclosure	$S_{g,0}$	F_s (separate enclosure)	¼ hour 1 day	low N	Puckridge & Ratkowski 1971–7
Wheat (Australian var)	Australia	Oct 4	4·6	enclosure	$S_{g,0}$	F_s (separate enclosure)	¼ hour 1 day	high N	Puckridge & Ratkowski 1971–8
Wheat (Australian var)	Australia	Oct 12	4·3	enclosure	$S_{g,0}$	F_s (separate enclosure)	¼ hour 1 day	high N	Puckridge & Ratkowski 1971–9
Wheat (Australian var)	Australia	Oct 8	1·7	enclosure	$S_{g,0}$	F_s (separate enclosure)	¼ hour 1 day	low N	Puckridge & Ratkowski 1971–10
Wheat (Mexican var)	Australia	Sep 5	0·8	enclosure	$S_{g,0}$	F_s (separate enclosure)	¼ hour 1 day	low N	Puckridge & Ratkowski 1971–11
Wheat (Mexican var)	Australia	Sep 14	4·9	enclosure	$S_{g,0}$	F_s (separate enclosure)	¼ hour 1 day	high N	Puckridge & Ratkowski 1971–12
Wheat (Mexican var)	Australia	Sep 11	1·2	enclosure	$S_{g,0}$	F_s (separate enclosure)	¼ hour 1 day	low N	Puckridge & Ratkowski 1971–13

Crop	Location	Date	Ratio	Method	Symbol	Flux	Duration	Condition	Reference
Wheat (Mexican var)	Australia	Sep 20	5·0	enclosure	$S_{g,0}$	F_s (separate enclosure)	¼ hour / 1 day	high N	Puckridge & Ratkowski 1971–14
Wheat (Mexican var)	Australia	Sep 17	2·2	enclosure	$S_{g,0}$	F_s (separate enclosure)	¼ hour / 1 day	low N	Puckridge & Ratkowski 1971–15
Wheat (Mexican var)	Australia	Sep 27	2·0	enclosure	$S_{g,0}$	F_s (separate enclosure)	¼ hour / 1 day	low N	Puckridge & Ratkowski 1971–16
Wheat (Mexican var)	Australia	Oct 16	4·2	enclosure	$S_{g,0}$	F_s (separate enclosure)	¼ hour / 1 day	high N	Puckridge & Ratkowski 1971–17
Wheat (Mexican var)	Australia	Oct 22	1·4	enclosure	$S_{g,0}$	F_s (separate enclosure)	¼ hour / 1 day	low N	Puckridge & Ratkowski 1971–18
Winter wheat	Kansas, US	May 16–25 ↑	peak 4·5	enclosure	Q_{abs}	F_n (subt.)	9 days	N–S orientated	Wall & Kanemasu 1990
Winter wheat	Kansas, US	April 20–June 19	?	enclosure	Q_{abs} (calcul.)	F_n (subt.)	1 day / 2 months		Wall et al. 1990
Cucumber	Australia control chamber	June 14–16 88	3·4	enclosure	Q_0	no	10 min / 3 days	glasshouse	Warren Wilson et al. 1992
Wheat	Australia	?	5·3–7·7	enclosure	$S_{g,int}$	no	1 day	irrigated, fertilized	Whitfield 1990

C4 crops

Crop	Location	Date	Ratio	Method	Symbol	Flux	Duration	Condition	Reference
Sorghum	Nebraska, US	→	3·7	eddy corr.	Q_0	no	several days		Anderson & Verma 1986–1
Sorghum	Nebraska, US	↑	2·7–3·6	eddy corr.	Q_0	no	several days		Anderson & Verma 1986–2
Sorghum	Nebraska, US	→	3·5	eddy corr.	Q_0	no	several days		Anderson & Verma 1986–3
Corn	Oregon, US	↑	1·8	eddy corr.	Q_0	F_n (not subt.)	1 day	water stress	Baldocchi 1994–2
Corn	?	Jul 17	?	eddy corr.	Q_{int}	no	1 hour / 1 day		Desjardins et al. 1984–1
Corn	?	Aug 15	?	eddy corr.	Q_{int}	no	several days	no water stress	Desjardins et al. 1984–2
Corn	Florida, US	May 10 ↑	3·26	enclosure	Q_0	no	several days	low N, irrigated	Jones et al. 1986–1

Appendix 1—*cont.*

Vegetation class	Location	Phenology date	L	Flux measurements	Radiation measurements	Respiration measurements	Time-step, duration	Treatment, conditions	Reference
Corn	Florida, US	May 10 ↑	4·02	enclosure	Q_0	no	several days	high N, irrigated	Jones et al. 1986–2
Corn	Florida, US	May 10 ↑	2·76	enclosure	Q_0	no	several days	low N, water stress	Jones et al. 1986–3
Corn	Florida, US	May 10 ↑	3·70	enclosure	Q_0	no	several days	high N, water stress	Jones et al. 1986–4
Corn	New York, US	Jul 29–Aug 8	?	enclosure	$S_{g,0}$	F_s (enclosure,) F_n	1 hour 1 day	CO_2 510 ppmv	Moss et al. 1961–1
Corn	New York, US	Jul 29–Aug 8	?	enclosure	$S_{g,0}$	F_s (enclosure,) F_n	1 hour 1 day	CO_2 270 ppmv	Moss et al. 1961–2
Corn	New York, US	Jul 29–Aug 8	?	enclosure	$S_{g,0}$	F_s (enclosure,) F_n	1 hour 1 day	CO_2 155 ppmv	Moss et al. 1961–3
Corn	New York, US	Jul 29–Aug 8	?	enclosure	$S_{g,0}$	F_s (enclosure,) F_n	1 hour 1 day	CO_2 30 ppmv	Moss et al. 1961–4
mixed vegetation types									
Mixed conifer, deciduous	Ottawa, Canada	summer	?	plane mounted eddy corr.	$S_{g,0}$	no	different passes several days		Desjardins et al. 1985–1
White pine	Ottawa, Canada	summer	?	plane mounted eddy corr.	$S_{g,0}$	no	different passes several days		Desjardins et al. 1985–2
Wheat fields	Manitoba, US	July 14 84	?	plane mounted eddy corr.	$S_{g,0}$	no	1·5 hour 1 day		Desjardins 1991
Various crops	Nebraska, US	June–Oct 72	?	flux gradient	$S_{g,0}$	F_n (not subt.)	daily means, growing season		Verma & Rosenberg 1976

APPENDIX 2: STATISTICS ON THE RELATIONSHIP BETWEEN CO$_2$ FLUX (F) AND PPFD (Q), ON INDIVIDUAL DATA SETS

In some cases two or more data sets have been aggregated to form bigger data sets, when original data sets are too small or present the same relationship. This is indicated in the reference column: for example the data set Desjardins *et al.* 1985–3 (1 + 2) results from the aggregation of data sets Desjardins *et al.* 1985–1 and Desjardins *et al.* 1985–2 of Appendix 1. *n* is number of data points; "means" indicates that reported data sets are averaged per radiation class. A rectangular hyperbolic (H) and a linear model (L) are tested; best fit is defined by the relationship having the highest coefficient of determination (r^2). Statistical parameters are not given when there is no statistical relationship, i.e. $r^2 < 0.4$ ("N" in column best fit). The parameters of the statistical models are initial slope (α), intercept on the y axis (R), and, for the hyperbolic model, value of F at saturating Q (F_∞). Derived parameters are departure from linearity (D) defined as r^2(H)–r^2(L), and photosynthetic capacity (F_m), defined as F at maximum Q ($1800\,\mu\text{mol m}^{-2}\text{s}^{-1}$ for instantaneous data sets, $80\,\text{mol m}^{-2}\text{d}^{-1}$ for daily data sets). $R = 0.00$ indicates that the intercept would be positive (therefore the value of R negative), but is constrained by the model.

2.1. Instantaneous data sets, F_∞, R and F_m are in $\mu mol\, m^{-2}\, s^{-1}$

References	n	Constraint	F_∞	α	R	r^2 (best fit)	D	F_m	Best fit
broadleaf forests									
Allen & Lemon 1976	5		/		/	−0·58	/	/	N
Baldocchi & Harley	277		43·80	0·040	4·74	0·60	0·05	22·49	H
Baldocchi et al. 1987	28		43·43	0·033	0·55	0·72	0·22	24·54	H
den Hartog et al. 1987	16		/	0·010	0·00	0·51	0·00	18·00	L
Denmead 1991 (Fig. 4A)	17		/	0·014	1·13	0·96	0·00	24·07	L
Fan et al. 1990	13		44·14	0·049	11·11	0·92	0·06	18·31	H
Hollinger et al. 1994−1	19	$R = 2·4$	7·64	0·009	2·40	0·78	0·00	3·62	H
Hollinger et al. 1994−2	47	$R = 4·5$	15·62	0·043	4·50	0·72	0·16	8·50	H
Hollinger et al. 1994−3	75	$R = 6·7$	16·05	0·042	6·70	0·50	0·09	6·54	H
Hollinger et al. 1994−4	57	$R = 5·2$	/		/	0·38	/	/	N
Hollinger et al. 1994−5	51	$R = 0·50$	24·37	0·031	0·50	0·51	0·01	16·46	H
Kelliher et al. 1989	19		40·33	0·048	4·48	0·93	0·01	23·02	H
Monteny 1989−1	7		30·58	0·076	12·20	0·98	0·25	12·79	H
Monteny 1989−2	7		52·26	0·089	11·75	0·99	0·16	27·66	H
Monteny 1989−3	7		81·70	0·076	11·16	0·99	0·09	39·99	H
Mordacq et al. 1991	45		41·10	0·073	0·39	0·90	0·30	30·92	H
Valentini et al. 1991	122		/	0·015	0·93	0·61	0·00	26·07	L
Verma et al. 1986−4 (1 + 2 + 3)	51		28·51	0·033	3·94	0·72	0·14	15·32	H
Wofsy et al. 1993−3 (1 + 2)	20		28·92	0·044	4·88	0·95	0·06	16·30	H
Wong & Dunin 1987−1 (Fig. 4B)	68		34·60	0·049	2·49	0·94	0·09	22·36	H
Wong & Dunin 1987−2	35		66·55	0·102	0·00	0·89	0·45	48·49	H
conifer forests									
Brunet et al. 1992	103		49·22	0·018	1·06	0·45	0·03	18·48	H
Jarvis 1994−1	4 (means)		59·95	0·174	12·73	0·99	0·07	37·59	H
Jarvis 1994−2	5 (means)		54·35	0·135	12·48	0·97	0·10	31·94	H
Jarvis 1994−3	4 (means)		42·98	0·049	7·53	0·99	0·04	21·37	H
Jarvis 1994−4	4 (means)		24·27	0·051	4·74	0·83	0·38	14·40	H
Jarvis 1994−5	5 (means)		26·91	0·023	5·04	0·99	0·02	11·27	H
Price & Black 1990−1	48		/		/	0·10	/	/	N

Study	n	R							N
Price & Black 1990–2	48		/	/	/	0·00	/	/	N
Price & Black 1990–3	48		40·86	0·020	3·58	0·72	0·02	14·87	H
C3 grasslands									
Belanger 1990–1	24	R = 0	44·60	0·112	0·00	0·94	0·57	36·52	H
Belanger 1990–2	27	R = 0	41·91	0·100	0·00	0·97	0·33	33·99	H
Belanger 1990–3	55	R = 0	17·52	0·029	0·00	0·90	0·77	13·12	H
Ripley & Saugier – 1	12	R = 3·22 (means)	13·65	0·039	3·63	0·94	0·21	7·80	H
Ripley & Saugier – 2	12	R = 2·54 (means)	15·79	0·012	2·54	0·88	0·17	6·58	H
Ripley & Saugier – 3	12	R = 1·45 (means)	4·13	0·010	1·45	0·98	0·18	1·91	H
Saugier 1986	13		62·42	0·061	0·86	0·98	0·21	38·94	H
Sheehy 1977–1	6	R = 5·05 (means)	33·85	0·126	5·05	0·97	0·19	24·40	H
Sheehy 1977–2	5	R = 5·62 (means)	40·08	0·224	5·62	0·99	0·21	30·78	H
Sheehy 1977–3	6	R = 3·78 (means)	40·38	0·099	3·78	0·99	0·12	29·14	H
Sheehy 1977–4	5	R = 6·94 (means)	41·84	0·190	6·94	0·98	0·12	30·34	H
Sheehy 1977–5	6	R = 3·78 (means)	39·53	0·119	3·78	0·99	0·12	29·59	H
Sheehy 1977–6	5	R = 6·94 (means)	43·65	0·220	0·94	0·98	0·12	38·36	H
Sheehy 1977–7	5	R = 0 (means)	71·74	0·206	0·00	0·99	0·49	60·11	H
Sheehy 1977–8	5	R = 0 (means)	61·87	0·186	0·00	0·99	0·53	52·22	H
Sheehy 1977–9	5	R = 0 (means)	69·82	0·177	0·00	0·99	0·41	57·27	H
Whiting et al. 1992–1	103		5·36	0·031	2·12	0·80	0·30	2·77	H
Whiting et al. 1992–2	73		2·48	0·033	1·14	0·80	0·37	0·89	H
Whiting et al. 1992–3	20		2·36	0·008	0·61	0·92	0·22	1·42	H
C4 grasslands									
Bartlett et al. 1990	31		37·79	0·031	3·45	0·69	0·00	52·35	L
Drake 1984	12		/	0·043	6·50	0·99	0·06	18·89	H
Houghton & Woodwell 1980	55		32·56	0·032	4·41	0·62	0·06	16·39	H
Kim & Verma 1990–1	77		53·61	0·035	7·99	0·89	0·07	20·97	H
Kim & Verma 1990–2	87		82·20	0·032	8·77	0·91	0·02	25·10	H
Kim & Verma 1990–3	36		76·24	0·017	7·60	0·91	0·01	14·24	H
Kim & Verma 1990–4	91		82·31	0·032	8·42	0·91	0·02	24·47	H
Kim & Verma 1990–5	71		73·61	0·030	10·25	0·81	0·02	20·90	H
Kim & Verma 1990–6	23		/	/	4·06	0·94	0·00	21·14	L
Kim & Verma 1990–7	50		18·56	0·014	4·85	0·68	0·06	7·40	H
Kim & Verma 1990–8	46		/	0·020	/	0·03	/	/	N

Appendix 2—cont.

References	n	Constraint	F_∞	α	R	r^2 (best fit)	D	F_m	Best fit
Leroux & Mordelet 1994–1	19		28·91	0·043	2·14	0·96	0·19	18·98	H
Leroux & Mordelet 1994–2	25		33·46	0·045	1·12	0·98	0·11	22·59	H
Leroux & Mordelet 1994–3	31		31·02	0·052	1·38	0·95	0·10	21·87	H
Morgan & Brown 1983–1	7 (means)		/	0·031	7·09	0·99	0·00	48·71	L
Morgan & Brown 1983–2	8 (means)		/	0·019	2·17	0·83	0·00	32·03	L
Verma et al. 1989–1	10		/	0·011	3·77	0·87	0·00	16·03	L
Verma et al. 1989–2	8		51·98	0·009	0·00	0·77	0·00	12·35	H
Verma et al. 1989–7(3 + 4 + 5 + 6)	17		/	/	/	0·33	/	/	N
C3 crops									
Anderson et al. 1984	16		/	0·018	1·01	0·77	0·00	31·39	L
Baker et al. 1990–1	110	$R = 0$	67·18	0·055	0·00	0·96	0·11	39·01	H
Baker et al. 1990–2	8·5	$R = 0$	/	0·048	0·00	0·98	0·00	86·40	L
Baldocchi 1994–1	47		/	0·021	3·94	0·98	0·00	33·86	L
Baldocchi et al. 1981–a	8		62·68	0·055	5·54	0·68	0·09	32·84	H
Baldocchi et al. 1981–b	36		/	0·017	2·59	0·81	0·00	28·01	L
Biscoe et al. 1975–1	28		/	0·020	0·00	0·81	0·00	36·00	L
Biscoe et al. 1975–2	8		34·62	0·098	4·72	0·98	0·17	24·22	H
Biscoe et al. 1975–3	7		31·42	0·047	2·79	0·98	0·07	20·12	H
Biscoe et al. 1975–4	6		33·33	0·031	4·06	0·98	0·06	16·81	H
Biscoe et al. 1975–5	8		25·57	0·031	3·60	0·98	0·09	13·93	H
Biscoe et al. 1975–6	7		10·07	0·055	3·17	0·96	0·26	5·97	H
Biscoe et al. 1975–7	14		/	0·014	4·90	0·96	0·00	20·30	L
Biscoe et al. 1975–8	17		/	0·010	4·23	0·80	0·00	13·77	L
Denmead 1976–1	63		62·76	0·030	3·72	0·66	0·03	25·35	H
Denmead 1976–2	29		/	0·021	2·75	0·87	0·00	35·05	L
Jones et al. 1984–1	64		/	0·032	0·00	0·98	0·01	57·60	L
Jones et al. 1984–2	62		/	0·050	0·00	0·95	0·04	90·00	L
Ohtaki 1980	28		/	0·020	3·12	0·94	0·00	32·88	L
Pettigrew et al. 1990	22		57·54	0·021	0·00	0·58	0·12	22·81	H

Reference	n							H/L
Puckridge & Ratkowski 1971–1	15	19·06	0·163	8·32	0·84	0·09	9·58	H
Puckridge & Ratkowski 1971–2	15	53·31	0·094	5·52	0·97	0·21	35·02	H
Puckridge & Ratkowski 1971–3	15	22·21	0·048	1·70	0·97	0·40	15·97	H
Puckridge & Ratkowski 1971–4	17	/	0·033	0·00	0·94	0·02	59·40	L
Puckridge & Ratkowski 1971–5	19	17·64	0·068	2·03	0·80	0·10	13·39	H
Puckridge & Ratkowski 1971–6	11	32·00	0·120	4·37	0·87	0·19	23·50	H
Puckridge & Ratkowski 1971–7	19	20·27	0·037	1·17	0·97	0·22	14·37	H
Puckridge & Ratkowski 1971–8	14	26·77	0·042	1·09	0·94	0·36	15·00	H
Puckridge & Ratkowski 1971–9	16	51·68	0·046	0·82	0·94	0·19	30·40	H
Puckridge & Ratkowski 1971–10	18	30·69	0·038	0·00	0·89	0·48	21·18	H
Puckridge & Ratkowski 1971–11	14	17·57	0·090	5·07	0·94	0·57	10·78	H
Puckridge & Ratkowski 1971–12	10	64·24	0·311	28·25	0·98	0·11	29·38	H
Puckridge & Ratkowski 1971–13	18	19·98	0·081	0·00	0·84	0·12	17·57	H
Puckridge & Ratkowski 1971–14	14	52·78	0·051	0·14	0·96	0·22	33·37	H
Puckridge & Ratkowski 1971–15	17	34·55	0·097	7·84	0·95	0·51	21·00	H
Puckridge & Ratkowski 1971–16	15	52·73	0·098	4·88	0·97	0·27	35·19	H
Puckridge & Ratkowski 1971–17	10	57·61	0·040	4·24	0·98	0·07	27·76	H
Puckridge & Ratkowski 1971–18	17	17·57	0·090	5·07	0·94	0·57	10·78	H
Puckridge 1970–1	17	71·08	0·109	0·00	0·78	0·06	52·18	H
Puckridge 1970–2	18	55·47	0·076	6·09	0·98	0·07	33·38	H
Wall & Kanemasu 1990 (Fig. 8a)	157	62·30	0·465	32·89	0·59	0·11	25·09	H
Warren Wilson et al. 1992	24	/	0·018	0·00	0·87	0·02	30·91	L
C4 crops								
Anderson & Verma 1986–1	24	/	0·018	1·49	0·87	0·01	30·91	L
Anderson & Verma 1986–2	14	48·76	0·106	6·89	0·85	0·37	31·94	H
Anderson & Verma 1986–3	10	43·48	0·059	5·84	0·98	0·17	25·16	H
Baldocchi 1994–2	48	/	0·015	2·82	0·97	0·01	24·18	L
Desjardins et al. 1984–1	10	98·14	0·058	7·21	0·61	0·01	43·38	H
Desjardins et al. 1984–2	13	/	0·039	0·00	0·88	0·05	70·20	L

Appendix 2—cont.

Appendix 2—cont.

References	n	Constraint	F_∞	α	R	r^2 (best fit)	D	F_m	Best fit
Jones et al. 1984–1	14		75·62	0·057	7·82	0·99	0·04	35·71	H
Jones et al. 1984–2	11		/	0·028	2·32	0·95	0·03	48·08	L
Jones et al. 1984–3	13		/	/	/	0·23	/	/	N
Jones et al. 1984–4	14		40·24	0·091	7·55	0·68	0·23	24·75	H
Moss et al. 1961–1	10		43·56	0·029	2·39	0·98	0·02	22·36	H
Moss et al. 1961–2	10		38·82	0·026	2·54	0·99	0·01	18·68	H
Moss et al. 1961–3	6		39·85	0·020	1·60	0·99	0·01	17·31	H
Moss et al. 1961–4	6		3·46	0·018	1·93	0·99	0·59	1·19	H
		mixed vegetation types							
Desjardins et al. 1985–3 (1 + 2)	8		/	0·014	2·90	0·56	0·00	22·30	L
Desjardins 1991	16		/	0·017	10·63	0·71	0·00	19·99	L

2.2. Daily data sets, F_∞, R and F_m are in $\mu mol\,m^{-2}\,d^{-1}$

References	n	Constraint	F_∞	α	R	r^2 (best fit)	D	F_m	Best fit
		various vegetation types							
Coyne & Kelley 1975–1 (Fig. 6a)	20		/	/	/	0·34	/	/	N
Coyne & Kelley 1975–1 (Fig. 6b)	61		0·94	0·052	0·32	0·60	0·03	0·44	H
Hollinger et al. 1994–6	8		/	/	/	0·21	/	/	N
Verma & Rosenberg 1976	28		/	/	/	0·12	/	/	N
Wall et al. 1990 (Fig. 8b)	45		/	0·027	0·34	0·67	0·00	1·82	L
Whitfield 1990	21		/	0·013	0·33	0·67	0·00	0·71	L

The Ecology of Symbiotic Micro-organisms

I. Introduction 69
    A. Why Study Symbiotic Micro-organisms? 69
    B. The Scope of the Article 71
II. The Phyletic Diversity of Endosymbiotic Micro-organisms 74
    A. A Comparison of the Phyletic Diversity of Hosts and
       Endosymbionts 74
    B. Taxonomic Practice and the Phyletic Diversity of Microbial
       Endosymbiosis 75
    C. Origins and Radiations of the Partners in Endosymbioses . . . 76
III. The Distribution of Micro-organisms in Endosymbiosis 78
    A. Host-related Determinants of Endosymbiont Distribution . . 78
    B. Determinants of Endosymbiont Distribution External to the
       Symbiosis 87
IV. Endosymbiotic Micro-organisms in the Environment 88
    A. On the Rarity of Symbiotic Micro-organisms in the Environment 88
    B. Symbiosis as a Source of Symbionts in the Environment . . . 90
V. The Host as a Habitat 93
VI. Conclusions 94
Acknowledgements 95
References 95

I. INTRODUCTION

A. Why Study Symbiotic Micro-organisms?

There is an increasing appreciation within mainstream ecology of the significance of symbioses. The erroneous but widespread view of symbioses as unstable and transient "curiosities of nature" is being replaced by a recognition that symbioses are abundant and widely distributed, and that they may play a central role in ecological processes. This changed perspective is indicated by the several texts devoted exclusively to the ecology of symbioses (e.g. Boucher, 1985; Margulis and Fester, 1991; Kawanabe *et al.*,

ADVANCES IN ECOLOGICAL RESEARCH VOL. 26
ISBN 0-12-013926-X

1993) and the inclusion of symbiosis in a minority of ecology textbooks (e.g. Begon et al., 1990; Howe and Westley, 1988) (see also Keddy, 1990).

Symbioses involving micro-organisms underpin some biological communities, and in these systems they are important to the flux of energy and nutrients. For example, in the marine environment, the symbiosis between corals and algae is the architectural foundation for all shallow-water coral reef communities, renowned for both their high productivity (1–5 kg C m^{-2}y^{-1}) (Hatcher, 1990) and biodiversity (Connell, 1978; Jackson, 1991). On land, the roots of most plant species are associated with symbiotic mycorrhizal fungi (Brundrett, 1991), which both channel soil-derived minerals to plants (so contributing to tight nutrient cycling) (Wood et al., 1984; Sprent, 1986) and act as a sink for 10–20% of plant primary production (Allen, 1991), amounting to an estimated 10^{13} kg C y^{-1} globally. In terrestrial environments where vascular plants do not flourish (e.g. many alpine-arctic habitats and some deserts), the dominant primary producers are the symbiotic algae or cyanobacteria in lichenized fungi (Kappen, 1988). Symbiosis is also crucial to biological nitrogen fixation on land, with virtually all of the estimated 140×10^9 kg fixed y^{-1} mediated by symbiotic bacteria in plant roots. Also of major ecological importance are the many microbial associations which underpin herbivory by many animals (Clarke and Bauchop, 1977; Martin, 1991; Douglas, 1992a).

The ecology of the microbial partners in these associations is a neglected aspect of symbiosis research, which is dominated by research on the larger "host" partner and the association per se. It is now a particularly appropriate time to consider these micro-organisms, for two different reasons: technical advances in the study of microbiology; and the vulnerability of symbiotic micro-organisms to recent anthropogenic changes to the environment.

The technical advances relate to the intractability of most symbiotic micro-organisms to traditional methods in microbiology, which require isolation into axenic culture for identification and study. Symbiotic micro-organisms are either unculturable, or they lose their symbiotic characteristics when isolated from the association. Much of this difficulty has been circumvented by the advent of molecular methods, especially those using PCR-technology, which provide a basis to study the abundance and diversity of micro-organisms in natural habitats, in principle without any recourse to axenic cultivation (Olsen et al., 1994). To date, the primary focus of most molecular studies of symbiotic micro-organisms has been to establish their taxonomic affiliations and to construct molecular phylogenies of the major groups (see Section II), but these methods are also well suited to address the ecological issues of the distribution and abundance of these micro-organisms.

The impact of human activities on some symbioses is well known. For example, the local extinction of many lichens by sulphur dioxide and other pollutants has been used for decades in pollution monitoring (Rose, 1970; Richardson, 1992); and the abundance and distribution of many ectomy-corrhizal fungi in Central Europe have declined precipitously in recent decades, probably as a result of pollution (Jansen and van Dobben, 1987; Ruhling and Tyler, 1991) with uncertain (but potentially profound) conse-quences for the long-term vigour of both coniferous and hardwood forests in the region. In the tropics and subtropics, the coral reefs are facing multiple, anthropogenic threats, including eutrophication, increased sedi-mentation, and, almost certainly, a progressive rise in sea surface tem-perature (Roberts, 1993). These various insults are widely believed to be contributing to the breakdown of the algal symbiosis in corals and other reef invertebrates, resulting in mass expulsion of the algae and "bleaching" of the animals (Glynn, 1993). Extreme bleaching events have led to large-scale coral mortality and a shift in the community from one dominated by corals to one dominated by macroalgae (e.g. Hughes, 1994). The impact of the loss of symbioses on broad ecological processes has barely been con-sidered (but see Bond (1994) for discussion of other symbioses). The future management and conservation of symbioses requires an understanding of the ecology and basic biology of the microbial partners, and this is the subject of the present article.

B. The Scope of the Article

The central aims of this article are to review the abundance and distri-bution of symbiotic micro-organisms, both in symbiosis (Section III) and in isolation (Section IV), and to explore the processes which may underlie the observed patterns. Our grasp of these issues has been strengthened greatly by an improved understanding of the taxonomy and diversity of the micro-organisms in symbiosis, arising from the molecular "revolution" in microbial taxonomy (see above); and current understanding of the diver-sity of these organisms is reviewed first, in Section II.

The symbiotic micro-organisms considered in this article are listed in Table 1. They and their partners form a very diverse group, both ecologi-cally and taxonomically, but in terms of the symbioses, they have three common characteristics.

(i) *The associations are endosymbioses, i.e. micro-organisms (the sym-bionts) are located within the body of their partner (the host), and usually in direct contact with the cells or cellular contents of their host.* Many of the micro-organisms are intracellular (usually located in the cytoplasmic compartment of host cells), and others lie between closely-apposed cells of their host. This review will not consider: the

Table 1
A survey of endosymbiotic micro-organisms[1]

The micro-organisms	The hosts	References
Nitrogen-fixing bacteria		
Rhizobium,	Many legumes	Sprent and Sprent (1990)
Azorhizobium &		
Bradyrhizobium		
Frankia	Various dicots	
Cyanobacteria (e.g.	Few plants and lichens	
Nostoc)		
Mycorrhizal fungi		
Arbuscular mycorrhizal	Most terrestrial plants	Harley and Smith (1983);
Ectomycorrhizal	Various gymnosperms	Allen (1991)
	and perennial dicots	
Ericoid mycorrhizal	Various Ericaceae	
Orchid mycorrhizal	All Orchidaceae	
Endophytic fungi	Various grasses and	Clay (1990)
(Members of	sedges	
Clavicipitaceae)		
Photosynthetic algae		
Symbiodinium	Marine protists &	Reisser (1992)
	invertebrates	
Chlorella	Freshwater protists &	
	invertebrates	
Trebouxia	Lichens	Honegger (1994)
Chemoautotrophic	Marine invertebrates	Fisher (1990)
bacteria	(e.g. all Pogonophora,[2]	
(γ-Proteobacteria)	various bivalves, few	
	annelids)	
Luminescent bacteria	Various teleost fish &	Haygood (1993)
(*Vibrio, Photobacterium*	cephalopod molluscs	
& unculturable forms)		
Mycetocyte symbionts[3]	Insects (e.g. all	Douglas (1989)
(Various bacteria &	cockroaches &	
yeasts)	homopterans, many	
	beetles)	
Methanogenic bacteria	Anaerobic ciliates	Fenchel and Finlay (1991)

[1] The micro-organisms included here are the principal taxa considered in the text. Detailed lists of symbiotic micro-organisms are provided in the references indicated and in Buchner (1965), Douglas (1994).
[2] Comprising perviate and vestimentiferan pogonophorans (some authorities consider the two groups as separate phyla, Pogonophora and Vestimentifera, respectively).
[3] Intracellular micro-organisms restricted to specific insect cells (known as mycetocytes). In the associations that have been studied, the micro-organisms provide essential nutrients (essential amino acids or vitamins). Microbial symbionts with this function may occur in a variety of invertebrate animals (Buchner, 1965), but they have not been studied in detail.

microbiota in the gut lumen of many animals; the micro-organisms that inhabit the surfaces of other organisms; and the micro-organisms that are maintained in the immediate environs of animals through most of the host life cycle, and internalized for transmission (e.g. ectosymbiotic fungi of insects, symbiotic bacteria in entomopathogenic nematodes).

(ii) *One organism in the association possesses a metabolic capability which its partner (which lacks the capability) utilizes.* The usual "donor" of the metabolic capability is the microbial symbiont. For example, many aquatic invertebrates use photosynthetic compounds derived from their symbiotic algae as a source of energy and carbon. In some associations, the flux of photosynthetic carbon from the algal cells to the animal is sufficient to supply the entire respiratory needs of the host (Muscatine, 1990). Another example concerns the grasses (Poaceae), which as a group lack the capacity to synthesize alkaloids. Many grass species use the high alkaloid content of fungal endophytes, which ramify through their shoot tissues, to deter and poison potential herbivores (Clay, 1990).

(iii) *The association is predominantly biotrophic*, i.e. the organisms gain nutrients from living cells of their partners. In most associations, nutrients are translocated in both directions between host and symbiont cells, but nutrient flux is unidirectional in a few systems. These include the luminescent bacteria and endophytic fungi which presumably derive their entire nutritional requirements from the surrounding animal and plant cells, respectively, but make no nutritional contribution to their hosts. Nutrient flux in lichens and in orchid mycorrhizas may also be very unbalanced, dominated by the transfer of nutrients from the symbiotic algae/cyanobacteria to the fungal host in lichens (Honegger, 1994) and from the mycorrhizal fungus to the plant in orchid mycorrhizas (Smith, 1967; Alexander and Hadley, 1985).

Much of the literature on symbiosis presupposes that the associations are mutualistic, but this view is not fully supported by direct experimental study. The usual method to identify benefit or harm is to compare the performance (e.g. as measured by growth, survival, fecundity) of an organism in symbiosis and in isolation. If the organism performs better in symbiosis, it is said to benefit from the association; and if it performs better in isolation, it is harmed by the association. Benefit (defined in this way) is not an invariant property of an association, but has been shown to vary with developmental age and environmental conditions for several systems, e.g. animal hosts of symbiotic algae, mycorrhizal plants (Douglas, 1994). Other organisms are not amenable to study because they are not known to

74 A.E. DOUGLASA.E. DOUGLAS

occur in isolation from their partners (this could be interpreted as either benefit or dependence). Because of these difficulties, the significance of the symbiosis to the performance or fitness of the organisms will not be used as a general characteristic of the associations considered in this article.

As indicated in Section I.A, the focus of the article is the micro-organisms in symbiosis. Aspects of the ecology of symbioses (e.g. the distribution of associations, and their contribution to ecological processes in different ecosystems) will not be addressed. These issues have not been researched in detail, but some relevant information can be obtained from the references in Table 1.

II. THE PHYLETIC DIVERSITY OF ENDOSYMBIOTIC MICRO-ORGANISMS

A. A Comparison of the Phyletic Diversity of Hosts and Endosymbionts

It has long been known that many different micro-organisms occur in symbiosis. The list of microbial symbionts in Table 1 includes a wide range of Bacteria, some Archaea (the methanogens in ciliates) and, among the Eukarya, both algae and fungi. However, for each type of association, the diversity of micro-organisms appears to be low. As examples, the only widespread symbiotic algae are the dinoflagellate *Symbiodinium* in the marine environment, *Chlorella* in freshwaters, and *Trebouxia* on land; and of all terrestrial nitrogen-fixing bacteria, fewer than ten genera are known to form intimate symbioses with plants (they include the α-Proteobacteria *Rhizobium*, *Azorhizobium* and *Bradyrhizobium*, the actinomycete *Frankia*, and the cyanobacteria *Nostoc, Anabaena* and *Calothrix*).

The apparently restricted diversity of micro-organisms in each type of association can be contrasted with the very high taxonomic diversity of most groups of hosts. This pattern was first recognized by Law and Lewis (1983), who found that the number of host genera or species in most types of association was at least ten-fold greater than the number of symbiont genera or species. For example, nearly 1600 genera of ascomycete fungi are lichenized with fewer than 40 genera of algae/cyanobacteria (Honegger, 1992; Hawksworth, 1988); *c.* 10 species of rhizobia and brady-rhizobia are recognized in symbiosis with 17 500 species of legumes (Young, 1991; Sprent and Sprent, 1990); and *c.* 120 species of arbuscular mycorrhizal fungi have been identified associated with an estimated 200 000 species of plants. (These are current estimates of the diversity in these groups, and they differ slightly from the values published in Law and Lewis (1983).) The sole major exception to this generality is the ectomycorrhizal fungi, comprising an estimated 3500 species associated with *c.* 5000 plant species.

B. Taxonomic Practice and the Phyletic Diversity of Microbial Endosymbionts

Law and Lewis (1983) attributed at least part of the difference in diversity of hosts and symbionts to difficulties in identifying micro-organisms. Arguably, the ectomycorrhizal fungi appear extraordinarily diverse relative to other symbionts because their conspicuous and morphologically diverse fruiting bodies (toadstools, truffles, etc.) provide many more taxonomically useful characters than, for example, unicellular bacteria. For some taxa of microbial symbionts, the limitations of traditional methods in microbial taxonomy have been compounded by the use of symbiosis as a taxonomically important character. As examples, the family Rhizobiaceae was erected to accommodate the nitrogen-fixing bacteria in the root nodules of legumes (Jordan, 1984); the mycetocyte symbionts in insects have been described as "allied to rickettsias" because both they and rickettsias are intracellular and unculturable (Lamb and Hinde, 1967); and all the gymnodinoid dinoflagellate algae in benthic marine animals have been assigned to a single species, *Symbiodinium microadriaticum* (Freudenthal, 1962). The construction of these taxa was subsequently used as evidence for a low phyletic diversity of micro-organisms in symbiosis. This reasoning is circular.

DNA-based studies have amply confirmed the comments of Law and Lewis (1983) that traditional taxonomic methods had greatly underestimated the phyletic diversity of many symbiotic micro-organisms. Previously unsuspected diversity has been identified in several groups. Analyses of 16S rRNA genes in the unculturable luminescent bacteria in anomalopid and ceratioid fish have identified two distinct groups of bacteria, each as diverse as all the culturable symbionts in the genera *Vibrio* and *Photobacterium* (Haygood and Distel, 1993). For the algal symbiont *Symbiodinium*, painstaking ultrastructural studies had revealed morphological variants interpreted by Trench and Blank (1987) as distinct species, but the variation in 18S rRNA gene sequence among *Symbiodinium* isolates is as great as in an entire family of nonsymbiotic algae (Rowan and Powers, 1991; McNally *et al.*, 1994). Among the intracellular symbionts of insects, at least two groups of γ-Proteobacteria (in aphids and whitefly: Munson *et al.*, 1991; Clark *et al.*, 1992), one group of β-Proteobacteria (in mealybugs: Munson *et al.*, 1992), and flavobacteria (in cockroaches: Bandi *et al.*, 1994) have been identified.

Molecular techniques, however, have not (to date) resolved a second difficulty: that direct comparisons between the number of species of symbiotic micro-organisms and their hosts may not be valid because a "species" of micro-organisms and higher organisms (especially animals and plants) may not be biologically equivalent. The uncertainty about the relevance of species concepts developed for sexually-reproducing animals and plants to micro-organisms arises from two attributes of microbial reproduction. First, reproduction in many micro-organisms (both eukaryotic and bacterial) is predominantly or exclusively asexual. Second, sexual reproduction in bacteria usually involves the unidirectional transfer of genes from a small portion of the genome (Maynard Smith *et al.*, 1991). In eukaryotes, genetic exchange is bidirectional and involves all parts of the genome.

C. Origins and Radiations of the Partners in Endosymbioses

Despite the several factors described in Section II.B which tend to underestimate symbiont diversity, Law and Lewis (1983) and other authors consider the discrepancy between the diversity of host and symbiont taxa to be so great that it is unlikely to be entirely a taxonomic artefact.

Law and Lewis (1983) attributed the greater diversity of hosts than endosymbionts to evolutionary processes in the intact association. Symbiosis is widely accepted to have triggered adaptive radiations on many host taxa (Douglas, 1992b), but the endosymbionts would not necessarily exhibit a parallel adaptive diversification. Law and Lewis (1983) propose that the conditions in the host environment are uniform, and that the endosymbionts are under stabilizing selection and diverge slowly. However, the rate of adaptive diversification (i.e. diversification in biological function), as considered by Law and Lewis (1983), is not equivalent to the rate of phyletic diversification as assessed by DNA sequence divergence. The molecular phylogenies are constructed, in large part, from selectively neutral sequence changes that are independent of the overlying phenotype.

A further factor contributing to the greater diversity of hosts than symbionts is that the symbiotic micro-organisms have evolved relatively rarely, but have been acquired on multiple occasions by taxonomically diverse hosts. Several systems, which have been the subject of molecular analyses, illustrate this point. The nitrogen-fixing symbionts of the genus *Frankia* almost certainly have a single evolutionary origin (Nazaret *et al.*, 1991), and have been acquired independently by members of 8 different plant families (and possibly more than once within some of these families) (Sprent and Sprent, 1990). The symbiotic chemoautotrophic bacteria also form a single, coherent group distinct from all known nonsymbiotic bacteria in the γ-Proteobacteria (Distel *et al.*, 1988; Distel and Wood, 1992; Eisen *et al.*, 1992), and they have been acquired as endosymbionts by members of at least three animal phyla (see Table 1 and Cavanaugh, 1994).

Detailed analyses of the phylogenies of some symbiotic micro-organisms, however, have revealed multiple evolutionary origins, but commonly from a restricted phylogenetic base. Two systems in which this has been studied in particular detail are the nitrogen-fixing bacteria in the root nodules of legumes and endophytic fungi in grasses. For the legume nodule bacteria, analyses of variation in the 16S rRNA genes have revealed multiple groups of α-Proteobacteria in symbiosis (Hennecke et al., 1985; Dreyfus et al., 1988). They are usually described as the genera Rhizobium, Bradyrhizobium and Azorhizobium, although some authorities would subdivide Rhizobium into two or more genera (e.g. Chen et al., 1991; de Lajudie et al., 1994). Recent studies have, further, shown that certain strains of Bradyrhizobium and Rhizobium are more closely related to the nonsymbiotic bacterium Rhodopseudomonas and the parasite Agrobacterium, respectively, than to other Bradyrhizobium/Rhizobium strains (e.g. Young et al., 1991; Willems and Collins, 1993; Yanaki and Yamasato, 1993). This suggests that symbiosis has evolved (and possibly been lost) several times within these groups of bacteria. The phyletic diversity of endophytic fungi in grasses has been enhanced by several instances of interspecific hybridization. The endophytic fungi have almost certainly evolved from parasites (probably Epichloe) within the ascomycete family Clavicipitaceae (Schardl et al., 1991). Recent molecular studies suggest that the endophyte Acremonium lolii in perennial ryegrass Lolium perenne is the product of hybridization between other Acremonium species and Epichloe typhina (Schardl et al., 1994), and at least three hybridization events involving Epichloe have contributed to the diversity of endophytic fungi in tall fescue grass Festuca arundinacea (Tsai et al., 1994).

Molecular studies have also been instrumental in the recognition of multiple origins among the symbiotic Chlorella (Douglas and Huss, 1986), methanogenic bacteria in ciliates (Embley and Finlay, 1994) and mycetocyte symbionts in insects (Moran and Baumann, 1993b).

In summary, the principal conclusion from recent molecular studies on symbiotic micro-organisms is that, although the capacity to enter into symbiosis has evolved rarely, these micro-organisms are far more diverse than was previously appreciated. The total tally of symbiont taxa is likely to continue to rise, as more hosts in a wider variety of habitats are sampled. For example, the number of described Rhizobium species has been increased from eight to ten, through a single taxonomic survey of the bacteria associated with the tree legumes Sesbania and Acacia in Senegal, West Africa (de Lajudie et al., 1994).

III. THE DISTRIBUTION OF MICRO-ORGANISMS IN ENDOSYMBIOSIS

The various taxa within any group of endosymbiotic micro-organisms are not distributed randomly among host species. The patterns of their distributions suggest that, first, the source from which individual hosts acquire their complement of symbionts and, second, the specificity of the associations are important determinants of the distribution of symbionts. These host-related factors are considered in Section III.A. Overlying the host-related determinants of symbiont distribution, however, are factors external to the symbioses, such as temperature or water availability; and these are discussed in Section III.B.

A. Host-related Determinants of Endosymbiont Distribution

1. Modes of Transmission

Symbiotic micro-organisms may be transmitted among hosts in two ways: vertically, from parent host to offspring, and horizontally. Vertical transmission is the norm in asexually reproducing hosts as, for example, in many protist hosts, and some invertebrates and lichens (which have specialized asexual propagules, such as soredia and isidia), and it also occurs in some sexually reproducing animals (especially insects) and plants (e.g. some hosts of endophytic fungi). Otherwise, horizontal transmission is the norm in sexually reproducing hosts, especially in morphologically complex hosts whose symbionts are housed at a distance from the gametes (e.g. no root symbionts of plants are transmitted vertically). Many horizontally transmitted symbionts are released from one host into the environment, from where they colonize other hosts. Examples include rhizobia in soils, and many algae and chemosynthetic and luminescent bacteria in the sea. Mycorrhizal fungi are also horizontally transmitted, either via spores (or other perennating structures) or, in established vegetation, by hyphal links between the root systems of neighbouring plants (Newman, 1988).

Vertical transmission tends to restrict the distribution of symbionts among hosts. If this mode of transmission is sustained through many host generations and horizontal transmission is entirely absent, the symbiont will be limited to a single host lineage, and ultimately the phylogeny of the symbionts will come to mirror that of the host. As an example, a remarkable congruence between host and symbiont phylogenies has been demonstrated for the aphid-*Buchnera* association (Munson *et al.*, 1991); the co-incidence of molecular and fossil evidence suggests that these bacteria have been transmitted vertically for some 200 million years (Moran *et al.*, 1993a) (Fig. 1A). A counter-example is the alga *Symbiodinium*. The phylogeny of *Symbiodinium*, as determined from 18S rRNA gene sequence, shows no concordance with the phylogeny of its cnidarian hosts (Rowan and Powers, 1991) (Fig. 1B), even though the *Symbiodinium* is transmitted vertically at asexual reproduction in virtually all hosts and at sexual reproduction in up to half of the host species. Congruence between the cnidarian and *Symbiodinium* lineages is probably precluded by the evolutionary plasticity of reproductive traits, including symbiont transmission, in the cnidarian hosts (Fautin, 1991).

The uniformity of the symbionts in vertically transmitted systems is further enhanced by the process of sorting. Sorting can best be explained by considering a host with multiple symbiont taxa. If the different symbionts are transmitted at random to the host offspring, it is likely that not all offspring will receive every type of symbiont in the same proportion as in the parent. If an offspring failed to acquire one type of symbiont altogether, then the diversity of symbionts in that individual and all its descendants would be reduced irrevocably. More generally, over multiple host generations, an initially mixed population of micro-organisms would become sorted, ultimately such that any one host individual would contain a single type of symbiont.

One crucial aspect of obligately vertically-transmitted symbionts is that their abundance and distribution is indistinguishable from that of their host. Although conflicts of interest may arise between vertically transmitted micro-organisms and their hosts (Hurst, 1992; Law and Hutson, 1992), the principal means by which a vertically transmitted symbiont enhances its own ecological amplitude is by increased effectiveness (i.e. by promoting host survival, growth, reproduction, etc.). The ecology of horizontally transmitted symbionts is more complex because their abundance and distribution are promoted not only by their effectiveness, but also by their infectivity, i.e. the number and diversity of hosts that they can colonize. Key components of the infectivity of these symbionts are their host range and competitiveness, as discussed below.

2. The Host Range of Symbiotic Micro-organisms

The host range of symbiotic micro-organisms is assessed by two complementary methods, commonly known as the "experimental" and "ecological" methods. In the experimental approach, the capacity of a micro-organism to form an association with a panel of potential hosts is determined; and in the ecological approach, the diversity of hosts bearing the micro-organism in nature is quantified. Estimates of host range are usually greater by the experimental approach, mostly because some host/symbiont combinations, which fail to survive in the field, persist under laboratory conditions.

In relation to the distribution of symbiotic micro-organisms, the most

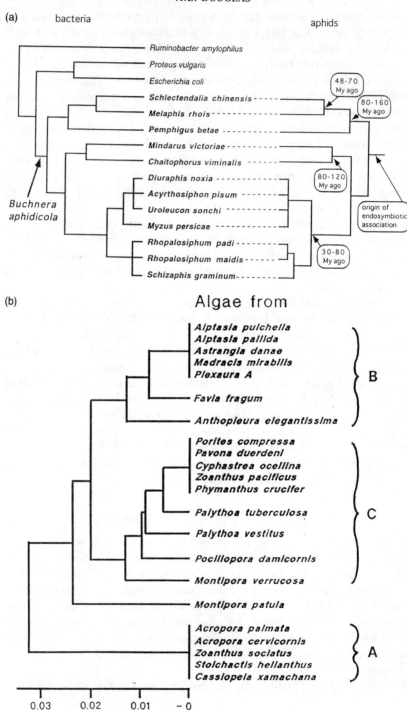

relevant information is the host range of the symbionts in natural conditions, i.e. as determined by the ecological approach (see review of Law, 1987). Unfortunately, virtually all research on the specificity of associations is conducted in the laboratory, and the focus of most studies is the specificity of the host for its microbial partners, and not (the concern here) that of the micro-organisms for their hosts. Relatively extensive data sets are available for only two groups of symbiotic micro-organisms: ectomycorrhizal fungi and rhizobia in soils. In both systems, the host range appears to vary widely among micro-organisms.

The host plant range of ectomycorrhizal fungi has been surveyed by Molina *et al.* (1992), and their data on more than 3000 fungal species are summarized in Table 2A. Just over a quarter of these fungi are restricted to a single plant genus, and a similar proportion have a very broad host range, utilizing many plant species including both angiosperms and gymnosperms. The remainder, nearly half of the total, have an intermediate host range. Examples of extremes in host range are, first *Suillus cavipes*, which utilizes only *Larix*; and second, *Cenococcum geophilum*, which has been described in association with most plant genera that form ectomycorrhizas and even some plants that typically form arbuscular mycorrhizas.

Variation in the host range among the rhizobia has been recognized for decades, but until recently the variation was perceived as a simple difference between *Bradyrhizobium* with a broad host range and *Rhizobium* with a narrow host range. With increasing information, it is becoming apparent that the host range of *Rhizobium* varies widely. Some isolates of *Rhizobium* from agricultural habitats can nodulate just a single legume genus, but other isolates, especially from plants in natural vegetation, have a broad host range. The host range of *Bradyrhizobium* isolates varies from several plant genera to "very broad" (e.g. Turk and Keyser, 1991; Souza *et al.*, 1994).

One further group of symbiotic micro-organisms, the arbuscular mycorrhizal fungi, warrants consideration because these symbionts are widely

Fig. 1. The impact of mode of transmission on the phylogenies of endosymbiotic micro-organisms.
a. The phylogeny of the vertically-transmitted bacteria of the genus *Buchnera* and their aphid hosts. The bacterial phylogeny is based on 16S rRNA gene sequences, and the aphid phylogeny is based on morphology. Dashed lines indicate associations, the taxa of *Buchnera* are indicated by their aphid hosts, and the dates are estimated from aphid fossils and/or biogeography.
(Figure reproduced from Moran and Baumann (1993b).
b. The phylogeny of the dinoflagellate algae *Symbiodinium* isolated from various Cnidarian hosts, as constructed from 18S rRNA gene sequences. Genetic distances (estimated average fraction of nucleotides substituted) are indicated. The algae are divisible into three groups (A, B, C), which are not congruent with host taxonomy.
(Figure reproduced from Rowan and Powers (1991)).

Table 2
The host plant range of basidiomycete ectomycorrhizal fungi[1]

A. *Host range of fungal species of different families*

Family of fungus	Number of fungal species		
	narrow host range	intermediate host range	broad host range
Amanitaceae	10	70	120
Boletaceae	156	142	67
Cortinariaceae	365	743	356
Gomphidiaceae	20	11	0
Hygrophoraceae	88	125	37
Russulaceae	257	345	200
Sclerodermataceae	8	15	8
Tricholomataceae	35	68	76
Total	939 (28%)	1519 (46%)	864 (26%)

B. *Host range of fungal species with different modes of spore dispersal*

Mode of dispersal	Number of fungal species		
	narrow host range	intermediate host range	broad host range
Epigeous	743 (25%)	1342 (46%)	864 (29%)
Hypogeous	196 (53%)	177 (47%)	0

[1] The values are calculated from the data in Molina *et al.* (1992). Narrow host range—restricted to a single plant genus; intermediate host range—colonizes members of one to several host families (but not both angiosperms and gymnosperms); broad host range—associated with a diversity of plants, including both angiosperms and gymnosperms.

cited as having a very broad host range. Glasshouse trials have routinely failed to detect specificity (Harley and Smith, 1983), leading to claims that arbuscular mycorrhizal fungi can colonize any plant capable of forming this type of association. However, it may not be appropriate to extrapolate these conclusions to the natural situation. Detailed studies of specificity in the field await the development of reliable molecular probes by which mycelium of the various fungi can be distinguished (Simon *et al.*, 1992; Clapp *et al.*, 1995), but an indication that some specificity may occur in natural conditions comes from the study of McGonigle and Fitter (1990) on the mycorrhizal community associated with herbaceous plants in a hay meadow in northern England. Three of the four dominant plant species bore mycorrhizal fungi that had coarse hyphae, but one, *Holcus lanatus*, was preferentially colonized by a fungus, possibly *Glomus tenue*, with distinctive fine hyphae.

THE ECOLOGY OF SYMBIOTIC MICRO-ORGANISMS

The indications from these few systems that the host range of symbionts may be variable raises the issue of the ecological determinants of host range. Do micro-organisms with broad and narrow host ranges occur in different ecological contexts? This question has barely been considered within the symbiosis literature, beyond the view that the host range of symbionts "should" be broad. The argument is that, for a mutualistic association, any mutation (in any partner) that restricts the variety of hosts that can be colonized by symbiotic micro-organisms would be selected against (Thompson, 1982; Law, 1987). This argument presupposes that symbioses are consistently mutualistic. In reality, symbioses are not invariably mutually beneficial (see Section I.B), and host resistance to some symbiotic micro-organisms may be promoted by antagonistic interactions, either between host and symbiont or between different hosts, as described below.

The selection pressure on the host to be resistant to ineffective symbionts may be an important factor limiting the host range of symbionts. This issue has not been explored, beyond the demonstration that some hosts are colonized by apparently ineffective endosymbionts in field situations (see Section III.A.3). Other factors may influence host susceptibility to symbionts. In associations where the complement of symbionts in one host may act as an inoculum for competing hosts, it may be in the first host's advantage to be colonized by symbionts with a narrow host range. The impact of inter-host competition on the symbiosis has not been studied in detail, but one possible example concerns the influence of ectomycorrhizas on the successional displacement of *Pseudotsuga menziesii* (Douglas fir) by *Tsuga heterophylla* (western hemlock) in temperate forests of western USA. Most of the fungi associated with *P. menziesii* are specialist to this species, while *T. heterophylla* predominantly uses fungi with a broad host range (Schoenberger and Perry, 1982). Kropp and Trappe (1982) suggest that it is advantageous for *P. menziesii* to be susceptible to fungi with narrow host range because this reduces the availability of fungal hyphae compatible with *T. heterophylla*, and so acts to retard successional displacement. In this context, the advantage of narrow host range to the fungal symbionts of *P. menziesii* is unclear.

The interaction between the host range of symbionts and competitive interactions among hosts deserves renewed attention, both in ectomycorrhizal vegetation and in other systems. The host range of the alga *Symbiodinium* in reef communities may be of particular interest because competitive interactions between corals are very intense (Jackson, 1991) and *Symbiodinium* is regularly expelled from corals (see Section IV.B.1) and, potentially, made available to the coral community as a whole.

In a strictly ecological context, the distribution of symbionts with different host ranges may often be determined primarily by the probability that symbionts contact hosts with which they are compatible. By definition, any symbiont which fails to encounter a compatible host will not persist in symbiosis. The variety of hosts encountered is determined, first, by the species diversity of hosts in a habitat and, second, by the characteristics of dispersal of symbiont propagules. The ectomycorrhizal fungi, again, provide good illustration of both these factors.

Many plant communities with ectomycorrhizal fungi are of low diversity, and ectomycorrhizal forests, such as the eucalypts in Australia and conifers in north America, support large guilds of highly specific ectomycorrhizal fungal species and smaller numbers of fungi with a broad plant range (Molina *et al.*, 1992). In disturbed habitats, where the composition of vegetation is very variable, either spatially or temporally, ectomycorrhizal fungi with broad plant ranges are commonly dominant. For example, in the arid forests of California, regular fires create a complex mosaic of successional stages between ericaceous shrubs (e.g. *Arctostaphylos uva-ursi*) and coniferous forest (especially of *Pseudotsuga menziesii*). The subterranean mycelia of the ectomycorrhizal fungi, which can survive the fires, are dominated by species (e.g. *Rhizopogon vinicolor*, *Suillus lakei*) that can utilize both ericaceous and coniferous plants and consequently persist, irrespective of the shifts in vegetation composition (Molina and Trappe, 1982). However, the relationship between host range of mycorrhizal fungi and species diversity of plants does not appear to hold for all types of vegetation. In particular, the ericoid mycorrhizal fungi isolated from *Calluna* heaths (of very low plant diversity) have a remarkably broad host range within the Ericales (Pearson and Read, 1973).

Among ectomycorrhizal fungi, the characteristics of propagule dispersal are correlated with the location of the fruiting bodies (Molina *et al.*, 1992). In fungi with epigeous (aboveground) structures, such as toadstools and puffballs, the spores are discharged to the air and dispersed randomly (with respect to host plant distribution) by air currents; but the fungal species with hypogeous (belowground) structures, such as truffles, are dispersed by mycophagous animals, with an increased probability of transmission between conspecific plants. As shown in Table 2B, a higher proportion of fungi with the hypogeous than epigeous habit have a narrow host range, and all the species with a very broad host range are epigeous.

3. Competition among Endosymbionts

Competition between micro-organisms for a host arises inevitably from the broad host range of many microbial symbionts and the finite capacity of hosts to accommodate micro-organisms. Experimentally, competition among symbionts can be identified as a reduction in colonization by an endosymbiont, when a host is challenged with several different symbionts.

Competitive interactions have been demonstrated among root symbionts, especially rhizobia and mycorrhizal fungi (Triplett and Sadowsky, 1992), and among animal symbioses with algae (Provasoli *et al.*, 1968) and luminescent bacteria (Lee and Ruby, 1994a). Most of the research has addressed two issues: the relationship between the competitiveness and effectiveness of various symbionts; and the host resources for which the symbionts may compete.

The first animal association in which competition among symbionts was identified was the symbiosis between the intertidal flatworm *Convoluta roscoffensis* and its symbiotic algae of the genus *Tetraselmis* (= *Platymonas*). A single algal species populates each animal, and the algal symbionts are derived from free-living cells, ingested by the juvenile (algafree) animal. The juvenile animal feeds indiscriminately on *Tetraselmis* species, and all but one species is discarded via the epithelial mucus glands (Douglas, 1980). Provasoli *et al.* (1968) and Douglas (1985) have demonstrated a hierarchy of competitiveness among *Tetraselmis* species, that is correlated with their effectiveness (as determined by the growth rate of the host) (Table 3). The resource for which the symbionts compete is unknown, but competitiveness and effectiveness are both correlated with the extent of photosynthate release to the animal tissues (Table 3).

In contrast to the *C. roscoffensis* association, the competitiveness of symbionts in the rhizobia-legume and mycorrhizal associations does not vary consistently with effectiveness, as defined by plant growth and yield (e.g. Triplett and Sadowsky, 1992; Pearson and Jackson, 1993). In other words, the host plant does not necessarily become infected with the symbiont strain that would maximize its performance.

Among rhizobia, the indigenous populations in agricultural soils are often relatively ineffective symbionts of legume crops, but they are highly competitive, preventing the formation of a symbiosis between effective rhizobial strains inoculated into the soils with the crops. This effect has been demonstrated elegantly by Thies *et al.* (1991) with the rhizobia in Hawaiian soils. A variety of legume crops was grown at 8 sites with widely divergent densities of indigenous rhizobia (which nodulated the legume crops very poorly). In soils with 10 or fewer indigenous rhizobia g^{-1} soil, inoculation increased the number and mass of nodules manyfold; at sites with 10–100 rhizobia g^{-1}, nodulation was enhanced two–threefold; and in soils with >100 indigenous rhizobia g^{-1} soil, inoculation did not significantly increase nodulation (even though nodulation was not maximal under these conditions). The "competition problem" between indigenous and inoculated rhizobia is a common cause of low yield of legume crops on low-nitrogen soils.

The characteristics of microbial symbionts of plant roots which determine their competitiveness are not well understood. Among rhizobia, com-

Table 3
The competitiveness and effectiveness of algal symbionts in *Convoluta roscoffensis*.[1] The experiments were conducted on *C. roscoffensis* collected from Aberthaw, Cardiff. This population contains two taxa of the algal genus *Tetraselmis*, distinguishable at the light-microscope level: algae of the subgenera *Tetraselmis* and *Prasinocladia* (see Douglas (1985)).

A: Competitiveness of subgenera *Tetraselmis* and *Prasinocladia*
Alga-free 5-day-old juvenile animals were exposed to cultures of the algae (as indicated) for 12 h, washed free of external algae. When the animals were 30 days old, their algal symbionts were identified (each animal contained just one algal taxon).

Infecting algae (at day 5) No. algal cells ml^{-1} external medium		Algae in symbiosis (at day 30) No. animals containing algae (n = 30)	
Tetraselmis	*Prasinocladia*	*Tetraselmis*	*Prasinocladia*
1×10^4	0	30	0
5×10^3	5×10^3	30	0
1×10^3	9×10^3	30	0
1×10^2	1×10^4	21	9
0	1×10^4	0	30

B: Effectiveness of subgenera *Tetraselmis* and *Prasinocladia*
The lengths of *C. roscoffensis* are for adults collected from the field site, and juveniles experimentally infected with *Tetraselmis* and *Prasinocladia* and maintained in the laboratory

Alga in animal	Length of adult animals mm (mean ± s.e., n = 50)	Length of 30-day-old juvenile animals mm (mean ± s.e., n = 28)
Tetraselmis	$2 \cdot 97 \pm 0 \cdot 12$	$0 \cdot 42 \pm 0 \cdot 08$
Prasinocladia	$2 \cdot 14 \pm 0 \cdot 16$	$0 \cdot 30 \pm 0 \cdot 03$
	$d = 4 \cdot 17$, p $< 0 \cdot 001$	$t = 2 \cdot 38$, p $< 0 \cdot 05$

C: Photosynthate transfer in symbioses with subgenera *Tetraselmis* and *Prasinocladia*
Photosynthetic ^{14}C fixation and translocation were quantified for 60-day-old animals reared, after experimental infection, in the laboratory

Alga in animals	% ^{14}C translocated to animal tissues mean ± s.e. (n = 10)
Tetraselmis	$48 \pm 5 \cdot 8$
Prasinocladia	$12 \pm 0 \cdot 9$

[1] Douglas, A.E. unpub. results.

petition for infection sites may be important. This is indicated by the depressed competitiveness of nonmotile mutants of *Rhizobium meliloti* (e.g. Caetano-Anolles *et al.*, 1988) and positive correlation between speed of nodulation and competitiveness (e.g. Kosslak *et al.*, 1983; McDermott and Graham, 1990). However, in both rhizobial-legume associations and mycorrhizal systems, the competitive hierarchy is maintained in split-root systems, where the two competing symbionts cannot make contact (Sargent *et al.*, 1987; Pearson *et al.*, 1993). These data indicate that the limiting resource is not invariably the availability of infection sites. Pearson *et al.* (1994) suggest that, in the mycorrhizal system, the fungi are probably competing for plant carbohydrate.

The repeated demonstration that the most competitive root symbionts are often not the most effective begs the question why host plants are susceptible to relatively ineffective symbionts. The problem is potentially complex because the effectiveness of the symbionts is likely to vary with environmental conditions. As an example, Newsham *et al.* (1994) have identified an arbuscular mycorrhizal fungus that is both very infective and (under most conditions) ineffective in promoting phosphorus acquisition as a symbiont of the winter annual grass *Vulpia ciliata*. However, the fungus protects the plant roots from fungal pathogens, and so is very effective in soils bearing the fungal pathogen *Fusarium*. For rhizobial-legume associations, the relationship between effectiveness and infectivity is further complicated by the agricultural industry, which defines symbiont effectiveness in terms of plant growth rate and yield. As the mycorrhizal example (above) illustrates, promotion of plant growth may not be the most important attribute of symbionts to the plant under natural conditions. It is conceivable, for example, that the competitiveness of indigenous rhizobia is advantageous to plants in natural situations, where the competitors are bacterial or fungal pathogens, but not when the competitors are "elite" symbiont inoculants selected by twentieth-century agriculture.

B. Determinants of Endosymbiont Distribution External to the Symbiosis

The factors external to an association which may directly influence symbiont distribution are predominantly abiotic, and may include temperature, moisture availability, etc. The impact of such factors on the micro-organisms in symbiosis depends critically on the homeostatic capabilities of the host. For example, external temperature would have a far greater influence on the temperature experienced by the microbiota in poikilothermic insects than in homeothermic mammals; and the effect of the availability of external water on the water relations of symbionts would be greater in the poikilohydric lichens than in the homeohydric vascular plants.

The identification of external factors influencing symbiont distribution is greatly simplified if there is a robust spatial pattern in the distribution of different symbiont species (whether over a scale of metres or thousands of kilometres) that can be correlated with environmental variables. Once such a correlation is established, the causal relationship between these variables and symbiont distribution can be explored by experimental manipulation (e.g. transport experiments, laboratory studies). However, interpretations of such a pattern maybe confounded by host-related factors, if the abundance of host species with different specificities varies with the environmental variables under study (although, in many cases, these host-related factors can be controlled for by appropriate sampling procedures and methods of analysis).

These issues have rarely been addressed for any symbiotic micro-organisms, but one particularly detailed study is that of Koske (1987) on the distribution of arbuscular mycorrhizal fungi. The mycorrhizal fungal species recovered as spores from vegetation in the dune habitats along 400 km of the east US coast varied consistently with latitude, with *Gigaspora gigantea*, *Scutellospora persica* and *S. calospora* favoured at the north end of the transect and *S. dipapillosa* and *S. fulgida* at the south. Overall, there was no striking effect of plant species on the fungal distribution, and the latitudinal gradient in fungal composition could be attributed primarily to variation in temperature.

Temperature has also been implicated in the distribution of the three culturable luminescent symbionts in the light organs of fish. Most fish bearing *Photobacterium phosphoreum* occur in cold (usually deep) waters, whereas temperate fish have *Vibrio fischeri* and tropical species tend to have *P. leiognathi*; and these distributions correlate well with the temperature optima for growth of the three bacterial species in culture (reviewed in Haygood, 1993). However, there are exceptions to this pattern, indicative of the impact of host-specificity on symbiont distributions. For example, leiognathid fish (which invariably have *P. leiognathi*) occur in both temperate and tropical waters (McFall-Ngai, 1991).

IV. ENDOSYMBIOTIC MICRO-ORGANISMS IN THE ENVIRONMENT

A. On the Rarity of Symbiotic Micro-organisms in the Environment

Most symbiotic micro-organisms are unknown or apparently rare free-living in the environment (i.e. apart from their hosts). Symbionts that have never been described in the external environment include the chemoauto-trophic bacteria in marine invertebrates and the mycetocyte bacteria in insects. Free-living cells of other symbiont taxa are either only known in the immediate vicinity of their hosts (e.g. the alga *Trebouxia* of lichens: Bubrick *et al.*, 1984), or as apparently perennating structures (e.g. the chlamydospores of arbuscular mycorrhizal fungi). Exceptionally, the rhizo-bia and some orchid mycorrhizal fungi have substantial free-living popu-lations in the soil (Beringer *et al.*, 1979; Molina *et al.*, 1992) (although rhizobia probably cannot persist indefinitely in the absence of host plants, see Section IV–C). Some symbiotic strains of the luminescent bacteria *Photobacterium* and *Vibrio* may also maintain large populations in the water column (Ruby *et al.*, 1980).

There are, however, two caveats in our understanding of the condition of symbiotic micro-organisms in the environment: the inadequacy of methods used to identify micro-organisms in natural environment, and ignorance of the underlying physiological reasons for the (apparent) ex-clusion of most symbionts from the environment.

The principal evidence for the view that most symbiotic micro-organisms lack substantial free-living populations comes from the absence of these taxa from the micro-organisms isolated into culture from water, soil, etc. It is conceivable that symbiotic micro-organisms are present, but remain undetected because they fail to grow (or grow very slowly) on the standard microbiological media used. In other words, the studies fail to distinguish between ecological obligacy (i.e. absence of free-living populations) and unculturability. This potential difficulty is not unique to the study of sym-biotic micro-organisms. Through molecular analyses of the microbiota in a variety of habitats, it has become apparent that methods based on *in vitro* cultivation may under-estimate microbial diversity by several orders of magnitude (e.g. Britschgi and Giovanonni, 1991; Fuhrman *et al.*, 1993; DeLong *et al.*, 1994). The continuing application of these molecular methods to aquatic and terrestrial environments could reveal hitherto unsuspected free-living populations of symbiotic micro-organisms. In other words, "routine" microbial ecology, based on techniques independent of *in vitro* cultivation, should provide the data to establish definitively whether most symbiotic micro-organisms are specialized to symbiosis.

Most of the reasons given in the literature for the (apparent) inability of symbiotic micro-organisms to maintain substantial free-living populations can be reduced to the truisms that they suffer high mortality or are out-competed by other micro-organisms. The rarity of a few symbiotic taxa can plausibly be attributed to their very fastidious nutritional requirements. For example, most ectomycorrhizal fungi preferentially utilize simple sugars as carbon sources (Harley and Smith, 1983) and they are therefore likely to be at a competitive disadvantage with saprotrophic fungi in the soil, where the principal carbon sources are cellulose, lignin, etc. It is also plausible that some bacteria would be outcompeted by non-symbiotic forms because they are genetically incapable of growing fast. In particular, the mycetocyte bacteria in some insects and luminescent bacteria in anomalopid fish have a single copy of rRNA genes, while bacteria capable of growing rapidly have multiple copies (Untermann and Baumann, 1990; Wolfe and Haygood, 1993). However, there is no definitive evidence that the persistence of micro-organisms is correlated with (or dependent on) their growth rates. *Bradyrhizobium* grows considerably more slowly than *Rhizobium* in culture, but there is no indication that *Bradyrhizobium* maintain smaller free-living populations than *Rhizobium* in the soil; symbiotic *Chlorella* are unknown apart from the symbiosis but when cultured on basal media, they have intrinsic growth rates comparable to those of other *Chlorella* (Douglas and Huss, 1986).

The only certain conclusion of these considerations is an urgent need for research on symbiotic micro-organisms in the natural environment. To date, virtually all studies of culturable symbionts have been conducted on laboratory cultures, whereas unculturable forms have been ignored.

One further issue, however, arises: the source of those symbiotic micro-organisms which have been identified in the environment. If these micro-organisms are specialized, such that they cannot persist indefinitely apart from their hosts, then the free-living forms are presumably derived from symbiosis. The significance of symbioses as a source of free-living populations of symbiotic micro-organisms is examined in the remainder of this section.

B. Symbiosis as a Source of Symbionts in the Environment

It is widely accepted, at least implicitly, that most free-living populations of symbiotic micro-organisms are not self-sustaining, but maintained by input from symbioses. This position requires, first, that viable micro-organisms are regularly released from symbiosis; and, second, that the free-living populations of symbiotic micro-organisms decline to extinction in the absence of the association. These two issues are considered here.

1. The Release of Micro-organisms from Symbiosis

Viable cells of symbionts are released from some associations into the external environment, where they may persist, at least transiently. In some systems, symbionts are regularly expelled from the intact association; this phenomenon may be adaptive for the host, as a means to enhance the probability that offspring are infected or to "cull" the symbiont population. Viable symbionts may also be released from many associations when the host is stressed or dies.

Extracellular symbionts with access to the external environment are particularly likely to be shed from the association. In particular, luminescent bacteria are expelled from the light organs of animals at rates of 10^6–10^8 cells per day from monocentrid and anomalopid fish (Nealson *et al.*, 1984), and *c.* 5×10^8 cells from the Hawaiian squid *Euprymna scolopes* (Lee and Ruby, 1994b). These values are substantial, relative to the standing stock of 10^7–10^9 bacteria in the animal.

Most intracellular symbionts are not routinely eliminated from the intact association, and in many systems, few, if any, symbiont cells ever gain access to the external environment. For example, the mycetocyte bacteria in some insects, notably aphids, are maintained in cells, except for a fleeting extracellular phase during transmission to the eggs in the maternal ovaries, and any bacterial cells that "escape" from the host cells are lysed in the insect haemocoel (Hinde, 1971). Exceptionally among the intracellular symbionts, the alga *Symbiodinium* in Cnidaria is routinely expelled. The extent of expulsion is probably low in unstressed animals. For example, Hoegh-Guldberg *et al.* (1987) estimate that less than 5% of the algal population is lost, daily, from the coral *Stylophora pistillata*. However, when Cnidaria are exposed to deleterious conditions, such as low salinity, high light intensity or high temperature, algal expulsion may be massive, leading to "bleaching" of the animal host (see Section I.A). The fate of the expelled *Symbiodinium* cells has not been investigated, but it is generally assumed that they do not persist indefinitely in the water column.

2. Impact of the Symbiosis on the Abundance of Symbionts in the Environment

The luminescent bacteria and rhizobia are technically the most feasible systems to study the contribution of symbiosis-derived micro-organisms to free-living populations because these two groups of bacteria attain substantial densities in the environment.

To date, the relationship between populations of luminescent bacteria in symbiosis and the environment has been investigated in just one association, between the squid *Euprymna scolopes* and *Vibrio fischeri* at Kaneohe Bay, Hawaii. Lee and Ruby (1994b) have demonstrated, first, that symbiotic *V. fischeri* is the dominant luminescent bacterium in the water column and sediment at sites regularly frequented by *E. scolopes*; and, second, that the abundance of *V. fischeri*, but not other luminescent bacteria, declines progressively with distance from the *E. scolopes* habitat. These data can reasonably be interpreted in terms of the symbiosis as a primary source of free-living populations of *V. fischeri*, but an analysis of the abundance of *V. fischeri* in *E. scolopes* habitats from which all hosts were experimentally removed would provide a definitive test.

Unlike the luminescent bacteria in light organs of marine animals, rhizobia are not released from the functional symbiosis with roots of legumes. The occasion when they may gain access to the soil is at nodule senescence. The life span of nodules varies from several weeks to more than a year, depending on the plant species and environmental conditions (Sprent and Sprent, 1990).

There is extensive evidence that host plants promote free-living rhizobial populations. The abundance of rhizobia in soils supporting host plants is high (often in excess of 10^4 cells g^{-1} rhizosphere soil) and declines progressively with both distance from host plants and time since host plants were grown in the soil (Beringer *et al.*, 1979; Kuykendall *et al.*, 1982; Woomer *et al.*, 1988). However, there is a longstanding debate whether the plants promote free-living populations because viable cells are released from root nodules (e.g. Gresshoff *et al.*, 1977; Kucey and Hynes, 1989) or because root exudates from plants bearing nodules provide resources that enhance free-living populations (Soberon and Shapiro, 1989; Olivieri and Frank, 1994), i.e. whether or not passage through the symbiosis is required for maintenance of free-living rhizobial populations. This debate illustrates the very real complexities that can underlie the relationship between microbial populations in symbiosis and in the environment, and it is summarized here.

The rhizobial cells in legume nodules certainly have the potential to amplify the free-living rhizobial population. The density of rhizobia in nodules is *c.* 10^9 cells ml^{-1}, several orders of magnitude greater than that in soils; and, at the demise of a nodule, the enclosed rhizobia are uniquely positioned to exploit the plant nutrients from senescing nodule tissues and to proliferate rapidly. However, the number of nodule-derived rhizobia which enter the soil population may be low because, first, many of the rhizobial cells may be destroyed along with the plant cells at nodule senescence (Pladys *et al.*, 1991) and, second, the nitrogen-fixing bacteria (known as bacteroids) may be terminally-differentiated, unable to switch from biotrophic lifestyle in the nodule to life as a saprotroph in the soil (Quispel, 1988). At present, there are no realistic estimates of the survival of rhizobia at nodule senescence. The principal method to assess viability has been to disrupt nodules, usually mechanically, and score the number of culturable bacteria recovered. This approach gives variable estimates of viability, depending on method of isolation, rhizobial strain and plant species (Paau *et al.*, 1980; Sutton *et al.*, 1977; Tsien *et al.*, 1977), and arguably it is not relevant to the survival of nodule senescence.

Research has also been conducted on the components of host root exudates that may be utilized by free-living rhizobia. Olivieri and Frank (1994) have argued that the putative nutritional resources in the plant exudates are utilized preferentially, or exclusively, by rhizobia (if the symbiosis-derived resources were generally available to the soil microbiota, then the "benefit" would be dissipated). Consistent with this reasoning, exudates of pea roots contain the non-protein amino acid, homoserine, which is almost exclusively utilized by pea-nodulating strains of *R. leguminosarum* (van Egeraat, 1975). In some strains of *R. meliloti*, the differentiated bacteroids in the plant nodule synthesize an inositol derivative (generically known as a rhizopine) which is released from the nodule and catabolized exclusively by bacteria of the same strain in the soil. The genes for the synthesis of the rhizopine (*mos* genes) are expressed only in the bacteroid cells in the nodule, and the genes for its catabolism (*moc* genes) are expressed only in free-living cells of the same strain. To date, rhizopines have been identified in a few strains of *R. meliloti* and possibly *R. loti* (Murphy *et al.*, 1987; Murphy and Saint, 1991), and the general significance of this interaction is uncertain.

Implicit in all these considerations of the relationship between rhizobia in symbiosis and in the soil is the assumption of a positive correlation between the abundance of particular genotypes of rhizobia in nodules and in the surrounding soil. A recent study of Leung *et al.* (1994) on *R. leguminosarum* bv *trifolii* infecting agricultural fields of annual clover (*Trifolium subterraneum*) suggests that this assumption may not be valid. At least 13 antigenically distinct strains of the rhizobium were identified; one (AS6) was dominant, occupying 60–73% of all nodules tested. However, contrary to expectation, strain AS6 was not the most abundant strain in the rhizosphere. The density of free-living cells of strain AS6 was 6–7 times lower than that of strain AS21, which colonized only 5–30% of nodules. For the clover-rhizobium system, the relationship between abundance in symbiosis and in the soil appears to be far more complex than is suggested by various hypotheses considered in this section. Further studies are needed to establish the generality of the findings of Leung *et al.* (1994).

It is sobering that, even for the intensively studied symbiosis between rhizobia and legume crops, the relationship between the populations of symbionts in association and the environment is uncertain.

V. THE HOST AS A HABITAT

The apparent rarity of endosymbiotic micro-organisms in the external environment (Section IV) suggests that the life histories of these organisms may be dominated by the symbiotic phase, and that the host organism is their principal habitat. This section considers the usefulness of general ecological classifications of habitats (Southwood, 1988) to describe the host habitat.

Endosymbionts have many characteristics of organisms at the high adversity/low disturbance pole of the scheme of Greenslade (1983), equivalent to the "stress tolerator" strategy of Grime (1979). These characteristics are that they have evolved rarely, are specialized to the host habitat, and have low growth and reproductive rates (Hoffmann and Parsons, 1993). Moulder (1979, 1985) has also drawn explicit parallels between intracellular micro-organisms and the micro-organisms that inhabit physically harsh environments, e.g. desert soils and hot springs.

It is, however, premature to describe symbiotic micro-organisms definitively as "stress/adversity tolerators". The principal difficulty is our collective ignorance of the nature of the putative stresses in endosymbioses. There are some indications that symbiotic algae may be nutrient-limited (reviewed in Douglas, 1994), but virtually no information is available on the nutritional status of symbiotic bacteria or fungi. The conditions of pH, osmolarity and ionic composition encountered by most symbiotic micro-organisms are also unknown, beyond the demonstration that the pH surrounding the algae *Chlorella* and *Symbiodinium* in aquatic invertebrates is not more acidic than 5·5 units (Rands *et al.*, 1992, 1993). Of possible relevance is the finding that several intracellular bacterial symbionts have elevated levels of a "stress protein", the chaperonin GroEL (Choi *et al.*, 1991; Kakeda and Ishikawa, 1991); but detailed ecological interpretation of these observations awaits a clear understanding of the function of this protein in the symbiotic context.

In the absence of definitive information on the stresses in symbiosis, consideration of the relationship between micro-organisms and their symbiotic habitat can lead to circular reasoning. This can be illustrated by the general pattern of microbial colonization of the lumen and epithelial cells in insect guts. The microbiota in the gut lumen is more diverse, has higher growth rates and is more amenable to cultivation than that in the gut epithelial cells (Buchner, 1965). One might describe the gut lumen as a more "favourable" environment than the cells; but the principal way in which the habitat quality is assessed is by the taxonomic diversity of colonizing organisms, growth rates and culturability.

VI. CONCLUSIONS

Perhaps, the single most important issue that is illustrated by this article is the dearth of basic information on the ecology of most endosymbiotic micro-organisms. Only infrequently have ecological principles been used to explore the biology of these organisms, and their ecology has not, to date, developed into a coherent field of research. The situation should, however, change rapidly in the coming years, with the application of molecular techniques to microbial ecology and increasing realization of the ecological significance of microbial endosymbioses.

Despite the current paucity of information, two generalities emerge from the reviewed literature: that endosymbiotic micro-organisms have evolved rarely, and that most are specialized to the "host habitat". The rarity (and perhaps absence) of symbiotic micro-organisms that can persist indefinitely apart from the association confirms the view of these organisms as a coherent group, ecologically distinct from micro-organisms which do not enter into endosymbioses. These characteristics of endosymbionts can be linked to the complexity of their interactions with hosts, involving the sustained exchange of specific nutrients between the partners and the co-ordinated growth and proliferation of host and symbiont cells. The integration of symbionts into the host environment has involved dramatic (and therefore evolutionarily improbable) changes in key biochemical and cellular processes of the micro-organisms (Douglas, 1994); and these modifications to basic processes may impair the capacity of the micro-organisms to persist indefinitely in the external environment. At every level, the biology of symbiotic micro-organisms is shaped primarily by their intimate association with larger organisms, such as animals and plants.

ACKNOWLEDGEMENTS

I thank Professor A.H. Fitter, Dr. R. Law and Professor J.P.W. Young for their valuable comments on an earlier draft of this article, which was written while in receipt of a 1983 University Research Fellowship of the Royal Society of London.

REFERENCES

Alexander, C. and Hadley, G. (1985) Carbon movement between host and mycorrhizal endophyte during the development of the orchid *Goodyera repens* Br. *New. Phytol.* **101**, 657–665.
Allen, M.F. (1991) *The Ecology of Mycorrhizae*. Cambridge University Press, Cambridge.
Bandi, C., Damiani, G., Magrassi, L., Grigolo, A., Fani, R. and Sacchi, L. (1994) Flavobacteria as intracellular symbionts in cockroaches. *Proc. R. Soc. Lond. B* **257**, 43–48.
Begon, M., Harper, J.L. and Townsend, C.R. (1990) *Ecology*. Blackwell Scientific Publications, UK.
Beringer, J.E., Brewin, N., Johnston, A.W.B., Schulman, H.M. and Hopwood, D.A. (1979) The *Rhizobium*-legume symbiosis. *Proc. R. Soc. Lond. B* **204**, 219–233.
Bond, W.J. (1994) Do mutualisms matter? Assessing the impact of pollinator and disperser disruption on plant extinction. *Phil. Trans. R. Soc. Lond. B* **344**, 83–90.

Boucher, D.H. (1985) *The Biology of Mutualism*. Croom-Helm, London.

Britschgi, T. and Giovannoni, S.J. (1991) Phylogenetic analysis of a natural bacterioplankton population by rRNA gene cloning and sequencing. *Appl. Env. Microbiol.* **57**, 1707–1713.

Brundrett, M. (1991) Mycorrhizas in terrestrial ecosystems. *Advances in Ecological Research* **21**, 171–315.

Bubrick, P., Galun, M. and Frensdorff, A. (1984) Observations on free-living *Trebouxia* de Puymaly and *Pseudotrebouxia* Archibald, and evidence that both symbionts from *Xanthoria parietina* (L.) Th. Fr. can be found free-living in nature. *New Phytol.* **97**, 455–462.

Buchner, P. (1965) *Endosymbioses of Animals with Plant Microorganisms*. Wiley, London.

Caetano-Anolles, G., Wall, L.G., de Micheli, A.T., Macchi, E.M., Bauer, W.D. and Favelukes, G. (1988) Role of motility and chemotaxis in efficiency of nodulation by *Rhizobium meliloti*. *Plant Physiol.* **86**, 1228–1235.

Cavanaugh, C.M. (1994) Microbial symbiosis: patterns in diversity in the marine environment. *Amer. Zool.* **34**, 79–89.

Chen, W.X., Yan, H.L. and Li, J.L. (1991) Numerical taxonomic study of fast-growing soybean rhizobia and a proposal that *Rhizobium fredii* be assigned to *Sinorhizobium* gen. nov. *Int. J. Syst. Bacteriol.* **38**, 392–397.

Choi, E.Y., Ahn, T.S. and Jeon, K.W. (1991) Elevated levels of stress proteins associated with bacterial symbiosis in *Amoeba proteus* and soybean root nodule cells. *BioSystems* **25**, 205–212.

Clapp, J.P., Young, J.P.W., Merryweather, J.W. and Fitter, A.H. (1995) Diversity of fungal symbionts in arbuscular mycorrhizas from a natural community. *New Phytol* **130**, 259–265.

Clark, M.A., Baumann, L., Munson, M.A., Baumann, P., Campbell, B.C., Duffus, J.E., Osborne, L.S. and Moran, N.A. (1992) The eubacterial symbionts of whiteflies (Homoptera: Aleyrodoidea) constitute a lineage distinct from the endosymbionts of aphids and mealybugs. *Current Microbiol.* **25**, 119–23.

Clarke, R.T.J. and Bauchop, T. (eds) (1977) *Microbial Ecology of the Gut*. Academic Press, London.

Clay, K. (1990) Fungal endophytes of plants. *Annu. Rev. Ecol. Syst.* **21**, 275–295.

Connell, J.H. (1978) Diversity in tropical rainforests and coral reefs. *Science* **199**, 1302–1310.

De Long, E.F., Wuk-Y., Prezelli, B.B. and Jovine, R.V.M. (1994) High abundance of Archaea in Antarctic marine picoplankton. *Nature* **371**, 695–697.

Distel, D.L. and Wood, A.P. (1992) Characterisation of the gill symbiont *Thyasira flexuosa* (Thiasiridae: Bivalvia) by use of polymerase chain reaction and 16S rRNA gene sequence analysis. *J. Bacteriol.* **174**, 3416–3421.

Distel, D.L., Lane, D.J., Olsen, G.J., Giovannoni, S.J., Pace, B., Pace, N.R. and Stahl, D.A. (1988) Sulfur-oxidizing bacterial endosymbionts: analysis of phylogeny and specificity by 16S rRNA sequences. *J. Bacteriol.* **170**, 2506–2510.

Douglas, A.E. (1980) Establishment of the symbiosis in *Convoluta roscoffensis*. *J. mar. biol. Ass. UK* **63**, 419–434.

Douglas, A.E. (1985) Growth and reproduction of *Convoluta roscoffensis* containing different naturally occurring algal symbionts. *J. mar. biol. Ass. UK* **65**, 871–879.

Douglas, A.E. (1989) Mycetocyte symbiosis in insects. *Biol. Revs* **64**, 409–434.

Douglas, A.E. (1992a) Microbial brokers of insect-plant interactions. In *Insect–Plant Relationships* (Ed. by S.B.J. Menken, J.H. Visser and P. Harrewijn), pp. 329–336. Kluwer Academic Publishers, Dordrecht, The Netherlands.

Douglas, A.E. (1992b) Symbiosis in evolution. *Oxford Surveys in Environmental Biology* **8**, 347–382.

Douglas, A.E. (1994) *Symbiotic Interactions*. Oxford University Press, Oxford.

Douglas, A.E. and Huss, V.A.R. (1986) On the characteristics and taxonomic position of symbiotic *Chlorella*. *Arch. Microbiol.* **145**, 80–84.

Dreyfus, B., Garcia, J.L. and Gillis, M. (1988) Characterisation of *Azorhizobium caulinodans* gen. nov., sp. nov., a stem-nodulating nitrogen-fixing bacterium isolated from *Sesbania rostrata*. *Int. J. Syst. Bacteriol.* **38**, 89–98.

Egeraat, A.W.S. van (1975) The possible role of homoserine in the development of *Rhizobium leguminosarum* in the rhizosphere of pea seedlings. *Plant Soil* **42**, 381–386.

Eisen, J.A., Smith, S.W. and Cavanaugh, C.M. (1992) Phylogenetic relationship of chemoautotrophic bacterial symbionts of *Solemya velum* Say (Mollusca: Bivalvia) determined by 16S rRNA gene sequence analysis. *J. Bacteriol.* **174**, 3416–3421.

Embley, T.M. and Finlay, B.J. (1994) The use of small subunit rRNA sequences to unravel the relationships between anaerobic ciliates and their methanogen symbionts. *Microbiology* **140**, 225–235.

Fautin, D.G. (1991) Developmental pathways of anthozoans. *Hydrobiologia* **216/217**, 143–149.

Fenchel, T. and Finlay, B.J. (1991) The biology of free-living anaerobic ciliates. *Eur. J. Protistol.* **26**, 201–215.

Fisher, C.R. (1990) Chemoautotrophic and methanotrophic symbioses in marine invertebrates. *Rev. Aquat. Sci.* **2**, 399–436.

Freudenthal, H.D. (1962) *Symbiodinium* gen. nov. and *Symbiodinium microadriaticum* sp. nov., a zooxanthella: taxonomy, life cycle and morphology. *J. Protozoology* **9**, 45–52.

Fuhrman, J.A., McCallum, K. and Davis, A.A. (1993) Phylogenetic diversity of subsurface marine microbial communities from the Atlantic and Pacific Oceans. *Appl. Env. Microbiol.* **59**, 1294–1302.

Glynn, P.W. (1993) Coral reef bleaching: ecological perspectives. *Coral Reefs* **12**, 1–17.

Greenslade, P.J.M. (1983) Adversity selection and the habitat templet. *Amer. Nat.* **122**, 352–365.

Gresshoff, P.M., Slotnicki, M.L., Eadie, J.F. and Rolfe, B.G. (1977) Viability of *Rhizobium trifolii* bacteroids from clover root nodules. *Plant Sci. Lett.* **10**, 299–304.

Grime, J.P. (1979) *Plant Strategies and Vegetation Processes*. John Wiley, Chichester, UK.

Harley, J.E. and Smith, S.E. (1983) *Mycorrhizal Symbiosis*. Academic Press, London.

Hatcher, B.G. (1990) Coral reef primary productivity: a hierarchy of pattern and process. *Trends Ecol. Evol.* **5**, 149–155.

Hawksworth, D.L. (1988) Coevolution of fungi with algae and cyanobacteria in lichen symbiosis. In *Coevolution of Fungi with Plants and Animals* (Ed. by K.A. Pirozynski and D.L. Hawksworth), pp. 125–148. Academic Press, London.

Haygood, M.G. (1993) Light organ symbioses in fishes. *Crit. Rev. Microbiol.* **19**, 191–216.

Haygood, M.G. and Distel, D.L. (1993) Bioluminescent symbionts of flashlight fishes and deep-sea anglerfishes form unique lineages related to the genus *Vibrio*. *Nature* **363**, 154–156.

Hennecke, H., Kaluza, K., Thonyl, B., Fuhrmann, M., Ludwig, W. and Stackebrandt, E. (1985) Concurrent evolution of nitrogenase genes and 16S rRNA in *Rhizobium* species and other nitrogen-fixing bacteria. *Arch. Microbiol.* **142**, 342–348.

Hinde, R. (1971) The control of the mycetome symbiotes of the aphids *Brevicoryne brassicae, Myzus persicae* and *Macrosiphum rosae. J. Ins. Physiol.* **17**, 1791–1800.

Hoegh-Guldberg, O., McCloskey, L.R. and Muscatine, L. (1987) Expulsion of zooxanthellae by symbiotic cnidarians from the Red Sea. *Mar. Ecol. Prog. Ser.* **57**, 173–186.

Hoffmann, A.A. and Parsons, P.A. (1993) *Evolutionary Genetics and Environmental Stress.* Oxford University Press, Oxford.

Honegger, R. (1992) Lichens: mycobiont-photobiont relationships. In *Algae and Symbioses* (Ed. by W. Reisser), pp. 255–276. Biopress Ltd, Bristol.

Honegger, R. (1994) Developmental biology of lichens. *New Phytol.* **125**, 659–677.

Howe, H.F. and Westley, L.C. (1988) *Ecological Relationships of Plants and Animals.* Oxford University Press, Oxford.

Hughes, T.P. (1994) Catastrophes, phase shifts, and large-scale degradation of a Caribbean coral reef. *Science* **265**, 1547–1551.

Hurst, L.D. (1992) Intragenomic conflict as an evolutionary force. *Proc. R. Soc. Lond.* B **248**, 135–140.

Jackson, J.B.C. (1991) Adaptation and diversity of reef corals. *BioScience* **41**, 475–482.

Jansen, E. and van Dobben, H.F. (1987) Is the decline of *Cantharellus cibarius* in the Netherlands due to air pollution? *Ambio* **16**, 211–213.

Jordan, D.C. (1984) Family III. Rhizobiaceae. In *Bergey's Manual of Systematic Bacteriology*, vol. 1 (Ed. by N.R. Krieg and J.G. Holt), pp. 234–244. Williams and Wilkins, Baltimore.

Kakeda, K. and Ishikawa, H. (1991) Molecular chaparon produced by an intracellular symbiont. *J. Biochem.* **110**, 583–587.

Kappen, L. (1988) Ecophysiological relationships in different climatic regions. In *Handbook of Lichenology* II, 37–100. CRC Press, Boca Raton, Florida.

Kawanabe, H., Cohen, J.E. and Iwasaki, K. (1993) *Mutualism and Community Organisation.* Oxford University Press, Oxford.

Keddy, P. (1990) Is mutualism really irrelevant to ecology? *Bull. Ecol. Soc. Amer.* **71**, 101–102.

Koske, R.E. (1987) Distribution of VA mycorrhizal fungi along a latitudinal temperature gradient. *Mycologia* **79**, 55–68.

Kosslak, R.M., Bohlool, B.B., Dowdle, S.F. and Sadowsky, M.J. (1983) Competition of *Rhizobium japonicum* strains in early stages of soybean nodulation. *Appl. Env. Microbiol.* **46**, 870–873.

Kropp, B.R. and Trappe, J.M. (1982) Ectomycorrhizal fungi of *Tsuga heterophylla. Mycologia* **74**, 479–488.

Kucey, R.M.N. and Hynes, M.F. (1989) Populations of *Rhizobium leguminosarum* biovars phaseoli and viceae in fields after bean or pea rotation with nonlegumes. *Can. J. Microbiol.* **35**, 661–667.

Kuykendall, L.D., Devine, T.E. and Cregan, P.B. (1982) Positive role of nodulation on the establishment of *Rhizobium japonicum* in subsequent crops of soybean. *Curr. Microbiol.* **7**, 79–81.

de Lajudie, P., Willems, A., Pot, B., Dewettinck, D., Maestrojuan, G., Neyra, M., Collins, M.D., Dreyfus, B., Kersters, K. and Gillis, M. (1994) Polyphasic taxonomy of rhizobia: emendation of the genus *Sinorhizobium* and description of *Sinorhizobium meliloti* comb. nov., *Sinorhizobium saheli* sp. nov. and *Sinorhizobium teranga* sp. nov. *Int. J. Syst. Bacteriol.* **44**, 715–733.

Lamb, K.P. and Hinde, R. (1967) Structure and development of the mycetome in the cabbage aphid, *Brevicoryne brassicae*. *J. Invert. Pathol.* **9**, 3–11.

Law, R. (1987) Some ecological properties of intimate mutualisms involving plants. In *Plant Population Ecology* (28th Symposium of the British Ecological Society) (Ed. by A.J. Davy, M.J. Hutchings and A.R. Watkinson), pp. 315–341. Blackwell Scientific Publications, Oxford.

Law, R. and Hutson, V. (1992) Intracellular symbionts and the evolution of uniparental cytoplasmic inheritance. *Proc. R. Soc. Lond.* B **248**, 69–77.

Law, R. and Lewis, D.H. (1983) Biotic environments and the maintenance of sex—some evidence from mutualistic symbioses. *Biol. J. Linn. Soc.* **20**, 249–276.

Lee, K-H. and Ruby, E.G. (1994a) Competition between *Vibrio fischeri* strains during initiation and maintenance of a light organ symbiosis. *J. Bacteriol.* **176**, 1985–1991.

Lee, K-H. and Ruby, E.G. (1994b) Effect of the squid host on the abundance and distribution on symbiotic *Vibrio fischeri* in nature. *Appl. Env. Microbiol.* **60**, 1565–1571.

Leung, K., Yap, K., Dashti, N. and Bottomley, P.J. (1994) Serological and ecological characteristics of a nodule-dominant serotype from an indigenous soil population of *Rhizobium leguminosarum* bv. trifolii. *Appl. Env. Microbiol.* **60**, 408–415.

McDermott, T.R. and Graham, P.H. (1990) Competitive ability and efficiency in nodule formation of strains of *Bradyrhizobium japonicum*. *Appl. Env. Microbiol.* **56**, 3035–3039.

McFall-Ngai, M.J. (1991) Luminous bacterial symbiosis in fish evolution: adaptive radiation among the leiognathid fishes. In: *Symbiosis as a Source of evolutionary Innovation* (Ed. by L. Margulis and R. Fester), pp. 381–409. The MIT Press. Cambridge, MA, USA.

McGonigle, T.P. and Fitter, A.H. (1990) Ecological specificity of vesicular-arbuscular mycorrhizal associations. *Mycol. Res.* **94**, 120–122.

McNally, K.L., Govind, N.S., Thome, P.E. and Trench, R.K. (1994) Small-subunit ribosomal DNA sequence analyses and a reconstruction of the inferred phylogeny among symbiotic dinoflagellates (Pyrrophyta). *J. Phycol.* **30**, 316–329.

Margulis, L. and Fester, R. (eds) (1991) *Symbiosis as a Source of Evolutionary Innovation*. The MIT Press. Cambridge, MA, USA.

Martin, M.M. (1991) The evolution of cellulose digestion in insects. *Phil. Trans.* B **333**, 281–288.

Maynard Smith, J., Dowson, C. and Spratt, B.G. (1991) Localised sex in bacteria. *Nature* **349**, 29–31.

Molina, R. and Trappe, J.M. (1982) Patterns of ectomycorrhizal host specificity and potential among Pacific Northwest conifers and fungi. *Forest Science* **28**, 423–458.

Molina, R., Massicotte, H. and Trappe, J.M. (1992) Specificity phenomena in

100 A.E. DOUGLAS

mycorrhizal symbioses: community-ecological consequences and practical implications. In *Mycorrhizal Functioning* (Ed. by M.F. Allen), pp. 357–423. Chapman and Hall, New York.

Moran, N.A., Munson, M.A., Baumann, P. and Ishikawa, H. (1993a) A molecular clock in endosymbiotic bacteria is calibrated using the insect hosts. *Proc. R. Soc. Lond. B* **253**, 167–171.

Moran, N.A. and Baumann, P. (1993b) Phylogenetics of cytoplasmically inherited microrganisms of anthropods. *Trends in Ecology and Evolution* **9**, 15–20.

Moulder, J.W. (1979) The cell as an extreme environment. *Proc. Roy. Soc. B* **204**, 199–210.

Moulder, J.W. (1985) Comparative biology of intracellular parasitism. *Microbiol. Revs.* **49**, 298–337.

Munson, M.A., Baumann, P., Clark, M.A., Baumann, L., Moran, N.A., Voegtlin, D.J. and Campbell, B.C. (1991) Evidence for the establishment of aphid-eubacterium endosymbiosis in an ancestor of four aphid families. *J. Bacteriol.* **173**, 6321–6324.

Munson, M.A., Baumann, P. and Moran, N.A. (1992) Phylogenetic relationships of the endosymbionts of mealybugs (Homoptera: Pseudococcidae) based on 16S rDNA sequences. *Mol. Phylogen. Evol.* **1**, 26–30.

Murphy, P.J.N., Heycke, N., Banfalzi, Z., Tate, M.E., de Bruijn, F.J., Kondorosi, A., Tempe, J. and Schell, J. (1987) Genes for the catabolism and synthesis of an opine-like compound in *Rhizobium meliloti* are closely linked and on the *sym* plasmid. *Proc. Natl Acad. Sci. USA* **84**, 493–497.

Murphy, P.J. and Saint, C.P. (1991) Rhizopines in the legume-*Rhizobium* symbiosis. *Molecular Signals in Plant-Microbe Communication* (Ed. by D.P.S. Verma), pp. 378–390. CRC Press, Boca Raton, Florida.

Muscatine, L. (1990) The role of symbiotic algae in carbon and energy flux in reef corals. In *Coral Reefs* (Ed. by Z. Dubinsky), pp. 75–87. Elsevier, Amsterdam.

Nazaret, S., Cournoyer, B., Normand, P. and Simonet, P. (1991) Phylogenetic relationships among *Frankia* genomic species determined by use of amplified 16S rDNA sequences. *J. Bacteriol.* **173**, 241–247.

Nealson, K.H., Haygood, M.G., Tebo, B.M., Roman, M., Miller, E. and McCosker, J.E. (1984) Contribution by symbiotically luminous fishes to the occurrence and bioluminescence of luminous bacteria in sea water. *Microb. Evol.* **10**, 69–77.

Newman, E.I. (1988) Mycorrhizal links between plants: functioning and ecological significance. *Adv. Ecol. Res.* **18**, 243–270.

Newsham, K.K., Fitter, A.H. and Watkinson, A.R. (1994) Root pathogenic and arbuscular mycorrhizal fungi determine fecundity of asymptomatic plants in the field. *J. Ecol.* **82**, 805–814.

Olivieri, I. and Frank, S.A. (1994) The evolution of nodulation in *Rhizobium*: altruism in the rhizosphere. *J. Heredity* **85**, 46–48.

Olsen, G.J., Woese, C.R. and Overbeck, R. (1994) The winds of (evolutionary) change: breathing new life into microbiology. *J. Bacteriol.* **176**, 1–6.

Paau, A.S., Bloch, C.B. and Brill, W.J. (1980) Developmental fate of *Rhizobium meliloti* bacteroids in alfalfa nodules. *J. Bacter.* **143**, 1480–1490.

Pearson, J.N. and Jackson, I. (1993) Symbiotic exchange of carbon and phosphorus between cucumber and three arbuscular mycorrhizal fungi. *New Phytol.* **124**, 481–488.

Pearson, J.N., Abbott, L.K. and Jasper, D.A. (1993) Modification of competition between two colonising arbuscular mycorrhizal fungi by the host plant. *New Phytol.* **123**, 93–98.

Pearson, J.N., Abbott, L.K. and Jasper, D.A. (1994) Phosphorus, soluble carbohydrates and the competition between two arbuscular mycorrhizal fungi colonising subterranean clover. *New Phytol.* **127**, 101–106.

Pearson, V. and Read, D.J. (1973) The biology of mycorrhiza in the Ericaceae. I. The isolation of the endophyte and synthesis of mycorrhizas in aseptic cultures. *New Phytol.* **72**, 371–379.

Pladys, D., Dmitrijevic, L. and Rigaud, J. (1991) Localisation of a protease in protoplast preparations in infected cells of French bean nodules. *Plant Physiol.* **97**, 1174–1180.

Provasoli, L., Yamasu, T. and Manton, I. (1968) Experiments on the resynthesis of symbiosis in *Convoluta roscoffensis* with different flagellate cultures. *J. mar. biol. Ass. UK* **48**, 465–479.

Quispel, A. (1988) Bacteria-plant interactions in symbiotic nitrogen fixation. *Physiol. Plant.* **74**, 783–790.

Rands, M.L., Douglas, A.E., Loughman, B.C. and Hawes, C.R. (1992) The pH of the perisymbiont space in the green hydra-*Chlorella* symbiosis: an immunocytochemical investigation. *Protoplasma* **170**, 90–93.

Rands, M.L., Loughman, B.C. and Douglas, A.E. (1993) The symbiotic interface in an alga-invertebrate symbiosis. *Proc. R. Soc. Lond. B.* **253**, 161–165.

Reisser, W. (Ed.) (1992) *Algae and Symbioses*. BioPress Ltd, Bristol, UK.

Richardson, D.H.S. (1992) *Pollution Monitoring with Lichens*. The Richmond Publishing Co. Ltd., Slough, UK.

Roberts, C.M. (1993) Coral reefs: health, hazards and history. *Trends Ecol. Evol.* **8**, 425–427.

Rose, R. (1970) *Lichens as pollution indicators. Your Environment* 5.

Rowan, R. and Powers, D.A. (1991) A molecular genetic classification of zooxanthellae and the evolution of animal-algal symbioses. *Science* **251**, 1348–1351.

Ruby, E.G., Greenberg, E.P. and Hastings, J.W. (1980) Planktonic marine luminous bacteria: species distribution in the water column. *Appl. Env. Microbiol.* **39**, 302–306.

Ruhling, A. and Tyler, G. (1991) Effects of simulated nitrogen deposition to the forest floor on the macrofungal flora of a beech forest. *Ambio* **20**, 261–263.

Sargent, L., Hunag, S.Z., Rolfe, B.G. and Djordjevic, M.A. (1987) Split-root assays using *Trifolium subterraneum* show that *Rhizobium* infection induces a systemic response that can inhibit nodulation of another invasive *Rhizobium* strain. *Appl. Env. Microbiol.* **53**, 1611–1619.

Schardl, C.M., Liu, J.S., White, J.F., Finkel, R.A., An, Z. and Siegel, M.R. (1991) Molecular phylogenetic relationships of nonpathogenic grass mycosymbionts and clavicipitaceous plant pathogens. *Plant Syst. Evol.* **178**, 27–41.

Schardl, C.L., Leuchtmann, A., Tsai, H-F., Collett, M.A., Watt, D.M. and Scott, B.D. (1994) Origin of a fungal symbiont of perennial ryegrass by interspecific hybridization of a mutualist with the ryegrass choke pathogen, *Epichloe typhina*. *Genetics* **136**, 1307–1317.

Schoenberger, M.M. and Perry, D.A. (1982) The effect of soil disturbance on growth and ectomycorrhizae of Douglas-fir and western hemlock seedlings: a greenhouse bioassay. *Can. J. Forest Res.* **12**, 343–353.

Simon, L., Lalonde, M. and Bruns, T.D. (1992) Specific amplification of 18S fungal ribosomal genes from VA endomycorrhizal fungi colonising roots. *Appl. Environ. Microbiol.* **58**, 291–295.

Smith, S.E. (1967) Carbohydrate translocation in orchid mycorrhizas. *New Phytol.* **66**, 371–378.

Soberon, J. and Shapiro, E. (1989) Population dynamics of a *Rhizobium*-legume interaction. *J. Theor. Biol.* **140**, 305–316.

Southwood, T.R.E. (1988) Tactics, strategies and templets. *Oikos* **52**, 3–18.

Souza, V., Eguiarte, L., Avila, G., Cappello, R., Gallardo, C., Montoya, J. and Pinero, D. (1994) Genetic structure of *Rhizobium etli* biovar phaseoli associated with wild and cultivated bean plants (*Phaseolus vulgaris* and *Phaseolus coccineus*) in Morelos, Mexico. *Appl. Env. Microbiol.* **60**, 1260–1268.

Sprent, J.I. (1986) *The Ecology of the Nitrogen Cycle*. Cambridge University Press, Cambridge, UK.

Sprent, J.I. and Sprent, P. (1990) *Nitrogen-fixing Organisms: Pure and Applied Aspects*. Chapman and Hall, London, UK.

Sutton, W.D., Jepsen, N.M. and Shaw, B.D. (1977) Changes in number, viability, and amino-acid-incorporating activity of *Rhizobium* bacteroids during lupin nodule development. *Plant Physiol.* **59**, 741–744.

Thies, J.E., Singleton, P.W. and Bohlool, B.B. (1991) Influence of the size of indigenous rhizobial populations on establishment and symbiotic performance of introduced rhizobia on field-grown legumes. *App. env. Microbiol.* **57**, 19–28.

Thompson, J.N. (1982) *Interaction and Coevolution*. John Wiley and Sons, New York.

Trench, R.K. and Blank, R.J. (1987) *Symbiodinium microadriaticum* Freudenthal, *S. goreauii* sp. nov., *S. kawaguti* sp. nov. and *S. pilosum* sp. nov.: gymnodinoid dinoflagellate symbionts of marine invertebrates. *J. Phycol.* **23**, 469–481.

Triplett, E.W. and Sadowsky, M.J. (1992) Genetics of competition for nodulation of legumes. *Annu. Rev. Micobiol.* **46**, 399–428.

Tsai, H-F., Liu, J-S., Staben, C., Christiensen, M.J., Latch, G.C.M., Siegel, M.R. and Schardl, C.L. (1994) Evolutionary diversification of fungal endophytes of tall fescue grass by hybridization with *Epichloe* species. *Proc. Natl Acad. Sci.* **91** 2542–2546.

Tsien, H.C., Cain, P.S. and Schmidt, E.L. (1977). Viability of *Rhizobium* bacteroids. *Appl. env. Microbiol.* **34**, 854–856.

Turk, D. and Keyser, H.H. (1991) Rhizobia that nodulate tree legumes: specificity of the host for nodulation and effectiveness. *Can. J. Microbiol.* **38**, 451–460.

Untermann, B.M. and Baumann, P. (1990) Partial characterisation of the ribosomal RNA operons of the pea-aphid endosymbionts: evolutionary and physiological implications. In *Aphid Plant Genome Interactions* (Ed. by R.K. Campbell and R.D. Eikenbary), pp. 329–350. Amsterdam, Elsevier.

Willems, A. and Collins, M.D. (1993) Phylogenetic analysis of rhizobia and agrobacteria based on 16S rRNA ribosomal DNA sequences. *Int. J. Syst. Bacteriol.* **43**, 305–313.

Wolfe, C.J. and Haygood, M.G. (1993) Bioluminescent symbionts of the Caribbean flashlight fish (*Kryptophanaron alfredi*) have a single rRNA operon. *Mol. Marine Biol. Biotechnol.* **2**, 189–197.

Wood, T., Borman, F.H. and Voight, G.K. (1984) Phosphorus cycling in a northern hardwood forest, biological and chemical control. *Proc. Natl Acad. Sci. USA* **223**, 391–393.

Woomer, P., Singleton, P.W. and Bohlool, B.B. (1988) Ecological indicators of native rhizobia in tropical soils. *Appl. Env. Microbiol.* **54**, 1112–1116.

Yanaki, M. and Yamasato, K. (1993) Phylogenetic analysis of the family Rhizobiaceae and related bacteria by sequencing of 16S rRNA gene using PCR and DNA sequencer. *FEMS Microbiol. Lett.* **107**, 115–120.
Young, J.P.W. (1991) Phylogenetic classification of nitrogen-fixing organisms. In *Biological Nitrogen Fixation* (Ed. by G. Stacey, R.H. Burris and H.J. Evans), pp. 43–86. Chapman and Hall, New York.
Young, J.P.W., Downer, H.L. and Eardley, B.D. (1991) Phylogeny of the phototrophic rhizobium strain BTAi1 by polymerase chain reaction-based sequencing of a 16S rRNA gene segment. *J. Bacteriol.* **173**, 2271–2277.

Impacts of Disturbance on Detritus Food Webs in Agro-Ecosystems of Contrasting Tillage and Weed Management Practices

D.A. WARDLE

I.   Summary 105
II.   Introduction 106
III. Tillage as a Disturbance Regime 109
      A. Relative Responses of Different Trophic Levels 109
      B. Overall Food Web Responses 127
      C. Response of Species Assemblages to Tillage 132
IV.   Alternatives to Tillage 136
      A. Consequences of Herbicide Use 137
      B. Use of Mulches 143
      C. Alternative Agricultural Systems 147
V.   Influence of Weeds on Detritus Food Webs 148
VI.   Agricultural Disturbance and Food Web Theory 150
VII. Conclusions 156
Acknowledgements 158
References 158
Appendices 182

I. SUMMARY

Agro-ecosystems under conventional tillage and no-tillage management experience vastly different disturbance regimes which impact upon the detritus food web. Indices of change calculated from the published literature demonstrate that larger organisms are likely to be reduced by tillage more than smaller ones, although some groups of intermediate size (bacterial-feeding nematodes, astigmatid mites, enchytraeids) often benefit from tillage. Since larger organisms usually occupy higher trophic levels, a unit of biomass of organisms at a lower trophic level (e.g. microflora) appears capable of supporting greater biomass of organisms at higher levels in no-tillage systems, indicating that consumption of lower trophic

ADVANCES IN ECOLOGICAL RESEARCH VOL. 26
ISBN 0–12–013926–X

levels is likely to be more important under reduced disturbance. Little consistent evidence emerges from published data for differences in response of bacterial-based and fungal-based compartments following tillage, except in relation to associations with buried litter ("resource islands"). Evidence of both top-down and bottom-up effects of tillage on components of the detritus food web are presented and these are due to tillage-induced physical disturbance, changes in residue distribution and soil physical factors. Some evidence also exists for trophic cascades which regulate lower trophic levels and litter decomposition. One large gap in understanding such interactions results from the very poor understanding of the linkages between macrofauna and other (micro- and meso-) faunal groups. This literature synthesis also demonstrated that species diversity (determined using diversity indices) of microfaunal groups is usually largely unchanged by tillage while diversity of macrofaunal groups may be either substantially elevated or reduced by tillage depending on the study considered. This suggests that tillage-induced disturbance does not cause predictable reductions of soil biodiversity. Herbicides are used more widely in no-tillage systems and little evidence emerges for consistent direct impacts of herbicides on soil organisms, which at least partly reflects inconsistent methodology. Other practices sometimes undertaken in no-tillage systems, viz. the use of non-living and living mulches, and residue addition, may substantially enhance many functional groups. Incomplete weed control, e.g. through herbicide use, usually stimulates meso- and macrofaunal groups, although little is known about the direct interactions between weeds and the overall microflora and microfauna. Below-ground food webs in agro-ecosystems appear relatively stable despite containing many features traditionally assumed to be "destabilizing" i.e. a high incidence of omnivory, long food chains, within-habitat compartments and probably a high value for species number x connectance, indicating that food webs in the soil environment do not conform well to ecological theory developed for other systems.

II. INTRODUCTION

Food webs are an integral component of both natural and managed ecosystems, and feeding links between the species present are critical in regulating nutrient cycling and energy flow. Since the development of the Hairston, Smith and Slobodkin (HSS) hypothesis, which indicated that the importance of predation and competition may alternate between trophic levels in food webs (Hairston et al., 1960), there has been an increasing recognition of the importance of food web theory in understanding the functioning of ecosystems (Schoener, 1989; Power, 1992). Food webs are often highly responsive to disturbance (at least at the energy-flow level) and changes in the abundance of some individual (taxonomic or genetic) species may influence organisms as many as two or three trophic levels away (Kareiva and Sahakian, 1990; Power, 1990; Strong, 1992); this is especially apparent in the case of "keystone species" which have a disproportionate effect on food web properties (Brown and Heske, 1990; Kerbes *et al.*, 1990).

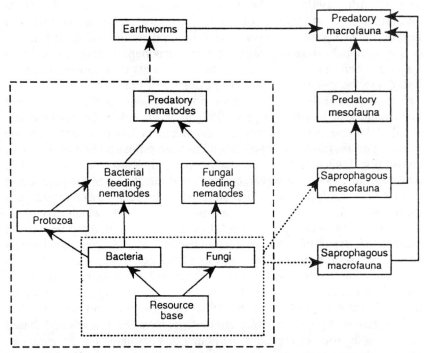

Fig. 1. Simplified functional food web for detritus-based systems. Organisms in the dotted and dashed boxes may be fed upon by organisms which consume substrate containing material of more than one trophic level.

The decomposer-based (detritus) food web (Fig. 1) is of particular importance in most terrestrial ecosystems, because it is closely linked to nutrient cycling and ecosystem productivity (De Angelis, 1992). It is also likely that the structure of the decomposer food web can itself regulate ecosystem level processes (Allen-Morley and Coleman, 1989; Kajak *et al.*, 1993; Bengtsson *et al.*, 1995). However, this food web is poorly understood and has thus contributed little to general ecological theory; soil-based food webs are virtually absent from the recent food web syntheses of Cohen *et al.* (1990) and Pimm *et al.* (1991). However, there is evidence that this food web may be quite different to corresponding above-ground food webs, through supporting longer food chains, distinct (bacterial- and fungal-based) within-habitat compartments, a higher incidence of omnivory and a much greater species richness (Coleman, 1985; Ingham *et al.*, 1986; Moore and Hunt, 1988; Wardle and Yeates, 1993). These features would all supposedly confer greater instability on food web structure (Pimm, 1982) although it has been suggested that the highly buffered three-dimensional nature of the soil matrix may enhance the stability of below-ground food-webs (Wardle, 1991). It is becoming increasingly appreciated that these "destabilizing" features are also probably more abundant in food webs in general than has often been acknowledged (Hall and Raffaelli, 1993).

Conventional tillage has been a major component of many agricultural systems, where it has been undertaken for a variety of purposes, including improving the physical and structural quality of the seed-bed, increasing soil aeration and drainage, incorporation of residues and fertilizers, and management of weeds and pests (Phillips *et al.*, 1980; Gebhardt *et al.*, 1985). Of these, weed control and management has probably been the most important (Sprague, 1986) and it has been estimated that over 50% of conventional tillage has been conducted to enable effective management of weeds (Ennis, 1979). However, conventional tillage also has several major and well-known drawbacks. Chief amongst these are increased erosion and topsoil loss, and associated deterioration of soil quality which may result from tillage practices (Phillips, 1984; Phillips *et al.*, 1980), a problem which has become apparent as mechanization has played an increasingly import-ant role in agriculture (Moorhouse, 1909; Faulkner, 1945). Repeated culti-vation may also reduce soil moisture retention (Unger and Cassel, 1991), water infiltration (Lal and van Doran, 1990) and soil organic matter con-tents (House *et al.*, 1984; Phillips, 1984). Because of the interest in reduc-ing tillage-based problems other alternatives have been developed, based largely on the use of herbicides. Although this has made the development of non-tillage systems feasible, herbicide-based systems also have per-ceived environmental drawbacks, relating usually to the fate of the herbi-cides used and potential contamination of other systems (Pimentel *et al.*, 1980; Torstensson, 1988). In recent years, there have been moves to reduce greatly or even eliminate use of synthetic herbicides (Morgan, 1989, 1992).

There is a clear gradient from high tillage/low herbicide to low tillage/ high herbicide management regimes (Triplett and Worsham, 1986). Mini-mum tillage systems, in which soil conditions are usually presumed to be less disturbed, are supposedly more closely related to natural systems, especially in terms of the organisms present and biotic processes occurring (House *et al.*, 1984). Agro-ecosystems that exhibit a higher similarity to natural systems supposedly have a higher diversity of soil-associated organ-isms and a greater capacity for self-regulation (Altieri, 1991). Because the detritus food web is based largely in the soil system, and is important in regulating nutrient cycling and energy flow, it is reasonable to expect that this food web may be a viable and meaningful indicator of disturbance exerted by tillage and alternative weed management strategies. Furthermore, the tillage : no-tillage comparison is of interest to ecologists because it enables comparison of food webs with structural simplicity (at least at the functional group level) and under vastly differing disturbance regimes (Beare *et al.*, 1992). As such, it provides a good opportunity for testing various aspects of food web theory. However, only a few studies have investigated the simultaneous response of several trophic levels of the detritus food web to such forms of agricultural management (e.g. Hendrix *et al.*, 1986; Paustian *et al.*, 1990), which reflects the poor understanding of the detritus food web which exists in general (Wardle and Yeates, 1993). The purpose of this review is to investigate how detritus-based food webs respond to various disturbance regimes occurring in conventionally tilled (CT) and non-tilled (NT) systems. Since the principal differences in disturbance regimes involve cultivation, herbicide application, residue management and manipulation of weed levels, emphasis will be placed on these factors. Much of this paper will be devoted to pooling the results of previous studies, determining how different components of the detritus food web respond to these disturbances relative to each other, and assessing the consequences of these disturbances for overall food web structure. Where applicable the findings of this study will be related to general food web theory.

III. TILLAGE AS A DISTURBANCE REGIME

A. Relative Responses of Different Trophic Levels

In assessing the response of detritus food webs to tillage it is necessary to consider how different functional groups respond to cultivation. In a quantitative literature survey (Appendix 1) I have compiled the results of 106 previous studies which have investigated the response to tillage of various components of the detritus food web, viz. the various functional groups of soil organisms, and soil carbon and nitrogen levels (indicative of the resource-base: Wardle, 1992). The list in Appendix 1 is intended to be representative. For each of these studies, data presented for the abundance or biomass of each group of organisms (or percent soil carbon and nitrogen) in CT and NT treatments was used to calculate the index V (Wardle and Parkinson, 1991b) for that group, i.e.

$$V = \frac{2M_{CT}}{M_{CT} + M_{NT}} - 1$$

110 D.A. WARDLE

where: M_{CT} and M_{NT} = abundance or mass of organisms, or percent carbon or nitrogen, under CT and NT respectively. This index V ranges from -1 (organisms occurring only under NT) to $+1$ (organisms occurring only under CT) with 0 indicating equal abundance under both CT and NT systems. In studies where a variety of sampling times and/or soil types were considered, the overall means for both treatments were used. Using this index, the following categories were constructed to express the degree of response to tillage:

extreme inhibition by tillage:	$V < -0.67$
moderate inhibition by tillage:	$-0.33 > V > -0.67$
mild inhibition by tillage:	$0 > V > -0.33$
mild stimulation by tillage:	$0 < V < 0.33$
moderate stimulation by tillage:	$0.33 < V < 0.67$
extreme stimulation by tillage:	$V > 0.67$

Two separate syntheses were performed. The first considered only those studies where the NT tillage treatment was under similar (mostly annual) cropping practices as the CT treatment (Fig. 2a); the second where the NT treatment was under a different management practice, i.e. usually a perennial crop, pasture or old-field (Fig. 2b). Studies where the NT treatment was under woody plants were not included. These syntheses were performed separately because it is often presumed that systems under perennial cover are more representative of natural environments (and more unlike a CT cropping system) than are NT annual cropping systems (Hendrix et al., 1986; Freckman and Ettema, 1993). It is noted that the taxonomic resolution of the studies used in compiling Fig. 2 is usually greater for higher trophic levels and this reflects the greater taxonomic difficulty in assessing organisms in lower trophic levels. This is a problem which is also apparent in other food webs syntheses (Hall and Raffaelli, 1993).

The remainder of Section III.A will involve constructing the detritus food web from the bottom-up, progressively adding larger organisms and those of higher trophic levels, and assessing the impact of tillage on each group (through the data in Fig. 2) and how they interact with each other group.

1. The Resource Base

The resource base, or the quantity of resource available to the detritus food web, comprises the organic matter available to micro-organisms in the soil system. Since the microflora is strongly influenced by soil carbon and nitrogen concentrations (Wardle, 1992), these variables are used here as an indicator of the potential resource base for the various organisms of the detritus food web. Different components of the resource base defined in this way may, however, differ vastly in microbial availability to microorganisms (Jenkinson and Rayner, 1977; McClaugherty, 1983).

The resource base carbon and nitrogen were lower under CT than NT for over 80% of the studies considered, regardless of whether the NT treatment was under similar or different management to the CT treatment (Fig. 2). However, moderate reduction of soil nitrogen by CT is apparent only in the comparison with NT perennial systems (Fig. 2b), where the index V was below -0.33 for 25% of all studies considered. The generally lower amounts of carbon and nitrogen in CT soils probably reflects depletion of organic matter resulting from greater microbial inefficiency associated with greater disturbance. Odum's (1969) theory of ecosystem succession predicts greater respiration per unit production in disturbed environments, and this has often been demonstrated for below-ground (microbially based) ecosystems (Anderson and Domsch, 1985; Insam and Domsch, 1988 but see Wardle, 1993). Coleman et al. (1976) estimated that production : respiration ratios may be around $1.2-1.3$ in agro-ecosystems compared with $1.8-2.0$ in natural systems; these differences probably stem largely from effects of tillage. If the detritivores in CT systems were operating inefficiently, then there would be a greater mineralization per unit biomass with long-term organic matter depletion (Blevins et al., 1977; Arshad et al., 1990; Wood and Edwards, 1992). Cultivation of previously untilled soils in particular induces a rapid initial reduction of organic matter, possibly in the order of 20–40% (Davidson and Ackerman, 1993).

2. Soil Microflora

The soil microbial biomass (defined as the mass of living microbial tissue in soil) was usually less in CT than NT systems when both systems were under similar cropping (Fig. 2a). However, any reductions, when they occurred, were relatively mild (V > -0.33) with a mean value for V of 0.10. Where bacteria and fungi were considered separately, they responded to tillage similarly. Data from most studies suggest that, under similar practices, bacteria and fungi in soil are often inhibited to similar degrees by tillage (e.g. Doran, 1980a; Linn and Doran, 1984; Wardle et al., 1993b) despite many fungal species being supposedly more K-selected than bacteria and thus more susceptible to reduction by disturbance resulting from cultivation (see Gerson and Chet, 1981). However, plant residues on the soil surface of NT systems tend to be dominated by fungi (Hendrix et al., 1986; Beare et al., 1992), which results from hyphal networks being able to form in the absence of disruption from cultivation, and the greater ability than bacteria to tolerate desiccation in surface residues (Holland and Coleman, 1987).

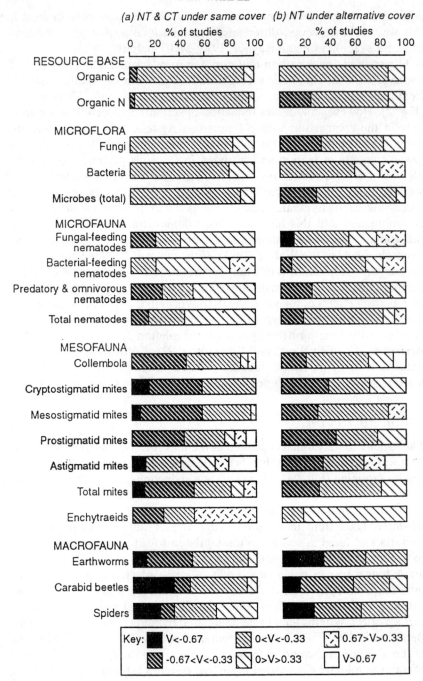

Although the microbial biomass is only mildly inhibited by CT relative to similarly cropped NT systems, it is often more strongly inhibited by tillage relative to perennial systems; the index V was below -0.33 in 29% of studies considered in Figure 2b (mean value of V = 0.30). In perennial NT systems, fungi may sometimes be strongly inhibited and bacteria strongly stimulated by CT (Fig. 2b), although the general applicability of this result is limited by the relatively small number of studies available on these groups, and largely appear to reflect the results of Gupta and Germida (1988). These effects in Figure 2b presumably represent differences in the effects of the NT practices between annual and perennial cropping systems.

Reduction of soil microbial biomass by tillage often parallels a decline in organic matter levels, especially in the top 5–10 cm soil layer (Saffigna et al., 1989; Angers et al., 1992). This reflects the well-established links between microbial biomass and soil organic carbon and nitrogen which have been reviewed elsewhere (Wardle, 1992). However, the extent of stimulation of microbial biomass by adoption of NT practices is often greater than that which would be expected by elevation of organic matter levels alone, and the ratio of microbial biomass carbon to organic carbon is frequently higher in NT situations (e.g. Powlson and Jenkinson, 1981; Carter and Rennie, 1982; Carter, 1991; Dalal et al., 1991). This may be in part a function of the quality of soil organic matter being superior in NT systems, with higher levels of microbially available carbohydrates and amino acids (Arshad et al., 1990). Microbial colonization of buried litter (e.g. following cultivation) tends to be much greater than that of surface-placed litter, resulting in accelerated litter decomposition and organic matter depletion (Beare et al., 1992; Wardle et al., 1993b).

The higher ratio of microbial biomass carbon to organic carbon in the surface soil layer of NT systems may be in part due to microclimate. In NT systems a significant residue layer often builds upon the soil surface, reducing aggregate disruption and sometimes pore sizes, all of which contribute

---

Fig. 2. Influence of tillage on the resource-base and various groups of organisms in the detritus food web, based on the pooled results of studies listed in Appendix 1. The proportion of each bar shaded in each hatching is equal to the proportion of published studies considered where the index V (calculated according to section III-A) was in the range specified in the key. V ranges from -1 to $+1$ and is increasingly negative or positive as the group considered is increasingly lesser or greater in conventional-tillage than no-tillage systems. Part (a) represents only those studies where the no-tillage treatment was under similar cropping or management practices as the conventional-tillage treatment; (b) includes only those studies where the non-tilled treatment is under an alternative cover, e.g. pasture, grassland, perennial ley.

114   D.A. WARDLE

to reduced water loss from soil (Adams, 1966; Lal et al., 1980; Phillips, 1984) and usually cooler temperatures (Blevins et al., 1984; Lal et al., 1990). The overall response of the soil microflora (or microbial biomass) to soil moisture and especially soil temperature is poorly understood and existing data conflict (Wardle, 1992). However, part of the microbial biomass is usually killed by extreme drying (Shields et al., 1974; Lund and Goksoyr, 1980), which is less likely to occur in NT situations. Further, in warmer (e.g. tropical) climates, extreme (microbicidal) heat levels are less likely under the residue layers in NT systems. Physical effects of soil tillage are reviewed in detail by Unger and Cassell (1991).

The soil microflora is likely to respond to differences in plant biomass between CT and NT systems, in terms of both quantities of rhizosphere exudates and above-ground litter following crop harvest. The relative crop yields between CT and NT systems are highly unpredictable (Lal et al., 1990); crop plant biomass may be greater in either system, depending mainly on soil type and drainage capacity (Dick and van Doran, 1985; Griffith et al., 1988; Dick et al., 1991). These influences are likely to affect residue levels and ultimately mass of soil microflora. Probably of more immediate (short-term) relevance to the soil microflora is the influence of tillage on plant rooting patterns and consequent distribution of root litter and rhizosphere exudates. Generally, plants in NT systems produce a greater proportion of roots nearer the soil surface compared with those in CT systems (Barber, 1971; Anderson, 1987; Cheng et al., 1990), although rooting in NT systems may occur to some depth in situations where earthworm burrows (which are more numerous in the absence of cultivation as discussed later) permit infiltration and aeration (Hargrove et al., 1988; Cheng et al., 1990). Rhizosphere effects usually greatly stimulate the soil microflora (see reviews by Newman, 1985; Smith and Paul, 1990) and distribution of soil microbes in different tillage regimes should reflect the influence that tillage has on plant rooting patterns. The often strong inhibition of microflora by tillage when compared with perennial NT systems (Fig. 2b) is likely to be due, in a large part, to sustained rhizosphere effects of perennial plants. CT and NT systems may also differ widely in weed biomass levels, which may have strong effects on the soil biota, but this depends on the nature and effectiveness of weed management adopted in NT systems, as is discussed in Section IV.

Arbuscular mycorrhizas (AM) are a major component of most agricultural soil-plant-microbial systems where they appear to act by enabling roots to exploit greater volumes of soil for immobile ions (e.g. phosphate) than otherwise possible (Allen, 1991; Read, 1991), possibly improving drought resistance, and encouraging resistance against pathogens (Anderson, 1992). Their linkage with the detritus food web, although probably important (Ingham et al., 1986; Moore et al., 1988) is poorly understood, and the influence of tillage on mycorrhizal spore densities and subsequent mycorrhizal infections has received little attention. However, it is likely that tillage may either enhance or reduce background levels of AM fungal spores in the soil depending on physical conditions (Kruckelman, 1975; Mitchell et al., 1980; Blevins et al., 1984). Tillage also disrupts the AM mycelial network (Evans and Miller, 1990), probably in a similar way to its influence on hyphal networks in general, and this appears to reduce the effectiveness of mycorrhizas (McGonigle and Miller, 1993).

3. Soil Microfauna

Depending upon soil conditions, a significant proportion of the soil pore space can be filled with water, providing an aqueous habitat for species assemblages of microfauna, (i.e. protozoa and nematodes) and microflora (Bamforth, 1988). The microfaunal groups are a major component of the detritus food web, feeding mainly on plant roots, microflora and other microfaunal groups, and playing an integral role in releasing plant-available nutrients (Stout, 1980; Ingham et al., 1985; Freckman, 1988). Plant-parasitic microfauna are only indirectly involved with the detritus food web and are outside the scope of this paper. Of those microfauna which feed on microflora, bacteria are fed upon mainly by protozoa and "bacterial-feeding" nematodes (i.e. nematodes adapted for engulfing prey) while fungi are fed upon mainly by "fungal-feeding" nematodes (equipped with stylets) (Swift et al., 1979). Some protozoa are also fungal-feeding although the importance of this interaction in the detritus food web has yet to be established (Bamforth, 1988).

The influence of tillage-related practices on protozoa has only occasionally been investigated, and insufficient data are available to present in Figure 2 or to arrive at any generalizations. Elliott et al. (1984) found no significant differences in protozoa populations between CT and stubble-mulched treatments and Parmelee et al. (1990) indicated that protozoa may be unaffected by tillage. Studies conducted in Sweden found that CT barley systems could contain either higher or lower protozoan numbers than corresponding NT perennial grass and lucerne leys (Andrén and Lagerlöf, 1983; Schnürer et al., 1986a; Paustian et al., 1990).

Total nematode numbers may either be enhanced or reduced by tillage, regardless of whether corresponding NT treatments are under a similar (annual) or perennial cropping regime (Figs. 2a, b). Sometimes these effects are reasonably strong: data of Andrén and Lagerlöf (1983), Sohlenius et al. (1988) and Yeates and Hughes (1990) suggest moderate inhibition by tillage while those of Hendrix et al. (1986) and Parmelee and Alston (1986) suggest moderate stimulation.

Essentially similar trends occur when bacterial-feeding and fungal-feeding nematode numbers are considered. Both these groups can demonstrate positive (e.g. Wasilewska, 1979; Parmelee and Alston, 1986) or negative (e.g. Thomas, 1978) responses to tillage. Bacterial-feeders may benefit more strongly than fungal-feeders from tillage—this is borne out strongly in data presented by Hendrix et al. (1986) and (to a lesser extent) Yeates and Hughes (1990), but not by Baird and Bernard (1984), Freckman and Ettema (1993) or Yeates et al. (1993b). The mean value of the index V from Figure 2a is $+0.08$ for bacterial-feeding nematodes and $+0.06$ for fungal-feeding nematodes when NT and CT cropping practices are similar. However, when the NT treatments are under perennial cover, the mean value of V is 0.00 and -0.02 for bacterial and fungal-feeders respectively (Fig. 2b). There therefore appears to be little real difference between bacterial and fungal feeders in their response to tillage. The ratio of bacterial-feeding : fungal-feeding nematode numbers may adequately reflect broad trophic shifts owing to tillage especially when the responses of the two groups are widely different (Hendrix et al., 1986; Parmelee and Alston, 1986) but may not be as sensitive to tillage-induced disturbance as other indices which can be extracted from nematode species data (Freckman and Ettema, 1993; Yeates et al., 1993b).

Data for predatory and "omnivorous" (= feeding at more than one trophic level) nematode numbers have been combined for the purposes of Figure 2 because many of the previously published studies considered combined data for the two groups together, not because they have any major functional similarity. Data for the sum of the predatory and omnivorous nematodes appear to parallel those of total nematode numbers in general (Fig. 2a,b). "Predators" are of particular importance in being the highest trophic level that actually occupies water-filled soil micro-pores. Few studies have specifically monitored the response of predators alone to tillage. Parmelee and Alston (1986) showed that predators may not respond to tillage, and Yeates and Hughes (1990) did not detect any significant difference in predatory Mononchus nematodes between CT and NT systems. Baird and Bernard (1984) indicated that predators were enhanced by tillage, whereas Yeates and Bird (1994) found cultivation to be detrimental to predators. Freckman and Ettema's (1993) data suggested that predator numbers in CT systems were intermediate between those of NT annual and NT perennial cropping systems. Clearly, this is a group requiring more attention, especially if, as suggested by Freckman and Ettema (1993), this group is responsive to human disturbance.

Nematodes show a highly variable response to tillage at all trophic levels considered here. This probably in part reflects the wide range of physical effects that cultivation may have on the soil system. Like bacteria, nematodes require moisture films for mobility, and are therefore less likely to be limited by moisture stress in no-tillage systems (Hendrix et al., 1986). The prevalence of anhydrobiotic coiling for withstanding moisture loss may enable the nematode fauna to tolerate soil drying to some extent (Freckman, 1978; Demure et al., 1979). Nematodes are also temperature-responsive (Freckman and Caswell, 1985) and since temperature appears to exert sufficient influence on microbial-feeding nematodes to justify identification as a niche axis (Anderson and Coleman, 1982), it is reasonable to expect that cooler NT systems may promote a different nematode fauna to warmer CT systems. The altered distribution of pore-sizes in CT systems may also influence the diversity of micro-habitats available for utilization by nematodes.

4. Microfauna–Microflora Interactions

Although tillage exerts important direct effects on soil microfauna through influencing soil physical factors, major indirect influences probably occur through tillage affecting lower trophic levels. Populations of bacterial- or fungal-feeding microflora can be strongly linked to levels of bacterial or fungal biomass (Clarholm, 1984; Sohlenius and Bostrom, 1984; Christensen et al., 1992). Frequently, however, there is no simple relationship between microfloral groups and their associated microfaunal group, and inverse (Coleman et al., 1978; Whitford et al., 1982) or neutral (Wardle and Yeates, 1993) relationships have been detected. The unpredictability of relationships between microbe-feeding nematodes and the microflora is reinforced by Figure 2a, which indicates that a significant proportion of studies show moderate inhibition (or in the case of bacterial-feeding nematodes, moderate stimulation) of various nematode groups under CT while none show moderate ($V = 0.33$ to 0.67 or $V = -0.33$ to -0.67) responses by microfloral groups. These differences may be in part due to microflora and microfauna responding differently to physical consequences of tillage but are more likely to reflect the complexity of trophic interactions and the involvement of other trophic levels. Two mechanisms appear to occur:

(i) Stimulation of microflora is likely to cause a rapid increase in associated microfauna (Christensen et al., 1992) but this increase in microfauna may in turn cause a (delayed) suppression of the microflora through grazing pressure (Ingham et al., 1986). This appears to be more likely to occur in relation to bacteria than fungi and potential increases of bacteria induced by organic matter return, plant productivity or favourable physical conditions are likely to be kept in check by grazing (Ingham et al., 1986; Wardle and Yeates, 1993). Bacteria are usually consumed at random (Freckman and Caswell,

118   D.A. WARDLE

1985) and are less well adapted than fungi at resisting grazing pressure. The ratio of the standing crop of bacterial-feeding nematodes to bacterial biomass is therefore usually substantially higher than the ratio of fungal-feeding nematodes to fungal biomass (Schnürer et al., 1986b; Ingham et al., 1989) and high grazing pressure on the total bacterial biomass may confer a substantial competitive advantage on total fungal biomass (Wardle and Yeates, 1993). Fungi are sometimes disadvantaged by nematode grazing pressure (Wasilewska et al., 1975), but those effects are not usually severe. Increased input of organic matter from plants is therefore likely to have a tri-trophic influence on the decomposer food web by stimulating microfloral (especially bacterial) feeding organisms, via the soil microflora. However, potential increases of microflora (especially bacteria) occupying the intermediate trophic level are in turn kept in check by enhanced grazing and this may explain in part why microfaunal responses to tillage are often more extreme than those of the microflora. Yeates (1979, 1987) demonstrated strong and consistent links between plant productivity and total nematode numbers in various pasture sites, and since a substantial proportion of these nematodes are likely to be microbial-feeders, this also confirms the existence of tri-trophic effects. Given the potential of microfauna to mineralize nutrients immobilized in microbial tissue (Darbyshire et al., 1994; Griffiths, 1994), these effects may be complicated by microbial-feeding nematodes simultaneously enhancing plant growth through enhancing nutrient cycling (Yeates, 1979; Ingham et al., 1985).

(ii) The top predatory nematode fauna, although poorly understood, is strongly linked to lower trophic levels (Freckman and Caswell, 1985) and the presence of anti-predation adaptations in many bacterial-feeding nematode species (Small and Grootaert, 1983) is presumably attributable to strong top-down effects of top predators. Wardle and Yeates (1993) demonstrated that enhancement of microbial biomass tended to be accompanied by a large increase in populations of predators but not by the microbe-feeding nematodes occupying the intermediate trophic level. This is indicative of a tri-trophic effect similar to that described above, with only every second trophic level responding to changes in resource quality or availability. Interactions which involve top predatory nematodes are poorly understood especially in field conditions, although ecosystem-level effects of top predators in regulating lower trophic levels may be an important factor in determining differences in these lower levels between CT and NT systems. There is, however, insufficient information regarding differences in predator numbers between these two systems to make meaningful predictions at this stage.

The tri-trophic effects mentioned in (i) and (ii) above enable testing of the applicability of certain ecological theories to below-ground ecosystems. The Hairston, Smith and Slobodkin (HSS) hypothesis (Hairston *et al.*, 1960) predicts that in plant-based terrestrial systems with three trophic levels the top and bottom levels are competition (resource) regulated while the intermediate level is predator-regulated. In contrast, the Menge and Sutherland (MS) theory originally developed for marine systems (Menge and Sutherland, 1976; Menge *et al.*, 1986) predicts that predation becomes more important at lower trophic levels. The information presented above, and that of Wardle and Yeates (1993) indicates that fungi tend to be less predator-regulated than bacteria, and that the bacterial-feeding and fungal-feeding nematodes tend to be regulated by predatory nematodes which are in turn resource regulated and likely to be influenced mainly by competitive interactions. This hypothesis is shown in Figure 3, and predicts that

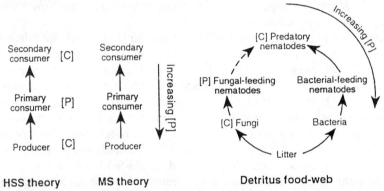

Fig. 3. Part of the detritus food web in relation to the Hairston, Smith and Slobodkin (HSS) theory and Menge and Sutherland (MS) theory. [C] and [P] indicates that competition (resource-limitation) and predation are likely to be the principle "regulatory forces" operating respectively.

bacterial-based food webs may conform more clearly to the MS theory while fungal-based food webs conform more to the HSS theory. Strong enhancement of bacteria by CT would not be detected because any increases are likely to be kept in check by higher trophic levels. Tillage-induced shifts in populations of microbial-feeding nematodes are also likely to be damped by predation although the ecosystem-level consequences of this remain largely unknown.

Various less-studied microfloral-microfaunal interactions may also be important. Fungal-feeding nematodes may consume arbuscular mycorrhizal fungi, possibly negatively influencing mycorrhizal infections (Ingham, 1988; Brundett, 1991) although the ecological consequences of this in relation to tillage remains unclear. Nematode-trapping fungi are likely to be important in moist situations with high organic matter (Dackman *et al.*, 1991); although their ecological significance in agroecosystems is poorly understood, they have been shown to regulate nematode populations in microcosms (Bouwman *et al.*, 1994), and their effects are likely to be more important in the uppermost soil and residue levels of NT systems. Testate protozoan predation on nematodes has recently been observed in forest litter but careful observations may lead to further reports (Yeates and Foissner, 1995).

5. Soil Microarthropods

The microarthropods, i.e. those which are less than 2 mm in length (Swift *et al.*, 1979), are a major component of the soil fauna. They consist of mainly the springtails (Collembola), mites (Acari), Protura and Pauropoda, and span a range of trophic levels, consuming litter, microflora, microfauna and other mesofauna.

Springtails are usually inhibited by cultivation (Fig. 2a,b) and when CT and NT treatments are under similar management practices, 44% of studies show moderate inhibition (V < −0·33). Data demonstrating moderately strong inhibition of springtails include those presented by Edwards (1975a, 1984), Winter *et al.* (1990) and Crossley *et al.* (1992). However, some previous studies also show tillage-induced stimulation of populations of springtails (e.g. Stinner *et al.*, 1986), while Hendrix *et al.* (1990) present data which suggest that CT systems have substantially higher populations of springtails than corresponding (non-tilled) old-field systems.

The total mite numbers show a wide range of responses to tillage (Fig. 2): both moderate (or extreme) reductions (e.g. Edwards, 1975a; Shams *et al.*, 1981; Stinner *et al.*, 1986, 1988; Hendrix *et al.*, 1990) and stimulations (e.g. van de Bund, 1970; Emmanuel *et al.*, 1985) have been detected. These effects are more extreme than those found for the microbial groups. Part of this variability in response may be due to differential responses to tillage among the various taxonomic groups of mites. This underscores the problem of reduced taxonomic resolution at lower trophic levels. When the CT and NT treatments are under similar management, over half of the studies show that cryptostigmatid and mesostigmatid mites are moderately or even extremely (e.g. Stinner *et al.*, 1988; Hendrix *et al.*, 1990) inhibited by tillage, but one study (Emmanuel *et al.*, 1985) shows moderate stimulation of the Mesostigmata by tillage, so these inhibitory effects are not universal. Prostigmatid mites are moderately inhibited by tillage in over 40% of studies in Figure 2, although moderate stimulation of this group sometimes occurs (Emmanuel *et al.*, 1985; Boles and Oseto, 1987; Perdue and Crossley, 1990). Astigmatid mites demonstrate a wider range of responses than the other microarthropod groups with reports of both extreme cases of tillage-induced inhibition (e.g. Andrén and Lagerlöf, 1983; Crossley et al., 1992) and stimulation (e.g. House and Parmelee, 1985; Perdue and Crossley, 1990).

As in the case of nematodes, shifts in microarthropod population size may reflect responses to the physical consequences of disturbance. Microarthropods are likely to be killed initially by tillage-induced abrasion and through being trapped in soil following soil inversion (Andrén and Lagerlöf, 1983). The relative abundance of microarthropods in CT and NT systems may also result from the nature of the physical environment created by tillage. However, these relationships are not always predictable. Springtails may demonstrate positive, negative or neutral relationships with soil moisture depending on species composition (Petersen and Luxton, 1982; Bardgett et al., 1993) resulting in unpredictable responses to tillage (Fig. 2). Litter accumulation often enhances populations of springtails (Steinberger et al., 1984). Mites also tend to prefer moister sites with higher organic matter although there is strong empirical evidence that the degree of this preference varies across the different orders. Cryptostigmatid and mesostigmatid mites prefer moister organic soils (Wallwork, 1982; Cepeda and Whitford, 1989) and have a low tolerance of physical disturbance (Holt, 1985) while prostigmatid mites have a lower requirement for soil moisture and organic matter, and can tolerate disturbance more easily (Loots and Ryke, 1967; Cepeda-Pizarro and Whitford, 1989). Astigmatid mites are less well understood but probably recover rapidly from disturbance (Andrén and Lagerlöf, 1980; Perdue and Crossley, 1990). Differential moisture and physical disturbance responses between the various soil mite orders may explain why cryptostigmatid and mesostigmatid mites may sometimes be more severely inhibited by tillage than prostigmatid and especially astigmatid mites. Modification of pore sizes and reduction of channel continuity by tillage (Unger and Cassell, 1991) also limits microarthropod populations (Winter et al., 1990). Microarthropods are capable of reinvading disturbed environments by migration (Whitford et al., 1981) and since their motility should enable them to recolonize tilled soils rapidly, long-term depressions of microarthropod numbers suggests that the nature of the soil environment in CT systems, rather than the direct one-off disturbance effects of tillage events themselves, may be at least partially responsible for regulating microarthropod numbers.

6. Interactions Between Microarthropods, Microfauna and Microflora

Microarthropods derive their nutrition from a variety of sources including all the trophic levels considered thus far (resource-base, microflora, and

122　　　　　　　　　　　　D.A. WARDLE

microbe feeding and predatory microfauna) and other microarthropods. It
is frequently assumed that the springtails, Cryptostigmata and Astigmata
(and some taxa of the Mesostigmata and Prostigmata) are "saprophagous", obtaining their nutrition from microflora and often plant litter,
while many components of the Prostigmata and Mesostigmata are predatory (see Petersen and Luxton, 1982). However, these distinctions are
highly blurred by omnivory, and "saprophagous" mites may feed on nematodes (Walter et al., 1986) and possibly protozoa (Walter and Kaplan,
1990) in addition to fungi. Predatory microarthropods also feed on nematodes, as well as other microarthropods (Ingham et al., 1986; Moore et al.,
1988). Therefore it is likely that microarthropod numbers reflect the biomass (or at least the productivity) of all the trophic levels occupied by the
microflora and microfauna. Also, in surface residues in NT systems,
greater hyphal networks are ultimately likely to result in greater microarthropod densities. However, depressions in microarthropod numbers as
caused by tillage are frequently well in excess of the negative impacts of
tillage on lower trophic levels, indicating either that negative impacts of
tillage may be amplified up the foodchain (as suggested by McNaughton
(1992) for above-ground food webs) or that microarthropods respond more
significantly to the changes in the physical factors resulting from tillage
than do the organisms they feed upon. Another possibility is that the
influences on lower trophic levels of grazing microarthropods prevents
standing crops of organisms in those levels increasing as much as they
otherwise might in NT (cf. CT) systems. Fungal productivity is probably
maximized at an intermediate grazing intensity and overgrazing of fungi
(under high microarthropod densities) may cause a decrease in fungal
biomass (Hanlon, 1981; Moore et al., 1988). Stimulation of fungi by microarthropods probably results from grazing of senescent hyphae, litter comminution and enhanced spatial dispersal (Visser, 1985; Moore et al., 1988)
and considerable compensatory fungal growth can occur, especially where
there is a high degree of habitat fragmentation (Bengtsson et al., 1993).
Enhanced comminution and dispersal may also enhance bacterial growth
(Hanlon, 1981; Lussenhop, 1992). These interactions are all likely to be
more prevalent in NT systems, due to higher densities of both microarthropods and fungi in the surface layers.

Microbe-feeding springtails and mites may also graze on arbuscular
mycorrhizal fungi, and thus reduce the effectiveness of mycorrhizas
(Finlay, 1985; Fitter and Sanders, 1992). This interaction is more likely to
be important in NT systems, which support a greater population of microarthropods and probably greater densities of AM fungal hyphae.

Predatory microarthropods that feed on nematodes have the potential to
reduce nematode populations, and tend to be negatively associated with
nematode densities (Santos and Whitford, 1981), a situation which is likely

to occur more strongly in NT systems. These effects are likely to be important, since they are sufficient to exert tri-trophic top-down (and possibly cascade) influences culminating in increases in bacterial populations (Santos et al., 1981; Seastadt, 1984; Moore et al., 1988). If predator pressure by microarthropods on nematodes is stronger in NT systems, then this may help explain why microbe-feeding nematodes are sometimes more abundant in CT systems (Fig. 2). This provides further evidence to suggest that bacteria are likely to be predator-limited, reflecting the findings of Wardle and Yeates (1993) in relation to systems dominated by predatory nematodes.

Top predatory microarthropods also have the potential to regulate populations of litter-feeding microarthropods although the consequences of this type of predator–prey interaction for the detritus food web remain largely unknown. This relationship is likely to be more important in NT systems, where microarthropods of most trophic levels occur in larger numbers than in CT systems.

7. Enchytraeids

The enchytraeids are a major, although often neglected, group of soil fauna, which are often abundant in both cultivated and undisturbed soil systems (Petersen and Luxton, 1982; Didden, 1993). Populations of enchytraeids may be either moderately strongly stimulated or inhibited by tillage, especially when CT systems are compared with NT systems under similar management (Fig. 2).

The influence of physical factors on enchytraeid populations is poorly understood. However, their comparative success in CT systems relative to most of the other mesofaunal groups considered in this study appears related to the ability of their populations to recover rapidly from disturbance (Lagerlöf et al., 1989), probably accompanied by a high metabolic activity (Golebiowska and Ryszkowski, 1977). Enchytraeids are dependent upon humidity, especially in warmer environments, and population recovery is slow after prolonged drought (Kasprzak, 1982). Given this moisture requirement, enchytraeids should be less abundant in CT systems. The lower abundance of enchytraeids under NT perennially cropped systems may be due in part to greater evapotranspiration when compared with conventionally tilled, annual-cropping systems (Lagerlöf et al., 1989).

Enchytraeids feed on microbes and detritus, and sometimes demonstrate preferential grazing of fungi (O'Connor, 1967). However, the influence of feeding by enchytraeids on lower trophic levels, or the dependency of enchytraeid populations on an abundant food-source, remains almost entirely unknown in either CT or NT systems.

8. Earthworms

Earthworms are a very important macrofaunal component of most agro-ecosystems, and are undoubtedly the most extensively studied group of soil animals. They are also highly sensitive to agricultural disturbance. Ninety-three percent of previous studies show greater numbers or biomass of earthworms in NT than CT systems when both systems are under equivalent cropping management, and 48% of these are moderate ($V < -0.33$) or extreme ($V < -0.67$) (Fig. 2a). In comparison with non-tilled systems under perennial cover, 100% of studies found fewer earthworms under CT, with 33% of studies showing moderate inhibition and a further 33% showing extreme inhibition by tillage, the highest for any group found in this study (Fig. 2b). Occasionally, however, mild positive effects of tillage have been suggested. Edwards and Lofty (1975) and Gerard and Hay (1979) note that some groups of earthworms are actually favoured by tillage and that modest cultivation may provide superior conditions for earthworms. More severe impacts of tillage on earthworm density may be related to greater intensities of cultivation (Barnes and Ellis, 1979; Haukka, 1988).

A principal factor influencing earthworms under CT is undoubtedly the physical disturbance associated with cultivation, since earthworms are usually large enough to be directly injured by machinery and thus suffer greater physiological stress (Rovira et al., 1987). Earthworms may also be killed by desiccation after being brought to the surface by ploughing, and by being preyed upon by birds when at the surface (Edwards, 1975a; Haines and Uren, 1990). Larger and deeper burrowing earthworms are particularly susceptible to cultivation (Edwards and Lofty, 1982; Rovira et al., 1987). Since earthworms aestivating below the tillage depth are not likely to be directly injured by cultivation, the nature of the physical environment following this disturbance is probably undesirable for these earthworms and this may reduce colonization following tillage. Reduced soil organic matter, surface residues and moisture levels, such as are more likely under conventional tillage, are less desirable for earthworm populations (Edwards, 1983; Rovira et al., 1987).

9. Interactions Between Earthworms and Other Trophic Groups

Earthworms ingest particles of soil and organic matter, containing the microfloral and microfaunal species assemblages found in the soil micropore network. As such, earthworms feed indiscriminately and obtain their nutrition from non-living organic matter (resource-base), fungi, bacteria and possibly microfaunal groups (Dash et al., 1980; Satchell, 1983). The greater abundance of food nearer the surface in no-tillage systems is likely to result in the stimulation of earthworm populations.

The influence of earthworms on the soil microflora results from modification of ingested material as it passes through the gut. It is well known that earthworm casts contain greatly modified species assemblages of micro-organisms (Ghilarov, 1963; Parle, 1963), and bacteria tend to be greatly enhanced, probably at the expense of fungi (Domsch and Banse, 1972; Daniel and Anderson, 1992). This appears to result from the tendency of the earthworm gut to stimulate some groups through modifying the chemical and physical nature of the soil environment, and simultaneously digest other (largely fungal) organisms (Edwards and Fletcher, 1988). The net result is that the composition of the microbial biomass is altered by earthworm activity while the total microbial biomass may be largely unaffected (Daniel and Anderson, 1992) or may respond unpredictably depending on soil and resource conditions (Wolters and Joergenson, 1992). Since earthworms are much more numerous under NT, their actions against fungi may dampen otherwise potential enhancement of fungi in non-cultivated systems. This is likely to be of considerable importance where they may, under high densities produce $2 \cdot 7 \times 10^5 \, \text{kg ha}^{-1} \text{yr}^{-1}$ of casts (Lee, 1985). Those organisms which are enhanced in the earthworm gut may enter a mutualistic relationship with the earthworm, facilitating breakdown of organic matter, particularly under warmer temperatures (Lavelle, 1995). Nematode populations appear to be reduced by high earthworm densities (Yeates, 1981; Hyvönen et al., 1994) suggesting that earthworms have a greater potential to displace (and possibly consume) nematodes in NT systems. This may dampen potential increases in nematode populations in the absence of cultivation. Earthworm burrowing activity in NT systems probably also affects lower trophic levels through increasing soil infiltration (Edwards et al., 1990), burying litter (and hence increasing microbial colonization: Wardle et al., 1993b) and dispersing microbial propagules (Satchell, 1983). Bacterial populations may also be enhanced in the walls of the burrows themselves (Loquet et al., 1977), resulting in a larger microbial biomass at greater depths. Distribution of soil mesofauna according to depth is also enhanced by earthworms through an improved network of soil pores (Marinissen and Bok, 1988). Earthworms can substantially alter plant community structure (Thompson et al., 1993) and earthworm-induced nutrient mineralization often stimulates plant growth (Edwards and Lofty, 1980) which is likely to stimulate other soil organisms through enhanced rhizosphere activity.

10. Saprophagous Mesoarthropods

Saprophagous mesoarthropods consist of a wide range of taxonomic sub-groups, including members of the Crustacea, Diplopoda, Isopoda, Coleoptera, Formicidae, Hymenoptera and (in the tropics) Isoptera. They are all probably highly omnivorous non-selective feeders, obtaining their nutrition from both litter and micro-organisms. This group has received much less attention than the other groups considered here (at least in relation to detritus food webs in agro-ecosystems) but they can demonstrate either negative or neutral responses to tillage (House and Stinner, 1983; Stinner *et al.*, 1988; House and Alzugaray, 1989; Wardle *et al.*, 1993a). Although saprophagous macroarthropods are probably susceptible to the physical disturbances associated with tillage, many are highly mobile and able to reinvade tilled sites.

Mesoarthropods can be extremely abundant in agricultural systems (Wardle *et al.*, 1993a). Since passage of material through the gut of these animals may have direct and important consequences for population of soil microflora (Anderson *et al.*, 1983; Ineson and Anderson, 1985; Wolters, 1989) and microfauna (Tajovský *et al.*, 1992), the consumption patterns of mesoarthropods need to be much better understood before a more complete understanding of food web differences between CT and NT systems can be developed.

11. Predatory Mesoarthropods

The top predators of this food web consist mainly of the Arachnomorpha, Chilopoda and Coleoptera, which consume mainly meso- and other macro-faunal groups. Two components of this group, i.e. the ground beetles (Carabidae) and spiders (Araneae), have been reasonably intensively studied in agro-ecosystems. Carabid beetles are often influenced adversely by tillage (Fig. 2): when CT and NT systems are under similar cropping management, populations are less under CT for 92% of studies, with 33% of studies showing extreme (V < −0·67) reduction by tillage. This pattern does not change appreciably when CT systems are compared with perennial NT systems. Extreme negative effects of tillage on carabid beetle populations are demonstrated by data of House and Stinner (1983), House and Parmelee (1985) and House and Alzugaray (1989). Occasional instances also occur when cultivation appears to stimulate some carabid species (Stassart and Gregorie-Wibo, 1983; Wardle *et al.*, 1995a).

The soil-associated Araneae are generally also disrupted by tillage when the CT and NT systems are under similar cropping practice (Fig. 2a) and although these effects can be quite severe (V < −0·67 in 22% of studies) only mild reductions by tillage were detected in over half the studies considered. Spider numbers or biomass respond considerably more dramatically to tillage when compared with NT systems under perennial cover (Fig. 2b) with 63% of studies showing moderate or extreme inhibition by tillage, indicating that NT annual systems may inhibit spider populations relative to NT perennial systems. It should be noted that many of the studies for which NT perennial systems were used involved crop headlands rather than perennial crops, which are less likely to be prone to disturbance. Extremely negative impacts of tillage are presented in data of House and Parmelee (1985), Hendrix *et al.* (1986) and Desender *et al.* (1989). Few studies have investigated other predatory groups, although the available studies on staphylinid beetles suggest that they are not strongly influenced by tillage (House, 1989; House and Alzugaray, 1989; Brust and House, 1990; Wardle *et al.*, 1993a).

The frequent extreme inhibition of carabid beetles and spiders by tillage may reflect in part their responses to physical disturbance, as previously discussed for earthworms. However, these organisms are often highly mobile, and their tendency not to reinvade tilled sites suggests that such systems are less suitable than NT systems, probably because of the absence of residues on the soil surface. Spiders and carabid beetles are largely active at or near the soil surface and often need surface-associated prey. Frequencies of these predators therefore may simply reflect populations of saprophagous meso- and macro-fauna associated with residues accumulated in the absence of tillage. The intensive nature of competitive interactions that can exist between carabid beetle species indicate that prey availability is an important regulatory force of this group (Niemalä, 1993).

The predatory meso-arthropod guild almost certainly represents the point where the detritus food web meets the above-ground (foliage-based) food web since many beetles and spiders are probably equally well adapted to feeding on Collembola, mites and foliage-associated insects of similar sizes. However, this linkage of above-ground and below-ground food webs has received negligible attention to date and is thus poorly understood (Stinner and House, 1990), although top-down regulation of litter-feeding fauna by these predators may potentially have important (and possibly cascade-type) effects on lower trophic levels, especially in non-tilled systems. Data presented by Kajak *et al.* (1993) indicate that excluding predatory macrofauna in a peat grassland induces a cascade effect down three trophic levels, enhancing mesofauna, microflora, and plant litter decomposition.

B. Overall Food Web Responses

When the food web as a whole is considered, it appears that there are mild responses to tillage experienced by those organisms associated with micro-

128   D.A. WARDLE

scopic (often water-filled) soil pores (i.e. microflora and microfauna) and
somewhat more dramatic responses by those of larger size (i.e. meso- and
macro-fauna). This may be in part a result of greater taxonomic resolution
of lower trophic levels (a problem recognized by Hall and Raffaelli (1993)
for other food web syntheses) but aggregation of components of higher
trophic levels does not appear to affect this conclusion significantly. The
relationship between organism size and response to tillage is demonstrated
in Figure 4, where the mean and variance of the index V values for all the

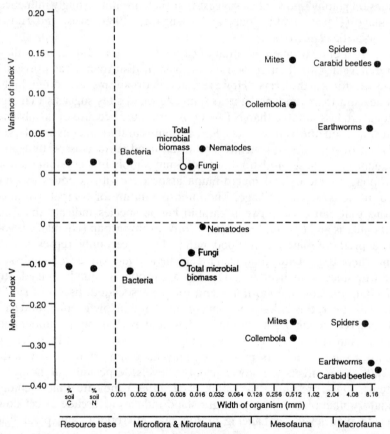

Fig. 4. Values for the variance and mean of index V (representing the level of
inhibition by conventional tillage vs. no-tillage systems and calculated according to
section III-A) for all the studies used to compile Figure 2a for various groups of
organisms, plotted against mean organism width (derived from Swift et al., 1979).
Mean and variance of index V values for resource-base (soil) carbon and nitrogen
are included for comparative purposes. The "microbial biomass" point was placed
assuming a bacterial:fungal ratio of 25:75 (Anderson and Domsch, 1975).

studies used in Figure 2a were calculated for each major taxonomic group. There is a distinct negative relationship between mean organism width (following log-transformation) and mean value of index V ($r = -0.86$, $p = 0.006$, $n = 8$), and a positive relationship between organism width and the variance of V ($r = 0.77$, $p = 0.024$, $n = 8$). This strongly suggests that responses to tillage are scale-dependent, and that organisms which are capable of occupying larger soil pores are more responsive to tillage-associated disturbance than smaller ones. Larger organisms are more prone to physical disruption, abrasion by tillage practices and, probably, modification of habitat. But they are also able to migrate readily to recently tilled sites from nearby undisturbed habitats and their tendency not to is probably reflective of the reduction of litter levels, and hence litter-associated organisms of lower trophic levels. They may, in fact, be demonstrating amplification of disturbance effects on lower trophic levels (McNaughton, 1992). Since large organisms usually occupy higher trophic levels than smaller ones, this indicates that a unit of biomass of organisms at a lower trophic level (e.g. microflora) appears capable of supporting a higher biomass of organisms in higher trophic levels in the absence of cultivation and that consumption of lower trophic levels is likely to be more important under reduced disturbance. The higher variance of the index V for larger organisms also indicates that their responses are less predictable than those of smaller organisms. While smaller organisms were similarly affected by tillage in most of the studies considered (i.e. slightly negatively), larger organisms appear to demonstrate rather different responses in different studies, ranging usually from slight inhibition to extreme inhibition (and some cases almost complete eradication). The lower resistance of larger organisms to disturbance is in contrast to what has sometimes been found in above-ground studies, particularly those on vertebrates (Pimm, 1991).

The large responses to tillage of larger organisms in Figure 4 may be of particular ecological importance if the background (temporal) variability of these organisms is otherwise low. Bengtsson (1994a) found, in surveying the literature on soil fauna in temperate forest soils, that population constancy was comparatively high, and if the same is true in agricultural soils, the consequences of tillage on the soil biota may be particularly significant.

The trends shown in Figure 4 also suggest that following tillage, bottom-up forces regulating larger organisms (i.e. tillage reducing small organisms, thus reducing food supply of larger organisms and reducing them too) are unlikely to be the only factors operating—the reduction of larger organisms is vastly in excess of what could be expected in terms of food availability alone. Top-down influences are also likely to be important (Bengtsson et al., 1995), and many instances of where higher trophic levels influence lower trophic levels (either positively or negatively) have already been presented in Section III.A. Some of these effects probably follow down the food chain for several trophic levels (see Abrams, 1993), and may in some cases originate from outside the below-ground food web. For example, birds feed preferentially in agricultural systems with reduced disturbance probably as a result of higher surface arthropod and earthworm densities (Tucker, 1992) although they may also reduce earthworm numbers in cultivated fields shortly after cultivation (Rovira et al., 1987) and therefore could conceivably indirectly affect lower trophic levels. Similar effects of birds on above-ground food webs, culminating in cascade effects down three other trophic levels, have been documented by Tscharntke (1992). Top-down effects may also dampen potential changes in lower trophic levels. Some large organisms which are more abundant under no-tillage (e.g. earthworms, springtails) tend to stimulate bacteria, and their activities may counteract what might otherwise be more positive effects of tillage on bacterial populations. Greater predation by top predatory nematodes on microbe-feeding nematodes (Wardle and Yeates, 1993) and macrofauna on mesofauna (Kajak et al., 1993) in NT systems have the potential to introduce trophic cascades, ultimately resulting in shifts in microbial populations.

Only eight taxonomic groups are presented in Figure 4 because insufficient data are available on other groups to provide meaningful estimation of the mean and variance of V. However, the Enchytraeidae may not conform to the trends identified in Figure 4. This group, and an order of the Acari (the Astigmata) appear to behave differently from the other taxonomic entities in the mesofaunal size-class, often being strongly positively influenced by tillage. The variance of V for these groups would appear to be in excess of the other groups depicted in Figure 4, although further data are required to verify this. Both of these groups (and to a lesser extent, the Prostigmata) are likely to have a rapid reproduction rate in soil and may be poor competitors although there is undoubtedly a wide variation of these properties within each group. Generally these are probably opportunistic groups that "colonize" early in the succession following tillage, and that would be reduced by competitive pressure from the other, slower-establishing mesofauna if tillage were delayed for sufficient time to allow the system to revert to a more natural state.

Few studies have investigated the simultaneous responses of more than one or two trophic levels to tillage. The results of three such studies are summarized in Table 1. The study of Paustian et al. (1990) conducted in Kjettslinge, Sweden (see also Andrén et al., 1990) followed changes in functional groups of organisms in tilled plots cropped to barley, and non-tilled plots containing perennial grass or lucerne leys. Although differences between tilled and non-tilled treatments may reflect in part differences in

Table 1

Biomass (g m^{-2}) of different taxonomic and functional groups in conventional-tillage and no-tillage systems

	Functional group	Study Paustian et al. (1990)[a] CT[d]	NT	V	Hendrix et al. (1986)[b] CT	NT	V	Beare et al. (1992)[c] CT	NT	V
Microflora	bacteria	80	90	−0·06				0·536	0·201	0·45
	fungi	190	170	0·06				0·687	0·611	0·06
Microfauna	protozoa	7·20	6·00	0·09				0·038	0·0068	0·70
	bacterial-feeding nematodes	0·075	0·090	−0·10	0·237	0·117	0·34	0·065	0·0094	0·75
	fungal-feeding nematodes	{ 0·050	0·060	−0·09	0·014	0·031	−0·38	0·013	0·0054	0·41
	predatory + omnivorous nematodes				0·017	0·040	−0·40	0·003	0·0004	0·76
Mesofauna	springtails	{ 0·020	0·035	−0·27	{ 0·118	0·303	−0·44	{ 0·213	0·072	0·49
	litter-feeding mites									
	predatory mites	0·025	0·025	0·00	0·059	0·017	0·55	0·024	0·008	0·50
	enchytraeids	0·037	0·019	0·32	3·13	20·31	−0·73			
Macrofauna	earthworms	1·24	3·22	−0·44	0·001	0·014	−0·87			
	spiders	{ 0·50	1·30	−0·44[e]	0·006	0·030	−0·67			
	carabid beetles									

[a] Data derived from Table 1 of Paustian et al. (1990). CT values are means of the two perennial ley treatments. Values apply to 27 cm soil depth. NT treatments are means of the two barley-cropped treatments.

[b] Data derived from Table 2 of Hendrix et al (1986).

[c] Data derived from Fig. 8 of Beare et al. (1992), and apply only to the biomass of each component on litter of Secale cereale L.

[d] CT, NT = conventionally-tilled and non-tilled systems. Index V calculated as according to section III-A.

[e] Total biomass of predatory macroarthropods.

the plants present and fertilization regimes adopted in that study, at least some of the differences may reflect responses of soil organisms to cultivation. Use of the index "V" indicates that at least some groups of larger organisms (Collembola and the macrofauna) are more susceptible to tillage than are the microfloral and microfaunal groups, and that the enchytraeids may benefit from tillage. The direct CT : NT comparison conducted at the "Horseshoe Bend" site (Georgia, USA) by Hendrix et al. (1986) (summarized in Table 1) again demonstrates that macrofaunal groups are the most susceptible to tillage. Enchytraeids and bacterial-feeding nematodes in contrast appear to benefit from tillage. Beare et al. (1992) presented data, determined from a litter-bag study at the same Horseshoe Bend site, which suggest that all groups except fungi are substantially enhanced in the litter of CT systems compared with litter of NT systems, mainly because much of the litter under CT is buried and in close contact with the soil system. Since NT systems usually have a higher overall biomass of mesofauna than CT systems (in contrast to the data presented by Beare et al. (1992) for litter only), enhancement of organisms by NT would therefore result from them being stimulated in the mineral matrix itself. Also, larger (macrofaunal) organisms that are not intimately associated with the "resource-islands" provided by the burial of litter are less likely to benefit as directly from litter burial as are the smaller organisms assessed by Beare et al. (1992), possibly amplifying differences in tillage-response between macrofaunal groups and other groups of soil biota.

C. Response of Species Assemblages to Tillage

This paper has so far considered only the response of functional groups (or "functional species") and major taxonomic entities to tillage. However, each functional grouping contains a wide range of taxonomic species, which may show differential responses to tillage (Yeates and Hughes, 1990; Freckman and Ettema, 1993), and species-level analysis may yield responses not detectable at the guild level (Bernard, 1992).

Ploughing favours organisms with short generation times and high metabolic activity (Hendrix et al., 1990). Microfloral species assemblages may be altered, with some bacterial groups (especially nitrifying and denitrifying bacteria: Broder et al., 1984; Linn and Doran, 1984) and certain fungal species (Wacha and Tiffany, 1979) being enhanced at the expense of others in NT systems. In relation to microfauna, shifts in species assemblages due to tillage are indicated in data of Domurat and Kozlowska (1974), Yeates and Hughes (1990) and Freckman and Ettema (1993) even though overall species diversity is not greatly affected. These changes are reflected in tillage-induced reductions in the nematode "maturity index", reflective of greater ecosystem-level disturbance (Yeates (1994) using modification of index of Bongers (1990)). Strong species-level effects of tillage are also apparent in most studies on meso- and macro-fauna, including those studies listed in Table 2. Earthworm species assemblages in particular exhibit strong differential responses to tillage, with ploughing reducing the deeper burrowing (rather than shallow burrowing) species (Edwards and Lofty, 1982). The effect is a shift to an earthworm fauna consisting mainly of smaller species (Rovira et al., 1987). Understanding changes in species assemblages following tillage is important in assessing the response of soil food web structure at the sub-functional level, as will be discussed in Section VI.

Recently there has been a greatly increased interest in biodiversity of ecosystems at the species level, because shifts in species diversity can be highly reflective of human disturbance (Wilson, 1988) and may impact upon both ecosystem function (Vitousek and Hooper, 1993) and population stability (Tilman and Downing, 1994). To assess the influence of tillage on the diversity of species of genera of the various functional groups considered in this study, 20 previous studies were selected which either presented the Shannon-Wiener diversity index (H'; a function of species evenness and richness) for one or more taxonomic groups in CT and NT systems, or which presented species/genus composition data from which H' could be calculated (Table 2). The one study in Table 2 conducted on microflora (Wacha and Tiffany, 1979) and the six studies on nematodes suggest little effect of tillage on fungal species diversity of these groups. On a survey over a larger geographical scale, Wasilewska (1979) found that richness of nematode genera and species was only slightly greater in (non-tilled) grassland than in (regularly-tilled) arable systems. Of the three studies on microarthropods, two (Lagerlöf and Andrén, 1988; 1991) provide data which suggest little response of species diversity to tillage, whereas the third (Emmanuel et al., 1985) suggests a strong tillage-induced decline. Two further studies (not in Table 2 because H' could not be calculated) presented microarthropod species richness data, of which one (Loring et al., 1981) found no tillage response while the other (van de Bund, 1970) detected a strong reduction in species number attributable to tillage. Ten studies have provided species-level data on various macroarthropod groups, and calculation of diversity indices for these studies suggests either positive or negative consequences of tillage may occur.

There appears to be a linkage between the size of organisms in each taxonomic group and the influence of tillage on the species diversity of that group. This can be demonstrated by calculating the index U (similar to the index V calculated earlier and that of Wardle and Parkinson, 1991b), given

Table 2

Shannon-Wiener species (or genus) diversity indices calculated for a range of studies in conventionally-tilled (CT) and non-tilled (NT) agro-ecosystems

Reference	Location	Taxonomic or functional group	Shannon-Wiener diversity index CT	NT (Same crop as NT)	Under grass or ley	Old-field or headland[a]
Wacha and Tiffany (1979)	Iowa, USA	Fungi	3·1	3·0		
Bostrom and Sohlenius (1986)	Central Sweden	Nematodes	4·1		4·3	
Domurat and Kozlowska (1974)	Poland	Nematodes	2·5			3·0
Freckman and Ettema (1993)	Michigan, USA	Nematodes	2·4	2·4	2·5	
Sohlenius et al. (1987)	Central Sweden	Nematodes	3·7		3·7	
Yeates and Hughes (1990)	New Zealand	Nematodes	2·2	2·2	2·2	
Yeates et al. (1993)	New Zealand	Nematodes	2·4	2·4[b]		
Lagerlöf and Andrén (1991)	Central Sweden	Springtails	1·5		1·7	
Emmanuel et al. (1985)	Northern Ireland	Mites	1·3	1·9		
Lagerlöf and Andrén (1988)	Central Sweden	Mites	3·2		3·3	
Blumberg and Crossley (1983)	Georgia, USA	Surface arthropods	2·1	2·7		2·6
Wardle et al. (1993a)	New Zealand	Surface beetles	0·5	0·4[b]		
Andersen (1987)	Denmark	Earthworms	1·2	1·6		
Barnes and Ellis (1979)	United Kingdom	Earthworms	1·0	1·0		
Haukka (1988)	Southern Finland	Earthworms	1·1	0·8		
Doane and Dondale (1979)	Saskatchewan, Canada	Spiders	2·3			3·1
Huhta and Raatikainen (1974)	Southern Finland	Spiders	1·9		2·7	
House and All (1981)	Georgia, USA	Carabid beetles	1·7	1·2	1·3	
Stassart and Gregorie-Wibo (1983)	Belgium	Carabid beetles	0·9	1·2		
von Klinger (1987)	Germany	Carabid beetles	1·7			1·4

[a] Studies with woody (non-herbaceous) plant cover not included.
[b] Reduced tillage systems.

by:

$$U = \frac{2H'_{CT}}{H'_{CT} + H'_{NT}} - 1$$

where: H'_{CT} and H'_{NT} = H' of taxonomic group in CT and NT systems respectively.

The index U ranges from −1 to +1 and is increasingly negative or positive if H' is increasingly lesser or greater in the CT system than the NT system. This index was calculated for each of the CT–NT comparisons

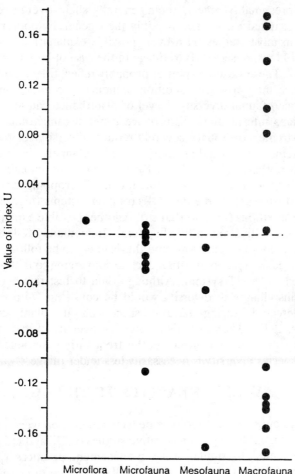

Fig. 5. Values for index U (calculated according to section III-C) comparing the Shannon-Wiener diversity index in conventional-tillage and no-tillage systems, using the data presented in Table 2. This index ranges from −1 to +1 and is increasingly negative or positive if the Shannon-Wiener index is increasingly lesser or greater in conventional-tillage than no-tillage systems.

presented in Table 2 (see Fig. 5) and demonstrates that tillage has a much more pronounced effect on macrofaunal than microfaunal groups at the species assemblage level. While diversity of microfauna is mostly non-responsive to tillage (U = -0·03 to +0·01 with one exception) the macrofaunal groups show a much wider range (U = -0·17 to +0·17), and there is no general trend. The much greater unpredictability of macrofauna than microfauna in their response to tillage is borne out by large differences in the variances of the index U for the two groups (F = 16·4, p < 0·001, n_1 = 12, n_2 = 8). The lack of response of the diversity of groups of smaller organisms presumably reflects their generally slight overall responses to tillage as discussed earlier and possibly their generally poorer taxonomic resolution in most studies. However, possible explanations for the more unpredictable responses of H' to tillage in the case of larger organisms are less obvious. These varied responses probably reflect the multiple effects of tillage, which may produce conditions conducive to either promoting or reducing macrofaunal diversity. Level of disturbance can also play a role here. Perhaps intermediate disturbance promotes macrofaunal diversity whereas extreme disturbance severely reduces diversity relative to undisturbed systems. This would be consistent with Connell's intermediate disturbance hypothesis (Connell, 1978) and may also explain why some groups of organisms are more abundant in NT cropping systems (intermediate disturbance) than either CT cropping systems (high disturbance) or old-field situations (no disturbance). Another possible explanation is the time-lag since cultivation in tilled systems. Immediately following tillage, diversity is likely to be very low, and this is likely to be followed over time by large increases, and an eventual decline converging to that which would be expected for a NT system. Although data to test this hypothesis are minimal, this change in diversity would be consistent with patterns frequently detected during natural succession in plant communities (Whittaker, 1975). The often higher species diversity of macroarthropods in CT than NT systems contradicts the frequently maintained generalization that species diversity is necessarily less under tillage.

IV. ALTERNATIVES TO TILLAGE

In no-tillage agriculture, there is a heavier reliance on weed management by other means. Although comparative studies of CT and NT systems are usually subjected to similar weed management practices (probably to reduce complicating factors), in reality NT practices often require a higher usage of other strategies, e.g. herbicide application, residue management and cover cropping.

A. Consequences of Herbicide Use

1. Response of Broad Taxonomic Groups

Herbicide application may influence organisms in the detritus food web either directly (through affecting their physiology) or indirectly (through influencing plants or other trophic levels). To determine the importance of these effects on the detritus food web, a literature survey was conducted which summarized the results of 110 previous studies that had investigated the impact of one or more herbicides on one or more taxonomic groups of soil organisms; these studies are listed in Appendix 2. For each study considered, the impact of each herbicide on each group of organisms was noted. In total 794 herbicide-organism combinations were found in these studies. Each of these combinations was given equal weighting, and the total distribution of these combinations is given in Tables 3 and 4.

Table 3 summarizes those studies conducted in laboratory incubation experiments. The majority of these were conducted in mineral soil, except for those on protozoa and nematodes, which were studied mainly in pure culture. Pure culture studies are useful in understanding the physiology of the organisms being investigated, but the response of organisms to herbicides in artificial media usually bears little resemblance to their response in soil (Greaves, 1987). For example, Wardle and Parkinson (1990b) demonstrated that most saprophytic fungal species were only stimulated by the herbicide glyphosate when it was added to soil, and only inhibited when added to agar. Bacteria may also be more tolerant of herbicides in soil than in artificial media (van Schreven et al., 1970). A significant proportion of studies in Table 3 utilized herbicide concentrations well in excess of those encountered in field conditions; herbicide concentrations in soil are usually less than $10\,\mu g$ a.i. (active ingredient) g^{-1} following spraying (see Domsch et al., 1983; Wardle and Parkinson, 1990a). Yet some studies have used concentrations of well over $1000\,\mu g$ a.i. g^{-1}. A breakdown of those organism–herbicide combinations tested in soil and using herbicide concentrations close to field levels reveals that the majority of combinations resulted in no detectable herbicide side-effects. However, bacteria (and to a lesser extent fungi) appear to be more susceptible to herbicides than the other groups considered here, with over 40% of combinations revealing inhibition. A small proportion of combinations also revealed stimulation of fungi, consistent with the possibility of fungi using easily degradable herbicides as resources (Wardle and Parkinson, 1990b). These results demonstrate that herbicides are probably more likely to affect smaller organisms directly than larger ones, although insufficient data points are available for a statistical test such as that conducted in Figure 3. It is reasonable to expect that herbicides may influence organisms which are more closely

Table 3

Summary of herbicide–organism combinations tested in laboratory studies, using studies tabulated in Appendix 2

Taxonomic group	Number of combinations using:		Number of combinations using:		% of combinations showing[a]		
	mineral soil	artificial media	realistic concentrations	excessive concentrations	+	0	−
Bacteria	45	34	48	31	3	56	41
Fungi	42	47	51	38	13	60	21
Total microflora	23	0	14	9	0	71	29
Protozoa	4	18	12	10	0	100	0
Nematodes	0	24	8	16	—	—	—
Springtails	8	12	10	10	0	83	17
Earthworms	64	53	52	65	0	83	17

[a] +, 0, −, indicates percentages of combinations where taxonomic group was stimulated, unaffected or inhibited by herbicide addition. Combinations included only those which were conducted in mineral soil at realistic concentrations.

Table 4

Summary of herbicide–organism combinations tested in field conditions, using studies listed in Appendix 2

Taxonomic group	Number of combinations		combinations with plants[b]			combinations without plants[c]		
	with plants	without plants[a]	+	0	–	+	0	–
Bacteria	29	2	17	52	31	50	50	0
Fungi	29	2	14	52	35	50	50	0
Total microflora	12	5	0	55	45	20	60	20
Nematodes	15	2	40	53	7	0	100	0
Enchytraeids	11	4	0	64	36	0	75	25
Mites	107	11	7	68	25	0	82	18
Springtails	46	8	7	54	39	0	63	38
Earthworms	11	6	0	9	91	0	83	17
Macroarthropods	68	5	3	63	34	0	80	20

[a] Studies without crop or weed plants in controls or treatments, and with hand-weeded controls.
[b] Percentages of herbicide–organism combinations in planted and/or non-weeded situations, which detected stimulatory (+), neutral (0) or inhibitory (–) effects of herbicides. Herbicide effects may be direct or indirect.
[c] Percentages of herbicide–organism combinations without plants (hand-weeded controls) which detected stimulatory (+), neutral (0) or inhibitory (–) effects of herbicides. Effects of herbicides are mostly direct.

linked with the resource-base (i.e. bacteria and fungi) since they are more likely to consume the organic matter containing adsorbed herbicide molecules. Organisms in higher trophic levels are often likely to consume herbicide molecules more indirectly. This may contradict the scheme of differential susceptibilities of soil faunal groups to pesticides in general as presented by Edwards (1989), in which mites, insects and millipedes are ranked as more susceptible than springtails, earthworms, nematodes and protozoa.

Laboratory incubation studies have one major drawback in that they do not easily enable comparison of herbicide side-effects on soil organisms with the natural (spatial and temporal) variability of the same organisms in the field. Although laboratory studies enable detection of even very small transient responses, most organisms naturally undergo very dramatic shifts in abundance or biomass, which would undoubtedly render small herbicide-induced responses as relatively unimportant (Domsch et al., 1983; Cook and Greaves, 1987). Field studies (Table 4) may partially overcome this problem—if a herbicide side-effect is still detectable against a background of field variability, it is more likely to be ecologically significant. But field studies have one major disadvantage: they do not easily enable separation between direct and indirect side-effects on soil organisms. Only

D.A. WARDLE

a relatively small proportion of studies have been conducted in the absence of plants (e.g. by using hand-weeded controls). When a study is conducted with plants present, the response of organisms to herbicides may be entirely indirect, due to herbicides killing existing plant cover and aiding organic matter return (Eijsackers and van de Bund, 1980; Mahn and Kastner, 1985), delaying weed emergence and thus subsequent effects of weed growth (Prasse, 1975; Sabatini et al., 1979), or stimulating crop growth through reduced weed competition, possibly modifying the nature of the rhizosphere. Indirect effects of herbicides on soil organisms may also occur in plant-free studies, although these mostly arise from the herbicide directly influencing another component of the soil biota, which then affects those organisms under investigation (Karg, 1964). Of those few herbicide–organism combinations tested in field situations (Table 4) the majority reveal no detectable herbicide side-effects. However, some studies reveal microbial stimulation and meso- and macro-faunal inhibition. These responses are not, however, large, especially when compared with the responses of these groups to tillage. Also, too few studies have been conducted in plant-free situations to enable any statistically valid comparison between taxonomic groups to be made.

In field studies that have significant plant (usually weed) biomass in their controls, responses to herbicide applications have been detected in a significant proportion of studies for all taxonomic groups considered (Table 4). Usually these effects are inhibitory, and probably stem from indirect effects via herbicides influencing plant cover. The response of all taxonomic groups is reasonably similar, except for (a) earthworms, which are significantly inhibited by herbicides in a significantly higher proportion of data sets than all other groups; and (b) nematodes, which are stimulated by herbicides in a higher proportion than most other groups. The inhibition of earthworms suggests that they may benefit by weedy conditions more than most other groups, possibly through the sustained input of organic matter increasing availability of food to lower trophic levels. Earthworms are highly sensitive to floristic changes (Fox, 1964) and this may help explain many of the apparent responses of earthworms to herbicides under field conditions outlined by Edwards and Bohlen (1992). Herbicide stimulation of nematodes (and, to a lesser extent, bacteria and fungi) in a significant proportion of studies appears to stem from studies in which herbicide use resulted in a return of organic matter by target plants present at the time of spraying (Sály and Staňová, 1983; Mahn and Kastner, 1985).

Tables 3 and 4 suggest that the direct effects of herbicides on the soil biota may be ecologically relatively unimportant. Indirect effects may be stronger, and most organisms (especially earthworms) are susceptible. The impact of herbicides in NT systems (and in CT systems wherever used) is likely to be largely dependent upon the level of weediness that would be expected in the absence of herbicide use. These indirect effects appear to apply more-or-less equally to all size classes of organisms (except for earthworms). Only larger organisms are often severely affected by tillage, and therefore, relative to the impacts of tillage, herbicide side-effects are more likely to be important in influencing smaller rather than larger organisms. It should be emphasized that the level of weediness present in the controls of many of the studies used for compiling Table 4 would be intolerable in a commercial cropping system, but permitting some weediness by reducing herbicide use may still enhance much of the soil biota. This is discussed in more depth in Section V. No evidence emerges from the studies used to compile Tables 3 and 4 to suggest that any herbicide group is consistently more toxic to the soil biota than any other group. It has often been concluded that herbicides such as DNOC and various triazines have stronger negative effects on soil organisms than other herbicides, but these conclusions are based on studies where indirect effects have not been corrected for, or studies for which no replication or statistical analyses were used.

Herbicides may alter the soil biota through influencing the microbial colonization of plant litter. Thus, Hendrix and Parmelee (1985) found that grass residues treated with herbicides resulted in an increase in microbe-feeding microarthropods but a reduction in predatory microarthropods. In contrast, House et al. (1987) found that various herbicides applied to sorghum residues exerted little effect on microarthropod populations, probably because of the overriding importance of micro-climatic factors. Wardle et al. (1994) determined that microbial colonization of herbicide-killed plants may differ greatly from plants killed by other means. These studies indicate that herbicide applications in no-tillage systems may have the potential to influence the soil biota in "resource-islands", and possibly affect the rate of decomposition of surface residues (see House et al., 1987). However, no study has yet compared the soil biota associated with buried litter in CT systems with that of herbicide treated surface litter in NT systems, even though this is likely to reveal important differences between the two systems.

Herbicide-induced plant death increases available organic matter derived from dead roots with a corresponding increase in microbial biomass (Wardle et al., 1994). Although this aspect is undoubtedly of importance in most agro-ecosystems, it has received very little attention. Herbicides may also affect rhizosphere populations of both target and non-target plants, mainly by altering the forms of carbon secreted (Kaiser and Reber, 1970; Moorman and Dowler, 1991). Arbuscular mycorrhizal fungi are also affected by herbicides, mainly indirectly through influencing the physiology of the host plant (Trappe et al., 1984) although direct effects may also occur (Dodd and Jeffries, 1989). Furthermore, weed species may act as hosts of

AM fungi during non-cropping seasons, and successful herbicidal weed control can reduce mycorrhizal spore levels thereby delaying mycorrhizal infection of crop species (Sieverding and Leihner, 1984).

Whenever herbicides influence any of the groups of organisms that make up the detritus food web, this may influence other groups which interact with the affected group, in a manner similar to the tillage-induced multi-trophic effects discussed in Section III. Multi-trophic effects of herbicide applications are more likely in NT than CT systems, not only because herbicides are more likely to be widely used, but also because NT systems are inherently more stable (not having been exposed to tillage-related disturbance) and there is probably a greater dependence between adjacent trophic levels. However, few studies have investigated herbicide effects over several trophic levels simultaneously. Eijsackers (1978a,b,c) reported a detailed study on springtail-carabid beetle interactions, where springtails sprayed with 2,4,5-T had an adverse effect on the beetles that fed upon them, inducing the beetles to avoid contaminated prey. Food-consumption patterns of saprophagous springtails (Eijsackers, 1975), isopods (Eijsackers, 1985) and earthworms (Fayolle, 1979) have also been shown to be influenced by herbicides with obvious implications for lower trophic levels. In the studies of Karg (1964) and Gottschalk and Shure (1979), herbicidal control of weeds and subsequent enhanced return of litter induced microfloral and microfaunal build-up, with resultant increases in microarthropod populations of higher trophic levels.

2. Responses of Species Assemblages

Herbicides may exert differential influences on different species or groups of species within the broader taxonomic groups considered in Tables 3 and 4. This is especially well known within the soil microflora, where some components (notably the nitrifying bacteria) tend to be more susceptible to herbicides than other groups, with obvious consequences for nutrient cycling (Kreutzer, 1963). Repeated herbicide use may lead to greater densities of microbial species adapted to herbicide breakdown, resulting in "enhanced degradation" of the herbicide (Racke, 1990). Modifications in species assemblages following herbicide application have also been detected for protozoa (Gel'cer and Geptner, 1976), nematodes (Ishibashi et al., 1978), microarthropods (Fratello et al., 1985) and macroarthropods (Jaworska, 1981). However, many of these effects were probably indirect, through herbicides modifying the weed flora present.

Shifts in biodiversity of soil organisms may potentially reflect impacts of herbicides on the soil biota. To assess this, fourteen studies from the literature have been selected, in which species composition data were presented in herbicide-treated and untreated soil. For each study, the Shannon-Wiener diversity index (H') has been calculated for each group of organisms in each of the herbicide treatments considered, using the composition data provided. Four of these studies were laboratory-incubation investigations, monitoring the response of soil fungal species assemblages (Abdel-Fattah *et al.*, 1983; Abdel-Mallek and Moharram, 1986; Mekwantanakarn and Sivasithamparam, 1987; Wardle and Parkinson, 1990b). Calculation of H' for each of these studies revealed that fungal species diversity was unaffected by herbicide application at concentrations close to those experienced in the field. The remaining ten studies involved herbicide side-effects on species assemblages in field conditions, for fungi (Wacha and Tiffany, 1979), nematodes (Ishibashi *et al.*, 1978; Yeates *et al.*, 1993b), mites and/or springtails (Rapoport and Cangioli, 1963; Karg, 1964; Davis, 1965; Sabatini *et al.*, 1979; Purvis and Curry, 1981), Carabidae (Jaworska, 1981) and total soil beetle fauna (Wardle *et al.*, 1993b). Diversity indices calculated for only two of these studies indicated likely responses to herbicides. For the data of Rapoport and Cangioli (1963), where MCPA and 2,4–D were applied to grass turf, H' for springtails was 1·67 for the control and 1·63 for the herbicide treatment six days after spraying; however, after four months this diverged to 2·14 (control) and 1·40 (treatment), suggesting a long-term reduction of species diversity. This effect is likely to be mainly indirect and attributable to herbicide-induced alteration of floristic composition. Strong herbicide-induced effects were revealed in the analyses of data of Sabatini *et al.* (1979) in a maize system. In that study, H' for springtails was 1·68 (control) and 0·82– 1·11 (various atrazine concentrations) at 1½ months after spraying; at 6 months H' ranged from 1·35 (control) to 0·85 (highest atrazine concentration). These effects may also have been largely indirect. Although few studies exist for which responses of species diversity of various groups to herbicide applications can be assessed, it appears that herbicides do not influence biodiversity in agro-ecosystems in the same way that tillage does for larger organisms, except perhaps indirectly by modifying the weed flora.

B. Use of Mulches

The use of mulches to reduce weed populations has been studied considerably less than tillage and herbicide regimes. Although mulches are not always practical or economically viable, they nevertheless enable the simultaneous reduction of tillage and herbicide use. They may therefore become an important component of NT systems whenever efforts are made to reduce herbicide usage.

1. Dead Mulches

Surface residues, usually in the form of plant litter, may exert strong influences on the detritus food web, especially in relation to NT systems, as has been discussed for much of this paper. The benefits of NT agriculture for the soil biota are often directly related to the amounts of residue present (Havlin et al., 1990) and it may therefore be considered desirable actively to increase the levels of residues on the soil surface to amplify these benefits. Dead mulches (e.g. straw, sawdust) should have obvious benefits, but have been little studied, probably because such strategies are used mainly in intensive, small-scale systems. Addition of mulches can greatly increase infiltration, reduce evapotranspiration and reduce organic matter loss (Adams, 1966; Lal et al., 1980), all of which have obvious beneficial consequences for the detritus food web. Surface residues and mulches are thus usually observed to stimulate the microflora (Holland and Coleman, 1987; Wardle et al., 1993a), microfauna (Beare et al., 1989) and arthropod populations (Riechert and Bishop, 1990). However, immobilization of soil nitrogen may occur (due to the high carbon–nitrogen ratio of most mulching materials) and limitation of organisms by nitrogen deficiency is more likely in no-tillage systems and where mulches are present (Blevins et al., 1984; Wardle, 1992).

The impact of mulching on the soil biota may be considerable. In August 1990 a study was initiated in an asparagus (perennial cropping) and maize (annual cropping) system near Hamilton, New Zealand, to compare the impacts of disturbance associated with five different weed management strategies on the below-ground food web structure, viz. two herbicide treatments, repeated 10-cm depth inter-row cultivation, application of a 10-cm depth sawdust mulch and a hand-hoed (minimum disturbance) "control" (Wardle et al., 1993a,b; Yeates et al., 1993b). In the asparagus site, measurements of the various groups of organisms in the mineral soil, and macroarthropods on the soil (or mulch) surface revealed that of all the treatments applied, the only one which exerted any consistent influence over the first 12 months was the application of sawdust mulch, which caused dramatic enhancements of many components of the soil biota (Table 5). Since soil carbon and nitrogen levels were unchanged by application of the mulch, the benefits probably resulted from the capacity of the mulch to elevate soil moisture levels. Of importance is the dramatic extent of the soil microbial enhancements by mulching; while there were no differences between the NT (hand-hoed control) and repeated cultivation treatments, very large differences emerged between NT treatments with and without surface mulch. For fungi, this difference was also larger than that for any of the CT–NT comparisons depicted in Figure 2 or any of the herbicide effects outlined in Table 4. The beneficial effects of mulching are

Table 5

Response of various groups of organisms in mineral soil to placement of 10-cm depth sawdust on the soil surface in asparagus-planted plots[a]

	Response variable	Soil under 10-cm depth sawdust	Soil in control (hand weeded) plots	F[b]
Resource-base	% soil nitrogen	0·50	0·55	6·7 NS
	% soil carbon	8·73	8·66	0·1 NS
	% soil moisture (dry wt. basis)	56·6	47·6	138·2***
Soil microflora[c]	Active bacterial biomass	1·74	1·25	21·2***
	Active fungal biomass	5·97	2·99	106·3***
Soil microfauna (No./m^2)	Bacterial feeding nematodes ($\times 10^5$)	3·39	2·37	9·8**
	Fungal-feeding nematodes ($\times 10^4$)	5·24	9·12	3·5 NS
	Top predatory nematodes ($\times 10^3$)	10·0	0·34	20·3***
Soil mesofauna (No./m^2)	Total springtails ($\times 10^2$)	163·0	97·5	3·2 NS
	Total mites ($\times 10^2$)	69·2	33·2	7·4*
Surface-dwelling macrofauna[d]	Millipedes	1·95	0·75	6·3*
	Harvestmen	4·20	1·30	20·1***
	Wolf spiders	1·50	0·65	4·8*
	Staphylinid beetles	6·55	0·50	36·9***
	Tiger beetles	12·9	16·1	0·6 NS
Active microbial biomass[c]	Surface-placed litterbags	115·7	58·7	19·3***
	10-cm depth buried litterbags	119·1	72·9	11·6***
Shannon-Wiener diversity index (H′)	Nematodes	2·31	2·49	2·7 NS
	Surface beetles	0·72	0·18	73·5***

[a] Data are averaged over a one-year period and have been adapted from Wardle et al. (1993a,b) and Yeates et al. (1993b). All measurements are in the 0–5 cm depth layer.
[b] F-value for comparison between treatments. *, **, *** indicates that F is significantly different from 0 at p = 0·05, 0·01, and 0·001 respectively; NS = no significant difference.
[c] Based on substrate-induced respiration data (units = $\mu g\ CO_2$—C. g soil^{-1}. h^{-1}).
[d] Relative figures, based on pitfall trap data.

likely to be relatively more important for smaller organisms, which usually demonstrate only a mild response to tillage. These effects appear to follow through the food-chain to some extent: predatory nematodes were greatly enhanced under mulching while microbe-feeding nematodes were not, suggesting a tri-trophic effect in which the intermediate level (microbe-feeding nematodes) were regulated by the top predator level (Wardle and Yeates, 1993). The impacts of mulching are apparent not only at the functional group level but also in terms of the genus or species composition of two groups that were investigated in greater detail, viz. nematodes and macroarthropods. In the case of macroarthropods, a consistently higher species diversity was detected in the sawdust-mulched treatments.

The site used receives around 1200 mm rain yr^{-1} so positive impacts of thick non-living mulches may be even more considerable in drier environments. The degree of difference between CT and NT systems must rely in large part on the thickness of residues on soil, and their ability to reduce soil moisture fluctuations and impose stability on the soil system. This supports previous data which suggest that removal or burning of surface residues can have highly negative consequences for the soil biota (Edwards, 1984).

2. Living Mulches

In the absence of cultivation, weed management can sometimes be effectively achieved by growing "green" or "live" mulch species under the desired crop, which supposedly reduce competitive effects of invading weed species without themselves exerting significant competitive effects on the crop. Popular live mulches consist of various perennial grass species (especially in the inter-rows of orchard systems) and densely growing clover species (Enache and Ilnicki, 1990; Ilnicki and Enache, 1992). The benefits of live mulches to the soil system are probably similar to those of perennial cropping in the NT–CT comparisons in Figure 2b, i.e. the ability to sustain organic matter levels of soil, enhanced release of rhizosphere exudates and possible beneficial effects on soil moisture (Fleige and Baumer, 1974; Mulognoy, 1986; Trujillo-Arriaga and Altieri, 1990). Agricultural practice involving live mulches tends to promote diversity of the cropping system, by actively encouraging the growth of two dominant species rather than just one. Enhancement of diversity of litter input may influence components of the decomposer food web, such as has been shown for forested systems by Blair et al. (1990).

Few studies have investigated soil organisms or the detritus food web under living mulches. However, live mulches may greatly stimulate both soil microflora and fauna (Mulognoy, 1986; Paoletti et al., 1991) and may preferentially stimulate saprophagous arthropods (Favretto et al., 1992).

Probably the properties of the soil biota in systems under live mulches are similar to those of perennial cropping and grassland systems. NT systems which rely on herbicides often result in bare or sparsely vegetated ground except in the area occupied by the crop itself. Therefore substituting the herbicide for a continuous cover species may enable the system to more closely resemble natural ecosystems, complete with greater diversity and biomass (and possibly diversity) of soil organisms (Ferris and Ferris, 1974; Mulognoy, 1986).

C. Alternative Agricultural Systems

There has been a greatly increased recent interest in developing agricultural systems which reduce environmental problems and rely less on fossil-fuel based energy and material inputs (Pimentel et al., 1983). These include "low-input" and "organic" agricultural systems, which aim to largely or entirely eliminate the use of synthetic pesticides and fertilizers (Werner and Dindal, 1990) and, in some cases, tillage (Foissner, 1992). Such methods therefore rely on ecosystem self-regulation as a substitute for these inputs (House and Brust, 1989). Alternative agricultural systems frequently contain higher levels of soil biota than comparable conventionally farmed systems (e.g. El Titi and Epach, 1989; Heinonen-Tanski, 1990; Werner and Dindal, 1990) and may also support modified species assemblages (e.g. nematodes (Freckman and Ettema, 1993); carabid beetles (Hokkanen and Holopainen, 1986)). De Ruiter et al. (1993) found most components of the decomposer food web to be enhanced by integrated farming practices. Frequently, stimulations resulting from alternative agricultural practice are not large, and Foissner (1992) concludes upon reviewing the evidence that: "It is increasingly evident that generalisations like—Conventional farming destroys life in the soil—or—Ecofarming stimulates soil life—are only partially supported by the available data". Differences in soil biota between conventional and alternative farming systems, when they occur, may stem from a variety of possibilities, but are most likely to result from differences in cultivation, weed biomass, mulching and residue management practices, and quantities of organic matter input. Curiously, many alternative agricultural philosophies allow the use of naturally occurring chemicals at unnatural concentrations for weed control (e.g. fatty acid herbicides: James and Rahman, 1992). Yet there is little evidence that many naturally occurring toxins are any more environmentally desirable than most synthetic herbicides at comparable concentrations, at least in relation to the soil biota. There is also little evidence to suggest that any of the preparations endemic to biodynamic farming are uniquely beneficial to soil organisms; although Reganold et al. (1993) concluded that biodynamic farming had beneficial effects on quality of soil biota, recalculation of their data demonstrates that little difference existed between biodynamic and conventional farming once adjustments were made for statistical artefact (see Wardle, 1994; Reganold, 1994).

V. INFLUENCE OF WEEDS ON DETRITUS FOOD WEBS

Often, studies which investigate the comparative effects of CT and NT effects on the soil ecosystem apply identical herbicide treatments to each treatment, to reduce sources of variability other than the soil disturbance regimes under investigation. In such studies, weed biomass is frequently similar under different tillage regimes (e.g. Stinner et al., 1984; Cheng et al., 1990). However, in agricultural practice, the use of herbicides is likely to be less under CT than NT (Stinner and House, 1990), so that weed levels may vary between different tillage treatments depending on the relative weed management strategies adopted. Therefore the influence of the weeds themselves on the detritus food web may be highly important in understanding NT and CT agro-ecosystems.

The influence of weeds on the soil saprophytic microflora has received little attention, although colonization of bare areas by plants (usually weedy ruderals) is usually accompanied by an increase in microbial bio-mass (Wardle, 1992). The data in Table 4 suggest that reduced weed populations may potentially reduce the soil microflora. In a field study (Wardle et al., 1995a) it was shown that high levels of weed biomass in an asparagus crop in the autumn may significantly stimulate the microbial biomass in the following summer, even when overall levels of total soil carbon remain unchanged. A corresponding glasshouse study (D.A. Wardle and K.S. Nicholson, unpublished) demonstrated that weedy species were capable of inducing substantial stimulation in soil microbial biomass, probably well in excess of crop plants of similar size. Weedy tissues can decompose extremely rapidly (Wardle et al., 1994) at least in part because they have a low C:N ratio (Nicholson and Wardle, 1994) and therefore represent an important form of readily available organic matter. Relatively little is known about the influence of weed abundance on the microfauna, but microbe-feeding nematode populations may respond indirectly to weed management via changes in the soil microflora (Mahn and Kastner, 1985). Tri-trophic effects of plant growth on bacterial-feeding nematode and enchytraeid numbers were demonstrated for four grass (including weedy Poa) species by Griffiths et al. (1992). That study also demonstrated the ability of different plant species to regulate differentially various trophic levels of the below-ground food web, probably as a result of different carbon allocation patterns in the rhizosphere.

Soil arthropods appear to be highly responsive to weed levels. Numerous studies have demonstrated that the principal mesoarthropod groups tend to be stimulated under weeds, usually through monitoring populations in cropped fields and in corresponding weedy headlands, field margins or weed-mulched plots (e.g. spiders: (Desender et al., 1989; Thomas et al., 1992); carabid beetles (von Klinger, 1987; Desender and Alderweireldt, 1988; Cardwell et al., 1994); ants (Altieri et al., 1985); opilionids (Kromp and Steinberger, 1992)). Long-term weedy areas appear to act as refugia for some predatory groups, allowing potential recolonization of tilled land (Altieri and Letourneau, 1984), as well as ensuring the existence of over-wintering habitats (Thomas et al., 1992; Lys and Nentwig, 1994). Weedy systems may induce large shifts in arthropod species composition, often accompanied by increased species richness (Doane and Dondale, 1979; Wallin, 1985; Lagerlöf and Wallin, 1993). Enhanced diversity in weedy conditions may be in part linked to the tendency of different arthropod species to be strongly associated with different weed species (House, 1989), possibly enabling a greater range of taxa to co-exist.

Predatory soil-associated mesoarthropods gain their nutrition from both the above-ground and below-ground food webs and weedy conditions may stimulate these predators through enhancement of lower trophic levels in either food web. However, only the predatory mesoarthropod:herbivore interactions have been studied in much detail, mainly because these re-lationships are important in terms of biological control of insect pests (Stinner and House, 1990). Although predator:herbivore interactions are beyond the main scope of this paper, it is necessary to note that since some weeds increase predator activity against herbivores (e.g. Speight and Law-ton, 1976; Altieri and Schmidt, 1984; Brust and House, 1990) with prob-able positive consequences for plant growth, they undoubtedly increase predator activity against saprophagous fauna, with probable consequences for other trophic levels of the detritus food web.

Systems with higher levels of weed biomass are conducive to supporting a rich and diverse soil biota. In comparison with weed-free situations, weedy systems bear stronger resemblance to natural systems. Tolerance of at least a partial weed flora is consistent with the concept of maintaining an agro-ecosystem with a wider variety of plant species, which in turn leads to a greater capacity for self-regulation (Altieri, 1991). Yet, excessive weedi-ness is obviously detrimental to crop growth, and conditions for main-taining optimal crop productivity are clearly different from those for maintaining a superior soil biota. However, even weed levels which are sufficiently low to cause no detectable effect on crop yield may encourage the soil biota (Wardle et al., 1993a). Weed management practices that allow low levels of weeds may therefore confer major advantages on the detritus food web and associated processes, especially in comparison with exhaustive weed control practices. Reduced tillage and herbicide use may therefore assist the soil biota simply through incomplete weed control.

VI. AGRICULTURAL DISTURBANCE AND FOOD WEB THEORY

Paine (1980) categorized three different means of representing empirical food webs, viz. "connectedness" food webs (which link each organism to its food-source), "energy" food webs (which estimate energy flows) and "functional" food webs (which estimate dynamic properties based on experimental manipulations of the consumers). Most of the present paper has emphasized the energetic and functional aspects but a substantial proportion of studies in food web ecology have emphasized the connectedness aspect, probably because of the relative ease of compiling connectedness food webs. Much of current food web theory is based on compilations of connectedness food webs from a range of environments, which have been analysed for underlying trends (Cohen et al., 1990) and which generally severely under-represent the detritus food web (Wardle, 1991). However, some doubt has been cast on many of the theories developed from such approaches, fuelled in part by an important paper by Paine (1988), appropriately entitled "Food webs: Road Maps of Interactions or Grist for Theoretical Development?". This section will evaluate the applicability of connectedness food web theory to the detritus food web and the response of this food web to tillage-related disturbance.

The taxonomy and feeding habits of most groups of soil organisms are poorly understood, and construction of below-ground connectedness food webs necessitates lumping taxonomic species together into broad functional (or feeding) groups. Although the impact of tillage and weed management may severely influence the success of many functional groups as evidenced by Figure 2 and Tables 3 and 4, no group is usually entirely eliminated. Thus, the connectedness food web remains largely the same, at least as far as the various functional groups and density of links between them is concerned. Meanwhile, energy and nutrient fluxes (and henceforth the energy food web) can differ widely between different systems (Paustian et al., 1990; Beare et al., 1992). Since the connectedness food web remains little altered by the disturbance regimes considered in this paper, so does food chain length. This is in contrast to the aquatic food chains studied by Havens (1994), which were shortened by chemical perturbation. The non-responsiveness of soil food chain length to tillage does not support the results of theoretical studies which suggest that food chains should be longer in ecosystems exempt from large perturbations (Pimm and Lawton, 1977) but may lend support to the findings of Briand and Cohen (1987) that food chain length does not vary between constant and fluctuating environments. Although long food chains have been presumed to be dynamically unstable and thus usually prevented from acquiring more than four trophic levels (Cohen *et al.*, 1990; Pimm *et al.*, 1991), detritus-based food chains may be considerably longer than this. Connectedness food webs presented by Ingham *et al.* (1986) and Hunt *et al.* (1987) suggest that detritus-based food chains may have up to seven links even when only plants, microflora and micro- and meso-fauna are considered. Since mesofauna are themselves consumed by macrofauna, which are in turn often consumed by other macrofauna and birds, and ultimately mammalian predators, nine or ten trophic levels may not be uncommon. This may be further enhanced when organisms at each of these levels themselves die and enter the decomposition subsystem. Yet, while tillage may drastically influence many of the organisms in the detritus food web, the food chain may still contain nine or ten trophic levels in spite of such intensive perturbation. Since detritus food webs are an important component of most terrestrial food webs, it appears that food webs are probably not as short as often maintained. The terrestrial food webs used by Briand and Cohen (1987) and Cohen *et al.* (1990) for assessing food chain length usually assume that predators draw their energy largely or entirely from the grazing food web, but if they actually derive their nutrition from both the grazing and detritus food web, food chain lengths may be much longer in most of these food webs than their results would suggest. It is becoming recognized that most other food webs have also been oversimplified and that larger food webs may contain much longer food chains than previously thought (e.g. Polis, 1991; Martinez, 1992; Hall and Raffaelli, 1993).

Omnivory has often been presumed to be rare in food webs because its presence in theoretical models has been shown to be destabilizing (Pimm, 1982) although some studies (e.g. Sprules and Bouwman, 1988; Polis, 1991) suggest that omnivory is commoner than frequently believed. Omnivory is widespread in the detritus food web (as discussed in Section 2) and probably becomes more important at the higher trophic levels (although it is unclear as to whether the incidence of omnivory differs from that which would be expected by chance alone: see Hall and Raffaelli, 1993). Only a proportion of microfauna are omnivores and others are specifically adapted for feeding only on hyphae or only on bacteria. Most meso- and macrofauna, however, feed more generally, and both saprophages and predators may gain nutrition from at least two (and sometimes several) trophic levels. Extreme reductions in meso- and macro-faunal groups due to tillage are therefore likely to result in a reduced incidence of omnivory, which is likely to be detectable in food webs based on energy flux but possibly not those based on connectedness. High omnivory probably makes concepts such as the scale-invariance of predator:prey species ratios (Briand and Cohen, 1984) largely inapplicable to below-ground systems, since systems with large numbers of species which may be regarded as both predators and prey can result in many species being double-counted (Closs et al., 1993). Any complete terrestrial food web requires a specific below-ground component, making such theories inapplicable to terrestrial food webs as a whole.

The high incidence of omnivory may lead to greater connectance in the detritus food web. Enhanced connectance is presumed to be linked to greater food web instability. May (1973) suggested that randomly assembled food webs were likely to be locally stable if:

$$\beta(SC)^{\frac{1}{2}} < 1$$

where: β = average strength of all direct interactions

S = number of species (in the detritus food web, this has usually been taken to mean the umber of functional groups)

C = connectance, i.e. fraction of all species pairs which interact directly.

This means that if β remains invariant, and stability remains unchanged, the product S × C should remain constant (Pimm, 1982) and the relationship between S and C should therefore be hyperbolic. This was supported by earlier studies (e.g. Rejmanek and Stary, 1979; Briand, 1983) although more recent analyses indicate that this relationship may not necessarily hold (Winemiller, 1989; Bengtsson, 1994b; Warren, 1994), and that it may in part reflect a spurious self-correlation between S and C (Peters, 1991). There is some evidence that connectance can be independent of species number (Warren, 1994), and Wardle et al. (1995b) found connectance to remain invariant in a successional study of decomposer organisms, even though species richness increased. Little is known about the connectedness properties of soil food web, and although Moore et al. (1988) estimated connectance for the detritus food web of a short-grass prairie (based on functional groups) to be of approximately the same magnitude as that of other food webs with a similar (functional) species number (e.g. those presented by Yodzis, 1981), connectedness may actually be considerably higher than this, for the following reasons:

(i) Insufficient may be known about feeding interactions to construct a complete connectedness food web. Interactions which have not usually been incorporated into detritus food webs (probably because they are very poorly understood) include nematode-trapping by fungi, mycophagy and nematophagy by protozoa, and the possible ingestion of bacteria by microarthropods. Predatory nematodes may feed on a wider range of other groups (e.g. oligochaetes, rotifers, other predatory nematodes: Small, 1987) than is often acknowledged in food web studies. Another interaction often not considered is the prospect of microarthropods gaining nutrition direct from plant litter, a realistic supposition considering the high levels of plant material frequently found in microarthropod gut analyses (Anderson and Healey, 1972; Visser, 1985).

(ii) Most connectedness food webs, including those of Moore *et al.* (1988), contain only some of the organisms found in the below-ground food webs, i.e. frequently only up to microarthropod level. Inclusion of predatory macroarthropods (some of which probably prey on microarthropods of all trophic levels) and saprophagous macroarthropods and earthworms (some of which may be highly omnivorous and gain their nutrition from several trophic levels) occupying the same habitat could substantially elevate the estimated level of connectance for the detritus food web.

(iii) As mentioned earlier there is a need to aggregate organisms into functional groups because the taxonomy and feeding biology of soil organisms is little known. For example, probably as few as 1% of bacterial and 3% of nematode species worldwide have been described (Klopatek *et al.*, 1992), a situation which is unlikely to change rapidly especially considering the current state of systematic biology (Feldman and Manning, 1992). But the concept of grouping several species together and treating them as one "trophic species" is problematic. If trophic species are defined according to Cohen *et al.* (1993), i.e. "a largest set of organisms with identical sets of predators (if any) and identical sets of prey (if any)" aggregation of taxonomic species is clearly inappropriate since this would often imply near-identical niche dimensions. In fact, even within taxonomic (genetic) species individuals interact with different predators and prey throughout different stages of their life-cycles, which would suggest that a younger instar of a carabid beetle larva feeding on saprophagous collembolans must act as a different trophic species from a later instar feeding on predatory spiders. This tendency to lump taxa together especially at lower trophic levels must severely alter (and possibly reduce) estimated connectance (Paine, 1988), and miss detection of critical links in the detritus food web (Walter *et al.*, 1991). For example, some predatory nematodes gain a substantial proportion of their nutrition from other predatory nematodes (Small, 1987), suggesting important internal structures which would be overlooked if they were all treated as one trophic species. A connectedness food web for below-ground systems based on individual species (if one were possible to construct) would also be more likely to reflect local disturbances than one based on species aggregates—a taxonomic species is more likely to be wiped out by disturbance than an entire functional group. Interestingly, it is now being suggested that the degree of connectance in other food webs is likely to be much greater than previously thought (Warren, 1990; Polis, 1991; Martinez, 1992).

Since the connectedness food web is unlikely to be strongly altered by tillage and other related disturbances at the trophic species level, the degree of connectance is also unlikely to be significantly affected. The same is probably true about the other main component affecting stability, namely "trophic species" number—whereas taxonomic species diversity (especially as estimated by H') may be influenced by tillage (Table 2), local extinctions of trophic (and probably taxonomic) species are likely to be uncommon, and overall species number may not be greatly affected. Regardless of disturbance regime, below-ground food webs have a much larger number of taxonomic species than corresponding above-ground systems. When taxonomic species rather than "trophic species" are considered (Paine, 1988) the product S.C in the detritus food web is almost certainly substantially higher than for most other systems, supposedly increasing complexity, and by implication instability of the below-ground system. Yet the below-ground system manages to remain locally stable in instances of agricultural disturbance at the connectedness level, and appears to show little difficulty in adapting to regular tillage while still maintaining a high species diversity and food web connectance.

Many of the features of the below-ground food web, viz. high omnivory, long food chains, high connectance and high taxonomic species diversity, all lead to supposedly greater instability. The apparent stability of the connectedness detritus food web suggests that there are probably fundamental differences between this web and other, more widely studied webs. One possible reason for this is that interaction strength (β) may be very weak, especially at the taxonomic species level: most species of soil fauna are highly non-specific in their feeding habits. Low values for β may offset increases in S and C while still maintaining stability according to May (1973). Wardle et al. (1995b) present empirical evidence from a primary succession of sawdust that most interactions involving nematodes in the decomposer food web are weak while a few are extremely strong, resulting in a highly skewed distribution of interaction strengths. This supports the results obtained by Paine (1992) in an intertidal ecosystem, and might therefore be a widespread trend. Another possible explanation is dimensionality of habitat. Briand and Cohen (1987) demonstrated that food webs that were able to arrange themselves spatially in three dimensions contained longer food chains than those in two dimensional habitats, although it is unclear as to whether this would remain the case if an explicit below-ground component was included in each of the terrestrial food webs they considered. In the case of the below-ground food web, the soil acts as a highly buffered three-dimensional matrix which may be able potentially to insulate the detritus food web against disturbance (Wardle, 1991). This is supported by the results of studies conducted on microarthropod–fungal interactions in two- and three-dimensional matrices of artificial beads, which reveal that trophic interactions may stabilize with increasing dimensionality of the substrate (Anderson and Ineson, 1983; Leonard, 1983). There is also theoretical evidence that detrital-based systems are inherently more stable than other systems (De Angelis, 1992) and the detrital pool has been suggested as a buffer against disturbance in aquatic ecosystems (Closs and Lake, 1994). The low temporal variability of forest soil animal populations compared with that of other animal taxa (Bengtsson, 1994a) may reflect greater stablity of detritus-based systems.

One aspect of below-ground food webs which possibly makes them unique is the presence of within-habitat compartments, maintained by the morphological differences between bacteria and fungi (Wardle and Yeates, 1993). Pimm (1982) and Pimm and Lawton (1980) concluded that there was little theoretical or empirical evidence for compartments in other systems, although an analysis of mostly aquatic food webs by Raffaelli and Hall (1992) suggests that compartments may exist in some cases. Although bacteria and fungi in soil separate themselves to some extent along the habitat niche axis (because the bacterial-based compartment is often more moisture dependent than the fungal-based compartment: Moore and Hunt, 1988), there is still habitat overlap between these compartments as indicated by their ability to displace each other under different conditions (Wardle and Yeates, 1993). Therefore, these two compartments may respond differently to disturbance as is evidenced by their response to tillage: the bacterial compartment benefits by the burial of organic matter relative to the fungal compartment (Hendrix et al., 1986; Beare et al., 1992). This results in changes in relative importance of different energy channels (Moore and Hunt, 1988). The existence of two trophically equivalent compartments enables potential comparison of properties of subsets of the entire food web. Bacterial-feeding microfauna, which engulf their prey, generally show little discrimination between bacterial species, whereas fungal-feeding organisms may prefer some fungal species above others (e.g. Parkinson et al., 1979; Newell, 1984), possibly by actively seeking preferred food items (Bengtsson et al., 1994). This presumably stems from fungi being morphologically more complex than bacteria and thus capable of showing a greater range of adaptations (and susceptibilities) to grazing (Wardle and Yeates, 1993). In other words, even though bacterial-feeding and fungal-feeding organisms in detritus food webs are both generalist feeders, especially in comparison with consumers in other food webs, fungal-feeders may be much less generalized than bacterial feeders. This would suggest that, independent of species number, connectance in fungal-based compartments is probably less than in bacterial-based compartments

(Fig. 6) which reveals the prospect that even between two compartments in the same food web the product S.C could be different. To confer equal stability on each compartment (as according to May, 1973), interaction strength would need to be correspondingly greater in the fungal-based compartment.

While the connectedness food web has limited applicability for understanding the consequences of agricultural disturbance, energy and carbon-based food webs, and food webs based on other resources such as nitrogen (Hunt *et al.*, 1987) can enable detection of perturbations, even when species are aggregated into broad functional groups. This has been demonstrated for many of the interactions outlined in Section III. Although some potential influences of agricultural disturbance (and possible cascade effects) become blurred at higher trophic levels because of the length of detritus-based food chains and the prevalence of omnivory, the below-ground food web clearly responds to such disturbance, and disturbance effects may manifest themselves over several trophic levels. Food web stability at the connectedness level of resolution may therefore be a poor indicator of actual food web responses to disturbance, especially when lower trophic levels are only aggregated according to kingdom level. Cohen *et al.* (1993) appropriately suggest that a better understanding of food webs may have potential in benefiting various practical problems, but such understandings in below-ground food webs are more likely to emerge from "energy" food webs and "functional" food webs (including studies aimed at identifying important interactions through manipulative experiments: Paine, 1992; Polis, 1994) than from compilations of "connectedness" food webs.

VII. CONCLUSIONS

The soil food web appears to respond to differences between CT and NT management mainly as a result of differences in three factors, viz. soil physical disruption, residue distribution and weed density. Physical disruption by tillage often results in severe inhibition of populations of larger (macrofaunal) organisms occupying higher trophic levels, with smaller groups being less affected, or in some cases stimulated. This may mean that a given unit of biomass of organisms at a lower trophic level supports a larger biomass of organisms higher up the food web, and that top-down influences by higher levels are likely to be more important under NT systems. Surface residues in NT systems enhance most groups of organisms and result in their concentration nearer the soil surface. Meanwhile, burial of litter following cultivation results in greater biomass of microflora and micro- and mesofauna around "resource-islands", resulting in accelerated

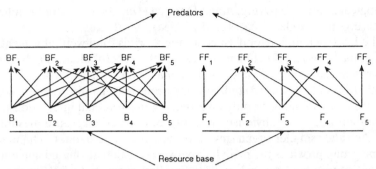

Fig. 6. Possible feeding links in the two main compartments of the detritus food web. The diagram is largely hypothetical but the important point is that, independent of species number, connectance is greater in the bacterial-based than the fungal-based compartment. B_1–B_5 and F_1–F_5 are bacterial and fungal species 1–5; BF_1—BF_5 and FF_1—FF_5 are bacterial- and fungal-feeding faunal species 1–5.

litter decay and consequent long-term loss of organic matter. Weed levels, which are most likely to be greater in NT systems in the absence of herbicides, often stimulate populations of soil organisms over and above crop-induced enhancement, through rhizosphere effects, addition of readily decomposed organic matter, and possibly by enhancing habitat diversity. These three factors interplay to create the physical and chemical environment for the soil biota. All three affect soil moisture levels which strongly influence the organisms at the lower levels of the detritus food web. Direct effects of herbicides on the soil biota may occur in some circumstances but little evidence exists to suggest they have long-term important effects in most soil systems.

While disturbances in CT systems may exert predictable negative effects on most groups of soil organisms, the responses of species assemblages are less predictable. Individual taxonomic species are undoubtedly a more sensitive indicator of ecosystem disturbance than are entire functional groups, and are therefore a more appropriate unit for biomonitoring purposes. The niche concept is most successfully applied to organisms at the genetic species level. However, the current state of systematic biology of most groups of organisms makes this extremely difficult. Furthermore, because so little is known about the feeding patterns of most organisms, the ecological consequences of shifts in abundances of individual species or genera following perturbation are likely to be estimated only by prediction. A more complete understanding of food web responses to perturbation is best able to be arrived at by studies emphasizing a finer taxonomic resolution than is normally used for investigating below-ground food webs (see Klopatek *et al.*, 1992) and by gathering and synthesizing information on major functional groups detailing, at the species or genus level, their ecological relationships (e.g. fungi: Domsch *et al.*, 1980) or trophic interactions (e.g. nematodes: Yeates *et al.*, 1993a).

The least well understood aspect of the detritus food web (at least at the functional level) is the linkage between the macrofauna and the other (microfloral and micro- and meso-faunal) components. This probably stems from soil-associated studies usually considering only organisms up to the size of mesofauna, or only investigating macrofaunal interactions. Yet both approaches are obviously incomplete, especially because mesofauna and the other smaller organisms all occupy the same habitat. The macrofaunal group provides the main link between the detritus-based and above-ground food webs and inclusion of this link suggests that there may be important influences of above-ground organisms on the detritus food web as demonstrated by Kajak *et al.* (1993). For example, CT and NT systems probably encourage vastly different levels of predation by birds on soil-associated macrofauna which may impact differently upon lower trophic level in the two systems (Tucker, 1992). But, the ecological consequences of these types of interactions (on corresponding bottom-up ones from the detritus food web) remain almost entirely unknown. However, any complete terrestrial food web has explicit above-ground and below-ground compartments, and a complete understanding of the effects of disturbance requires an understanding of both components and their level of interaction.

ACKNOWLEDGEMENTS

I thank G.M. Barker, A.H. Fitter, F.D. Panetta, D. Raffaelli, A. Rahman, J. Springett and G.W. Yeates for valuable comments on the manuscript, K.S. Nicholson for help with the references, P. Hunt for preparing the illustrations, P. Klaffenbach for help in interpreting German language literature, and J.P. Bruce for typing the manuscript. This work was funded by The National Foundation for Research, Science and Technology.

REFERENCES

Abdel-Fattah, H.M., Abdel-Kader, M.I.A. and Hamida, S. (1983). Selective effects of two triazine herbicides on Egyptian soil fungi. *Mycopathol.* **82**, 143–151.

Abdel-Mallek, A.Y. and Moharram, A.M. (1986). Effect of the herbicide ametryn on cellulose-decomposing fungi in Egyptian soil. *Folia Microbiol.* **31**, 375–381.

Abrams, P.A. (1993). Effect of increased productivity on the abundance of trophic levels. *Am. Nat.* **141**, 351–371.

Adams, J.E. (1966). Influence of mulches on runoff, erosion and soil moisture depletion. *Soil Sci. Soc. Am. Proc.* **30**, 110–114.

Allen, M.F. (1991). *The Ecology of Mycorrhizae.* Cambridge University Press, Cambridge.

Allen-Morley, C.R. and Coleman, D.C. (1989). Resilience of soil biota in various food webs to freezing perturbations. *Ecology* **70**, 1127–1141.

Altieri, M.A. (1991). How can we best use biodiversity in agroecosystems? *Outl. Agric.* **20**, 15–23.

Altieri, M.A. and Letourneau, D.K. (1984). Vegetation diversity and insect pest outbreaks. *Crit. Rev. Plant Sci.* **2**, 131–169.

Altieri, M.A. and Schmidt, L.L. (1984). Abundance and feeding activity of communities in abandoned, organic and commercial apple orchards in northern California. *Agric. Ecosyst. Environ.* **11**, 341–352.

Altieri, M.A., Wilson, R.C. and Schmidt, L.L. (1985). The effects of living mulches and weed cover on the foliage- and soil-arthropod dynamics in three crop systems. *Crop Prot.* **4**, 201–213.

Andariese, S.W. and Vitousek, P.M. (1988). Soil nitrogen turnover is altered by herbicide treatment in a North Carolina piedmont forest soil. *For. Ecol. Manage.* **23**, 19–25.

Andersen, A. (1987). Regnorme uddrevet med strom in forsog med saning og plojning. *Tidsskr. Planteavl.* **91**, 3–14.

Anderson, A.J. (1992). The influence of plant root on mycorrhizal formation. In: *Mycorrhizal Functioning* (Ed. by M.F. Allen), pp. 37–64. Chapman and Hall, New York.

Anderson, E.L. (1987). Corn root growth as influenced by tillage and nitrogen fertilisation. *Agron. J.* **79**, 544–549.

Anderson, J.M. and Healey, I.N. (1972). Seasonal and interspecific variations in major components of the gut content of woodland Collembola. *J. Anim. Ecol.* **41**, 359–368.

Anderson, J.M. and Ineson, P. (1983). Interactions between soil arthropods and microorganisms in carbon, nitrogen and mineral element fluxes from decomposing leaf litter. In: *Nitrogen as an Ecological Factor* (Ed. by J.A. Lee, S. McNeill and I.H. Rorison), pp. 413–432. Blackwell Scientific Publishers, Oxford.

Anderson, J.M., Ineson, P. and Huish, S.A. (1983). Nitrogen and cation mobilisation by soil fauna feeding on leaf litter and soil organic matter from deciduous woodlands. *Soil Biol. Biochem.* **15**, 463–467.

Anderson, J.P.E. and Domsch, K.H. (1975). Measurement of bacterial and fungal contribution to respiration of selected agricultural and forest soils. *Can. J. Micro.* **21**, 314–322.

Anderson, R.V. and Coleman, D.C. (1982). Nematode temperature responses: a niche dimension in populations of bacterial-feeding nematodes. *J. Nematol.* **14**, 69–76.

Anderson, T-H and Domsch, K.H. (1985). Determination of ecophysiological maintenance requirements of soil microorganisms in a dormant state. *Biol. Fertil. Soils* **1**, 81–89.

Andrén, O. and Lagerlöf, J. (1980). The abundance of soil animals (Microarthropoda, Enchytraeidae, Nematoda) in a crop rotation dominated by ley and in a rotation with varied crops. In: *Soil Biology as Related to Land Use Practices* (Ed. by D. Dindal), pp. 274–279. Environmental Protection Agency, Washington D.C.

Andrén, O. and Lagerlöf, J. (1983). Soil fauna (microarthropods, enchytraeids nematodes) in Swedish agricultural cropping systems. *Acta Agric. Scand.* **33**, 33–52.

Andrén, O., Lindberg, T., Paustian, K. and Rosswall, T. (Eds.) (1990). *Ecology of Arable Land*. Ecological Bulletins, Copenhagen.

Andrén, O., Paustian, K. and Rosswall, T. (1988). Soil biotic interactions in the functioning of agroecosystems. *Agric. Ecosyst. Environ.* **24**, 57–67.

Angers, D.A., Pesant, A. and Vigneux, J. (1992). Early cropping-induced changes in soil aggregation, organic matter and microbial biomass. *Soil Sci. Soc. Am. J.* **56**, 115–119.

Arshad, M.A., Schnitzer, M., Angers, D.A. and Rigmeester, J.A. (1990). Effect of till vs. no-till on the quality of soil organic matter. *Soil Biol. Biochem.* **22**, 595–599.

Arthur, M.F. and Frea, J.I. (1988). Microbial activity in soils contaminated with 2,3,7,8–TCDD. *Env. Toxicol. Chem.* **7**, 5–13.

Atlavinyte, O., Daciulyte, J. and Lugauskas, A. (1977). The effect of Lumbricidae on plant humification and soil organism processes under application of pesticides. *Ecol. Bull.* **25**, 222–228.

Baarschers, W.H., Donnelly, J.G. and Heitland, H.S. (1988). Microbial toxicity of triclopyr and related herbicides. *Tox. Assess.* **3**, 127–136.

Baird, S.M. and Bernard, E.C. (1984). Nematode population and community dynamics in soybean-wheat cropping and tillage regimes. *J. Nematol.* **16**, 379–386.

Bamforth, S.S. (1988). Interactions between protozoa and other organisms. *Agric. Ecosyst. Environ.* **24**, 229–234.

Barber, S.A. (1971). Effect of tillage practice on corn (*Zea mays L.*) root distribution and morphology. *Agron. J.* **63**, 724–726.

Bardgett, R.D., Frankland, J.C. and Whittaker, J.B. (1993). The effects of agricultural management on the soil biota of some upland grasslands. *Agric. Ecosyst. Environ.* **45**, 25–45.

Barnes, B.T. and Ellis, F.B. (1979). Effects of different methods of cultivation and direct drilling and dispersal of straw residues, on populations of earthworms. *J. Soil Sci.* **30**, 669–679.

Baudissin, F.G. (1952). Die wirking von pflanzenschutzmitteln auf Collembolen und milben in verschiedenen boden. *Zool. Jahnb.* **81**, 47–90.

Bauer, A. and Black, A.L. (1981). Soil carbon, nitrogen and bulk density in two cropland tillage systems after 25 years and in virgin grassland. *Soil Sci. Soc. Am. J.* **45**, 1166–1170.

Beare, M.H., Blair, J.M. and Parmelee, R.W. (1989). Resource quality and trophic responses to simulated throughfall: effects on decomposition and nutrient flow in a no-tillage agroecosystem. *Soil Biol. Biochem.* **21**, 1027–1036.

Beare, M.H., Parmelee, R.W., Hendrix, P.F., Cheng, W., Coleman, D.C. and Crossley, D.A. (1992). Microbial and faunal interactions and effects on litter nitrogen and decomposition in agroecosystems. *Ecol. Monogr.* **62**, 569–591.

Bengtsson, G., Hedlund, K. and Rundgren, S. (1993). Patchiness and compensatory growth in a fungus-Collembola system. *Oecologia* **93**, 295–302.

Bengtsson, G., Hedlund, K. and Rundgren, S. (1994). Food- and density-dispersal: evidence from a soil Collembolan. *J. Anim. Ecol.* **63**, 513–520.

Bengtsson, J. (1994a). Temporal predictability in forest soil communities. *J. Anim. Ecol.* **63**, 653–665.

Bengtsson, J. (1994b). Confounding variables and independent observations in comparative analysis of food webs. *Ecology* **75**, 1282–1288.

Bengtsson, J., Zheng, D.W. Ågren, G.I. and Persson, T. (1995). Food webs in soil: an interface between population and ecosystem ecology. In: *Liking Species and Ecosystems* (Ed. by C.G. Jones and J.H. Lawton). Chapman and Hall, New York, In press.

Bernard, E.C. (1992). Soil nematode biodiversity. *Biol. Fertil. Soils* **14**, 99–103.

Bhattacharya, T. and Joy, V.C. (1977). Effect of Banvel-D, a herbicide, on the microarthropod population of soil. *Ind. Biol.* **9**, 47–51.

Blair, J.M., Parmelee, R.W. and Beare, M.H. (1990). Decay rates, nitrogen fluxes and decomposer communities of single and mixed-species foliar litter. *Ecology* **71**, 1976–1985.

Blevins, R.L., Smith, M.S. and Thomas, G.W. (1984). Changes in soil properties under no-tillage. In: *No Tillage Agriculture: Principles and Practices* (Ed. by R.E. Phillips and S.H. Phillips), pp. 190–230. Van Nostrand Reinhold Co., New York.

Blevins, R.L., Thomas, G.W. and Cornelius, P.L. (1977). Influence of no-tillage and nitrogen fertilisation on certain soil properties after five years of continuous corn. *Agron. J.* **69**, 383–386.

Blevins, R.L., Thomas, G.W., Smith, M.S., Frye, W.W. and Cornelius, P.L. (1983). Changes in soil properties after 10 years non-tilled and conventionally tilled corn. *Soil Till. Res.* **3**, 135–146.

Blumberg, A.Y. and Crossley, D.A. (1983). Comparison of soil surface arthropod populations in conventional tillage, no-tillage and old-field systems. *Agro-Ecosyst.* **8**, 247–253.

Boiteau, G. (1984). Effect of planting date, plant spacing and weed cover on populations of insects, arachnids and entomophthoran fungi in potato fields. *Environ. Entomol.* **13**, 751–756.

Boles, M. and Oseto, Y. (1987). A survey of the microarthropod populations under conventional-tillage and no-tillage systems. *North Dakota Farm News* **44**, 17–18.

Bongers, T. (1990). The maturity index: an ecological measure of environmental disturbance based on nematode species composition. *Oecologia* **83**, 14–19.

Bostrom, S. and Sohlenius, B. (1986). Short-term dynamics of nematode communities in arable soil: influence of a perennial and an annual cropping system. *Pedobiol.* **29**, 345–357.

Bouwman, L.A., Bloem, J., van de Boogert, P.H.J.F., Bremer, F., Hoenderboom, G.H.J. and de Ruiter, P.C. (1994). Short-term and long-term effects of bacterivorous nematodes and nematophagous fungi on carbon and nitrogen mineralisation in microcosms. *Biol. Fertil. Soils* **17**, 249–256.

Breazeale, F.W. and Camper, N.D. (1970). Bacterial, fungal and actinomycete populations in soils receiving repeated applications of 2,4-Dichlorophenoxyacetic acid and trifluralin. *Appl. Microbiol.* **19**, 379–380.

Briand, F. (1983). Environmental control of food web structure. *Ecology* **64**, 253–263.

Briand, F. and Cohen, J.E. (1984). Community food webs have scale-invariant structure. *Nature* **307**, 264–267.

Briand, F. and Cohen, J.E. (1987). Environmental correlates of food-chain length. *Science* **238**, 956–960.

Broder, M.W., Doran, J.W., Peterson, G.A. and Fenster, C.R. (1984). Fallow tillage influences on spring populations of soil nitrifiers, denitrifiers and available nitrogen. *Soil Sci. Soc. Amer. J.* **48**, 1061–1067.

Brown, J.H. and Heske, E.J. (1990). Control of a desert-grassland transition by a keystone rodent guild. *Science* **250**, 1705–1708.

162   D.A. WARDLE

Brundett, M. (1991). Mycorrhizas in natural ecosystems. *Adv. Ecol. Res.* **21**, 171–313.
Brust, G.E. (1990). Direct and indirect effects of four herbicides on the activity of Carabid beetles (Coleoptera : Carabidae). *Pestic. Sci.* **30**, 309–320.
Brust, G.E. and House, G.J. (1990). Effects of soil moisture, no tillage and predators on southern corn cutworm (*Diatrobica undecimpunctata howardi*) survival in corn agroecosystems. *Agric. Ecosyst. Environ.* **31**, 199–216.
Buchanan, M. and King, L.D. (1992). Seasonal fluctuations in soil microbial biomass carbon, phosphorus, and activity in no-till and reduced-chemical-input maize agroecosystems. *Biol. Fertil. Soils* **13**, 211–217.
Bund, C.F. van de (1970). Influence of crop and tillage on mites and springtails in arable soil. *Neth. J. Agric. Sci.* **18**, 308–314.
Camper, N.D., Moherek, E.A. and Huffman, J. (1973). Changes in microbial populations in paraquat-treated soil. *Weed Res.* **13**, 231–233.
Cardwell, C., Hassall, M. and White, P. (1994). Effect of headland management on carabid beetle communities in Breckland cereal fields. *Pedobiol.* **38**, 50–62.
Carter, M.R. (1986). Microbial biomass as an index for tillage-induced changes in soil biological properties. *Soil Till. Res.* **7**, 29–40.
Carter, M.R. (1991). The influence of tillage on the proportion of organic carbon and nitrogen in the microbial biomass of medium-textured soils in a humid climate. *Biol. Fertil. Soils* **11**, 135–139.
Carter, M.R. (1992). Influence of reduced tillage on organic matter, microbial biomass, macro-aggregate distribution and structural stability of the surface soil in a humid climate. *Soil Till. Res.* **23**, 361–372.
Carter, M.R. and Mele, P.M. (1992). Changes in the microbial biomass and structural stability at the surface of a duplex soil under direct drilling and stubble retention in north-eastern Victoria. *Aust. J. Soil. Res.* **30**, 493–503.
Carter, M.R. and Rennie, D.A. (1982). Changes in soil quality under zero tillage farming systems: distribution of microbial biomass and mineralisable C and N potentials. *Can. J. Soil Sci.* **62**, 587–597.
Caseley, J.C. and Eno, C.F. (1966). Survival and reproduction of earthworm species and a rotifer following herbicide treatment. *Soil Sci. Soc. Amer. Proc.* **30**, 346–350.
Cepeda, J.G. and Whitford, W.G. (1989). The relationships between abiotic factors and the abundance patterns of soil microarthropods on a desert watershed. *Pedobiol.* **33**, 79–86.
Cepeda-Pizarro, J.G. and Whitford, W.G. (1989). Spatial and temporal variability of higher microarthropod taxa along a transect in a northern Chihuahuan Desert watershed. *Pedobiol.* **33**, 101–111.
Chakravarty, P. and Chatarpaul, L. (1990). Non-target effects of herbicides. 1. Effect of glyphosate and hexazinone on soil microbial activity, microbial population and *in vitro* growth of ectomycorrhizal fungi. *Pestic. Sci.* **28**, 233–241.
Chakravarty, P. and Sidhu, S.S. (1987). Effect of glyphosate, hexazinone and triclopyr on *in vitro* growth of ectomycorrhizal fungi. *Eur. J. For. Pathol.* **17**, 204–210.
Chandra, P. (1964). Herbicide effects of certain soil microbial activities in some brown soils of Saskatchewan. *Weed Res.* **4**, 54–63.
Cheng, W., Coleman, D.C. and Box, J.E. (1990). Root dynamics, production and distribution in agroecosystems on the Georgia piedmont using mini-rhizotrons. *J. Appl. Ecol.* **27**, 592–604.

Chiverton, P.A. and Sotherton, N.W. (1991). The effects on beneficial arthropods of the exclusion of herbicides from cereal edges. *J. Appl. Ecol.* **28**, 1027–1039.
Christensen, S., Griffiths, B.S., Ekelund, F. and Ronn, R. (1992). Huge increase in bacterivores on freshly killed barley roots. *FEMS Microbiol. Ecol.* **86**, 303–310.
Clarholm, M. (1984). Heterotrophic, free living protozoa: neglected microorganisms with an important task in regulating bacterial populations. In: *Current Perspectives in Microbial Ecology* (Ed. by M.J. Klug and C.A. Reddy), pp. 321–326. American Society of Microbiology, Washington DC.
Closs, G.P. and Lake, P.S. (1994). Spatial and temporal variation in the structure of an intermittent-stream food web. *Ecol. Monogr.* **64**, 1–21.
Closs, G.P., Watterson, G.A. and Donnelly, P.J. (1993). Constant predator-prey ratios: an arithmetic artefact? *Ecology* **74**, 238–243.
Cohen, J.E., Beaver, R.A., Cousins, S.H., De Angelis, D.L., Goldwasser, L., Heong, K.L., Holt, R.D., Kohn, A.J., Lawton, J.H., Martinez, N.D., O'Malley, R., Page, L.M., Batten, B.C., Pimm, S.L., Polis, G.A., Rejmanek, M., Schoener, T.W., Schoenly, K., Spurles, W.G., Teal, J.M., Ulanowicz, R.E., Warren, P.H., Wilmur, H.M. and Yodzis, P. (1993). Improving food webs. *Ecology* **74**, 252–258.
Cohen, J.E., Briand, F. and Newman, C.M. (1990). *Community Food webs: Data and Theory*. Springer-Verlag, New York.
Coleman, D.C. (1985). Through a ped darkly: an ecological assessment of root-soil-microbial-faunal interactions. In: *Ecological Interactions in Soil* (Ed. by A.H. Fitter), pp. 1–21. Blackwell Scientific Publications, Oxford.
Coleman, D.C., Anderson, R.V., Cole, C.V., Elliott, E.T., Woods, L. and Campion, M.K. (1978). Trophic interactions in soil as they affect energy and nutrient dynamics. IV. Flow of metabolic and biomass carbon. *Microb. Ecol.* **4**, 373–380.
Coleman, D.C., Andrews, R., Ellis, J.E. and Singh, J.S. (1976). Energy flow and partitioning in selected man-managed and natural ecosystems. *Agro-Ecosyst.* **3**, 45–54.
Connell, J.H. (1978). Diversity in tropical rain forests and coral reefs. *Science* **199**, 1302–1310.
Cook, K.A. and Greaves, M.P. (1987). Natural variability in microbial activities. In: *Pesticide Effects on the Soil Microflora* (Ed. by L. Somerville and M.P. Greaves), pp. 15–43. Taylor and Francis, London.
Cooper, S.L., Wingfield, G.I., Lawley, R. and Greaves, M.P. (1978). Miniaturised methods for testing the toxicity of pesticides to microorganisms. *Weed Res.* **18**, 105–107.
Crossley, D.A., Mueller, B.R. and Perdue, J.C. (1992). Biodiversity of micro-arthropods in agricultural soils: relations to processes. *Agric. Ecosyst. Environ.* **40**, 37–46.
Curry, J.P. (1970). The effects of different methods of new sward establishment and the effect of the herbicides paraquat and dalapon on the soil fauna. *Pedobiol.* **10**, 329–361.
Curry, J.P. and Purvis, G. (1981). Studies on the influence of weeds and farmyard manure on the arthropod fauna of sugar-beet. *J. Life Sci. Royal Dublin Soc.* **3**, 397–408.
Czulakow, S.A. and Zarasow, S.U. (1975). Microbiological activity of south Kazakhstan soils depending on herbicides. *Rocz. Glebozn. T.* **26**, 179–184.
Dackman, C., Jansson, H-B. and Nordbring-Hertz, B. (1991). Nematophagous

fungi and their activities in soil. In: *Soil Microbiology. Volume 7* (Ed. by G. Stotzky and J.M. Bollag), pp. 95–130. Marcel Dekker, New York.

Dalal, R.C., Henderson, P.A. and Glasby, J.M. (1991). Organic matter and microbial biomass in a vertisol after 20 years of zero tillage. *Soil Biol. Biochem.* **23**, 435–441.

Daniel, O. and Anderson, J.M. (1992). Microbial biomass and activity in contrasting soil materials after passing through the gut of the earthworm *Lumbricus rubellus* Hoffmaster. *Soil Biol. Biochem.* **24**, 465–470.

Darbyshire, J.F., Davidson, M.S., Chapman, S.J. and Ritchie, S. (1994). Excretion of nitrogen and phosphorus by the soil ciliate *Colpoda steinii* when fed upon the bacteria *Arthrobacter* sp. *Soil Biol. Biochem.* **26**, 1193–1199.

Dash, M.C., Senapati, B.K. and Mishra, C.C. (1980). Nematode-feeding by tropical earthworms. *Oikos* **34**, 322–325.

Davidson, E.A. and Ackerman, I.L. (1993). Changes in soil carbon inventories following cultivation of previously untilled soils. *Biogeochem.* **20**, 161–193.

Davis, B.N.K. (1965). The immediate and long-term effects of the herbicide MCPA on soil arthropods. *Bull. Entom. Res.* **56**, 357–366.

De Angelis, D.L. (1992). *Dynamics of Nutrient Cycling and Food webs*. Chapman and Hall, London.

Demure, Y., Freckman, D.W. and Gundy, S.D. van (1979). Anhydrobiotic coiling of nematodes in soil. *J. Nematol.* **11** 189–195.

De Ruiter, P.C., Moore, J.C., Zwart, K.B., Bouwman, L.A., Hassink, J., Bloem, J., de Vos, J.A., Marinissen, J.C.Y., Didden, W.I.M., Lebbink, G. and Brussaard, L. (1993). Simulation of nitrogen mineralisation in the below-ground food webs of two winter wheat fields. *J. Appl. Ecol.* **30**, 95–106.

Desender, K. and Alderweireldt, M. (1988). Population dynamics of adult and larval Carabid beetles in a maize field and its boundary. *J. Appl. Entom.* **106**, 13–19.

Desender, K., Alderweireldt, M. and Pollet, M. (1989). Field edges and their importance for polyphagous predatory arthropods. *Med. Fac. Landbouw. Rijksuniv. Gent.* **54**, 823–833.

De St. Remy, E.A. and Daynard, T.B. (1982). Effects of tillage on earthworm populations in monoculture corn. *Can. J. Soil Sci.* **62**, 699–703.

Dick, W.A. (1983). Organic carbon, nitrogen and phosphorus concentrations and pH in soil profiles as affected by tillage intensity. *Soil Sci. Soc. Am. J.* **47**, 102–107.

Dick, W.A. and Doran, D.M. van (1985). Continuous tillage and rotation combination effects on corn, soybean and oat yields. *Agron. J.* **77**, 459–465.

Dick, W.A., McCoy, E.L., Edwards, W.M. and Lal, R. (1991). Continuous application of no-tillage to Ohio soils. *Agron. J.* **83**, 65–73.

Didden, W.A.M. (1993). Ecology of terrestrial Enchytraeidae. *Pedobiol.* **37**, 2–29.

Doane, J.F. and Dondale, C.D. (1979). Seasonal captures of spiders (Araneae) in a wheat field and its grassy borders in central Saskatchewan. *Canadian Entomol.* **111**, 439–445.

Dodd, J.C. and Jeffries, P. (1989). Effects of herbicides on three vesicular-arbuscular fungi associated with winter wheat (*Triticum aestivum L.*). *Biol. Fertil. Soils* **7**, 113–119.

Domsch, K.H. and Banse, H.J. (1972). Mykologische untersuchungen am regenwurm exkrementen. *Soil Biol. Biochem.* **4**, 31–38.

Domsch, K.H., Gams, W. and Anderson, T-H (1980). *Compendium of Soil Fungi*. Academic Press, London.

Domsch, K.H., Jagnow, G. and Anderson, T-H. (1983). An ecological concept for the assessment of side-effects of agrochemicals on soil microorganisms. *Res. Rev.* **86**, 65–105.

Domurat, K. and Kozlowska, J. (1974). Comparative investigations on the communities of nematodes of cultivated fields and neighbouring barrens. *Zesz. Problem. Postepow Nauk Rolnicz.* **154**, 191–199.

Doran, J.W. (1980a). Microbial changes associated with residue management with reduced tillage. *Soil Sci. Soc. Am. J.* **44**, 518–524.

Doran, J.W. (1980b). Soil microbial and biochemical changes associated with reduced tillage. *Soil Sci. Soc. Am. J.* **44**, 765–771.

Doran, J.W. (1987). Microbial biomass and mineralisable nitrogen distributions in no-tillage and plowed soils. *Biol. Fertil. Soils* **5**, 68–75.

Duah-Yentumi, S. and Johnson, D.B. (1986). Changes in soil microflora in response to repeated applications of some pesticides. *Soil Biol. Biochem.* **18**, 629–635.

Dumontet, S. and Perucci, P. (1992). The effect of acifluorfen and trifluralin on the size of microbial biomass in soil. *Sci. Total Environ.* **123/124**, 261–266.

Edwards, C.A. (1970). Effects of herbicides on the soil fauna. *Proc. 10th British Weed Control Conf.*, 1052–1062.

Edwards, C.A. (1975a). Investigations into the influences of agricultural practice on soil invertebrates. *Ann. Appl. Biol.* **87**, 515–520.

Edwards, C.A. (1975b). Effects of direct drilling on the soil fauna. *Outl. Agric.* **8**, 243–244.

Edwards, C.A. (1983). Earthworm ecology in cultivated soils. In: *Earthworm Ecology* (Ed. by J.E. Satchell), pp. 123–137. Chapman and Hall, New York.

Edwards, C.A. (1984). Changes in agricultural practices and their impact on soil organisms. In: *Agriculture and the Environment* (Ed. by D. Jenkin), pp. 56–65. Lavenham Press, Suffolk.

Edwards, C.A. (1989). Impact of herbicides on soil ecosystems. *Crit. Rev. Plant Sci.* **8**, 221–257.

Edwards, C.A. and Bohlen, P.J. (1992). The effects of toxic chemicals on earthworms. *Rev. Environ. Contam. Toxicol.* **125**, 23–99.

Edwards, C.A. and Fletcher, K.E. (1988). Interactions between earthworms and microorganisms in organic matter breakdown. *Agric. Ecosyst. Environ.* **24**, 235–247.

Edwards, C.A. and Lofty, J.R. (1975). The influence of cultivations on soil animal populations. In: *Progress in Soil Zoology* (Ed. by J. Vanek), pp. 399–407. Academia, Prague.

Edwards, C.A. and Lofty, J.R. (1980). Effects of earthworm inoculation upon the root growth of direct drilled cereals. *J. Appl. Ecol.* **17**, 533–543.

Edwards, C.A. and Lofty, J.R. (1982). The effects of direct drilling and minimal cultivation on earthworm populations. *J. Appl. Ecol.* **19**, 723–734.

Edwards, C.A. and Stafford, C.J. (1979). Interactions between herbicides and the soil fauna. *Ann. Appl. Biol.* **91**, 132–137.

Edwards, W.M., Shipitalo, M.J., Owens, L.B. and Norton, L.D. (1990). Effect of *Lumbricus terrestris* L. burrows on hydrology of continuous no-till corn fields. *Geoderma* **46**, 73–84.

Ehlers, W. (1975). Observations on earthworm channels and infiltration on tilled and untilled loessial soils. *Soil Sci.* **119**, 242–249.

Eijsackers, H. (1975). Effects of the herbicide 2,4,5-T on *Onychiurus quadriocella-*

166 D.A. WARDLE

tus Gisin (Collembola). In: *Progress in Soil Zoology* (Ed. by J. Vanek), pp. 481–488. Academia, Prague.

Eijsackers, H. (1978a). Side-effects of the herbicide 2,4,5-T affecting the Carabid *Notiophilus biguttatus* Fabr., a predator of springtails. *Zeitschr. Angew. Entomol.* **85**, 113–128.

Eijsackers, H. (1978b). Side-effects of the herbicide 2,4,5-T on reproduction, food consumption and moulting of the springtail *Onychiurus quadriocellatus* Gisin (Collembola). *Zeitschr. Angew. Entomol.* **85**, 341–360.

Eijsackers, H. (1978c). Side-effects of the herbicide 2,4,5-T affecting mobility and mortality of the springtail *Onychiurus quadriocellatus* Gisin (Collembola). *Zeitzchr. Angew. Entomol.* **85**, 361–372.

Eijsackers, H. (1985). Effects of glyphosate on the soil fauna. In: *The Herbicide Glyphosate* (Ed. by E. Grossbard and D. Atkinson), pp. 151–157. Butterworths, London.

Eijsackers, H. and Bund, C.F. van de (1980). Effects on soil fauna. In: *Interactions Between Herbicides and the Soil.* (Ed. by R.J. Hance). pp. 225–305. Academic Press, London.

Elliott, E.T., Horton, K., Moore, J.C., Coleman, D.C. and Cole, C.V. (1984). Mineralisation dynamics in fallow dryland plots, Colorado. *Plant Soil* **76**, 149–155.

El Titi, A. and Epach, U. (1989). Soil fauna in sustainable agriculture: results of an integrated farming system at Lautenbach, F.R.G. *Agric. Ecosyst. Environ.* **27**, 561–572.

Emmanuel, N., Curry, J.P. and Evans, G.O. (1985). The soil Acari of barley plots with different cultural treatments. *Exper. Appl. Acarol.* **1**, 101–113.

Enache, A.J. and Ilnicki, R.D. (1990). Weed control by subterranean clover (*Trifolium subterraneum*) used as a living mulch. *Weed Technol.* **4**, 534–538.

Ennis, W.B. (1979). *Introduction to Crop Protection.* American Society of Agronomy, Madison.

Evans, D.G. and Miller, M.H. (1990). The role of the external mycelial network in the effect of disturbance upon vesicular-arbuscular mycorrhizal colonisation of maize. *New Phytol.* **114**, 65–71.

Everts, J.W., Aukema, B., Hengeveld, R. and Koeman, J.H. (1989). Side-effects of pesticides on ground-dwelling predatory arthropods in arable ecosystems. *Envir. Pollution* **59**, 203–225.

Faulkner, E.H. (1945). *Ploughman's Folly.* Michael Joseph, London.

Favretto, M.R., Paoletti, M.G., Caporali, F., Nannipieri, P., Onnis, A. and Tomei, P.E. (1992). Invertebrates and nutrients in a Mediterranean vineyard mulched with subterranean clover (*Trifolium subterraneum L.*). *Biol. Fertil. Soils* **14**, 151–158.

Fayolle, L. (1979). Consequences de l'apport de contaminants sur les Lombriciens. *Doc. Pedozool.* **1**, 34–65.

Fedorko, A., Kamionek, M., Kozlowska, J. and Mianowska, E. (1977). The effects of some carbamide herbicides on nematodes from different ecological groups. *Pol. Ecol. Stud.* **3**, 23–28.

Feldman, R.M. and Manning, R.B. (1992). Crisis in systematic biology in the "age of biodiversity". *J. Palaentol.* **66**, 157–158.

Ferris, V.R. and Ferris, J.M. (1974). Interrelationships between nematodes and plant communities in agricultural ecosystems. *Agro-Ecosyst.* **1**, 275–299.

Finlay, R.D. (1985). Interactions between soil micro-arthropods and endomycorrhizal associations of higher plants. In: *Ecological Interactions in Soil* (Ed. by

A.H. Fitter, D. Atkinson, D.J. Read and M.B. Usher), pp. 319–331. Blackwell Scientific Publications, Oxford.

Finlayson, D.G., Campbell, C.J. and Roberts, H.A. (1975). Herbicides and insecticides: their compatability and effects on weeds, insects and earthworms in the minicauliflower crop. *Ann. Appl. Biol.* **79**, 95–108.

Fischer, E. (1989). Effects of atrazine and paraquat containing herbicides on *Eisenia foetida* (Annelida, Oligochaeta). *Zool. Anz.* **223**, 291–300.

Fitter, A.H. and Sanders, I.R. (1992). Interactions with the soil fauna. In: *Mycorrhizal Functioning* (Ed. by M.F. Allen), pp. 333–354. Chapman and Hall, New York.

Fleige, H. and Baumer, K. (1974). Effect of zero tillage on organic carbon and total nitrogen content, and their distribution in different N-fractions in loessial soils. *Agro-Ecosyst.* **1**, 19–29.

Foissner, W. (1992). Comparative studies on the soil life in ecofarmed and conventionally farmed fields and grasslands in Austria. *Agric. Ecosyst. Environ.* **40**, 207–218.

Follett, R.F. and Schimel, D.S. (1989). Effect of tillage practices on microbial biomass dynamics. *Soil Sci. Soc. Am. J.* **53**, 1091–1096.

Fox, C.J.F. (1964). The effects of five herbicides on the numbers of certain invertebrate animals in grassland soil. *Can. J. Plant Sci.* **44**, 405–409.

Fratello, B., Bertolani, R., Sabatini, M.A., Mola, L. and Rassu, M.A. (1985). Effects of atrazine on soil microarthropods in experimental maize fields. *Pedobiol.* **28**, 161–168.

Freckman, D.W. (1978). Ecology of anhydrobiotic nematodes. In: *Dried Biological Systems* (Ed. by J.H. Crowe and J.S. Clegy), pp. 345–357. Academic Press, New York.

Freckman, D.W. (1988). Bacterivorous nematodes and organic matter decomposition. *Agric. Ecosyst. Environ.* **24**, 195–217.

Freckman, D.W. and Caswell, E.P. (1985). The ecology of nematodes in agroecosystems. *Ann. Rev. Phytopathol.* **23**, 275–296.

Freckman, D.W. and Ettema, C.H. (1993). Assessing nematode communities in agroecosystems of varying human disturbance. *Agric. Ecosyst. Environ.*, **45**, 239–265.

Frey, F. (1976). Untersuchungen uber dic wirkungen von im obstbau verwendeten herbiciden auf den testnematoden *Acreboloides buetschlii* (de Man 1884) Steiner und Buhrer 1933. *Zeitschr. Pflanzenkrank. Pflanzenschutz* **83**, 434–441.

Gebhardt, M.R., Daniel, T.C., Schweizer, E.E. and Allmaras, R.R. (1985). Conservation tillage. *Science* **230**, 625–630.

Gel'cer, Ju.G. and Geptner, V.A. (1976). O protistidnom dejstvii gerbicidov. *Pedobiol.* **16**, 171–184.

Gerard, B.M. and Hay, R.K.M. (1979). The effects on earthworms of ploughing, tined cultivation, direct drilling and nitrogen in a barley monoculture system. *J. Agric. Sci.* **93**, 147–155.

Gerson, U. and Chet, I. (1981). Are allochthonous and autochthonous soil microorganisms r- and K-selected? *Rev. Ecol. Biol. Sol* **18**, 285–289.

Gestel, C.A.M. van, Dirven-Bremen van, E.M., Baerselman, H.J.B., Janssen, J.A.M., Postuma, R. and Vliet, P.J.M. van. (1992). Comparison of sublethal and lethal criteria for nine different chemicals in standardising toxicity tests in the earthworm *Eisenia andrei*. *Ecotox. Environ. Safety* **23**, 206–220.

Ghabbour, S.I. and Imam, M. (1967). The effects of five herbicides on three Oligochaete species. *Rev. Ecol. Biol. Sol* **4**, 119–122.

168   D.A. WARDLE

Ghilarov, M.S. (1963). On the interrelationships between soil-dwelling invertebrates and microorganisms. In: *Soil Organisms* (Ed. by J. Doeksen and J.
van der Drift), pp. 255–259. North Holland Publishing Company, Amsterdam.

Ghilarov, M.S. (1975). General trends of changes in soil animal populations of
arable land. In: *Progress in Soil Zoology* (Ed. by J. Vanek), pp. 31–39. Academia, Prague.

Golebiowska, J. and Ryszkowski, L. (1977). Energy and carbon fluxes in soil
compartments of agro-ecosystems. *Ecol. Bull.* **25**, 274–283.

Gomez, M.A. and Sagardoy, M.A. (1985). Influencia del herbicida gliphosato
sobre la microflora y mesofauna de un arenoso de la region semiarida. *Revista
Latinoamer. Microbiol. Mexico* **27**, 351–357.

Gottschalk, M.R. and Shure, D.J. (1979). Herbicide effects on leaf litter decomposition processes in an oak-hickory forest. *Ecology* **60**, 143–151.

Granatstein, D.M., Bezdicek, D.F., Cochran, V.L. and Elliott, L.F. (1987). Long-
term tillage and rotation effects on soil biomass, carbon and nitrogen. *Biol.
Fertil. Soils* **5**, 265–270.

Greaves, M.P. (1987). Side-effects testing—an alternative approach. In: *Pesticide
Effects on the Soil Microflora* (Ed. by L. Somerville and M.P. Greaves), pp.
183–190. Taylor and Francis, London.

Griffith, D.R., Kladivko, E.J., Mannering, J.V., West, T.D. and Parsons, S.D.
(1988). Long-term tillage and rotation effects on corn growth and yield on high
and low organic matter, poorly draining soils. *Agron. J.* **80**, 599–605.

Griffiths, B.S. (1994). Soil nutrient flow. In: *Soil Protozoa* (Ed. by J.F.
Darbyshire), pp. 65–91. CAB International, Wallingford.

Griffiths, B.S., Welschen, R., Arendonk, J.J.C.M. van and Lambers, H. (1992).
The effect of nitrate-nitrogen on bacteria and the bacterial-feeding fauna in the
rhizosphere of different grass species. *Oecologia* **91**, 253–259.

Gupta, V.V.S.R. and Germida, J.J. (1988). Distribution of microbial biomass and
its activity in different soil aggregate size classes as affected by cultivation. *Soil
Biol. Biochem.* **20**, 777–786.

Haines, P.J. and Uren, N.C. (1990). Effects of conservation tillage farming on soil
microbial biomass, organic matter and earthworm populations in north-eastern
Victoria. *Aust. J. Exper. Agric.* **30**, 365–371.

Hairston, N.G., Smith, F.E. and Slobodkin, L.B. (1960). Community structure,
population control and competition. *Am. Nat.* **94**, 421–425.

Hall, S.J. and Raffaelli, D.G. (1993). Food webs: theory and reality. *Adv. Ecol.
Res.* **24**, 187–239.

Hanlon, R.D. (1981). Influences of grazing by Collembola on the activity of senescent fungal colonies grown on media of different nutrient concentrations.
Oikos **36**, 362–367.

Haque, A. and Ebing, W. (1983). Toxicity determination of pesticides to earthworms in the soil substrate. *Zeitschr. Pflanzenkrank. Pflanzenschutz* **90**, 395–
408.

Hargrove, W.L., Box, J.E., Radcliffe, D.E., Johnson, J.W. and Rothcrock, C.S.
(1988). Influence of long-term no-tillage on crop rooting in an Ultisol. *Proc.
1988 Southern Tillage Conf., Tupelo, Mississippi*, 22–25.

Hargrove, W.L., Reid, J.T., Touchton, J.T. and Gallagher, N. (1982). Influence of
tillage practices on the fertility status of an acid soil double-cropped to wheat
and soybeans. *Agron. J.* **74**, 684–687.

Hassall, M., Hawthorne, A., Maudsley, M., White, P. and Cardwell, C. (1992).

Effects of headland management on invertebrate communities in cereal fields. *Agric. Ecosyst. Environ.* **40**, 155–178.

Haukka, J. (1988). Effect of various cultivation methods on earthworm biomass and communities on different soil types. *Ann. Agric. Fenn.* **27**, 263–269.

Havens, K.E. (1994). Experimental perturbation of a freshwater plankton community: a test of hypotheses regarding the effects of stress. *Oikos* **69**, 147–153.

Havlin, J.L., Kissel, D.E., Maddux, L.D., Claasen, M.M. and Long, J.H. (1990). Crop rotation and tillage effects on soil organic carbon and nitrogen. *Soil Sci. Soc. Am. J.* **54**, 448–452.

Haynes, R.J. and Knight, T.L. (1989). Comparison of soil chemical properties, enzyme activities, levels of biomass N and aggregate stability in the soil profile under conventional and no-tillage in Canterbury, New Zealand. *Soil Till. Res.* **14**, 197–208.

Heimbach, F. (1984). Correlations between three methods for determining the toxicity of chemicals to earthworms. *Pestic. Sci.* **15**, 605–611.

Heinonen-Tanski, H. (1990). Conventional and organic cropping systems at Suitia. III Microbial activity in soils. *J. Agric. Sci. Finl.* **62**, 321–330.

Heinonen-Tanski, H., Siltanen, H., Kilpi, S., Simojoki, P., Rosenberg, C. and Makinen, S. (1986). The effect of the annual use of some pesticides on soil microorganisms, pesticide residues in soil and carrot yields. *Pestic. Sci.* **17**, 135–142.

Helmecke, K., Hickisch, B., Mahn, E-G., Prasse, J. and Sternkopf, G. (1977). Beitrage zur wirkung des herbizideinsatzes auf struktur und stoffhausholt von agro-okosystemen. 1. Zur beeinflussung von phytozonose und bodenorganismengemeinschaften nach mehrjahrigem herbizideinsatz. *Herc. N.F. Leipzig* **14**, 375–398.

Hendrix, P.F. and Parmelee, R.W. (1985). Decomposition, nutrient loss and microarthropod densities in herbicide-treated grass litter in a Georgia piedment agroecosystem. *Soil Biol. Biochem.* **17**, 421–428.

Hendrix, P.F., Crossley, D.A., Blair, J.M. and Coleman, D.C. (1990). Soil biota as components of sustainable agroecosystems. In: *Sustainable Agricultural Systems* (Ed. by C.A. Edwards, R. Lal, P. Madden, R.H. Miller and G.J. House), pp. 637–654. Soil and Water Conservation Society, Ankeny.

Hendrix, P.F., Parmelee, R.W., Crossley, D.A., Coleman, D.C., Odum, E.P. and Groffman, P.M. (1986). Detritus food webs in conventional and no-tillage agroecosystems. *Bioscience* **36**, 374–380.

Hickisch, B., Prasse, J. and Machulla, G. (1986). Nebenwirkung von herbiziden auf bodenorganismen nach ein-und mehrmaliger applikation. 4. Mitteilung: freilandversuch mit getreidehetonter fruchfolge. *Zentralbl. Mikrobiol.* **141**, 81–96.

Hokkanen, H. and Holopainen, J.K. (1986). Carabid species and activity in biologically and conventionally managed cabbage fields. *J. Appl. Entomol.* **102**, 353–363.

Holland, E.A. and Coleman, D.C. (1987). Litter placement effects on microbial and organic matter dynamics in an agroecosystem. *Ecology* **68**, 425–433.

Holt, J.A. (1985). Acari and Collembola in the littoral soil of three Queensland rain forests. *Aust. J. Ecol.* **10**, 57–65.

Honczarenko, J. (1969). Wstepne budania nad dzialaniem niektorych herbicidydow na owady glebowe in dzdzownice. *Pol. Pismo Entomol.* **39**, 567–578.

House, G.J. (1989). Soil arthropods from weed and crop roots of an agroecosystem

170   D.A. WARDLE

in a wheat-soybean rotation: impact of tillage and herbicides. *Agric. Ecosyst. Environ.* **25**, 233–244.

House, G.J. and All, J.N. (1981). Carabid beetles in soybean ecosystems. *Environ. Entomol.* **10**, 194–196.

House, G.J. and Alzugaray, M.D.R. (1989). Influence of cover-cropping and no-tillage practices on community composition of soil arthropods in a North Carolina agroecosystem. *Environ. Entomol.* **18**, 302–307.

House, G.J. and Brust, G.E. (1989). Ecology of low-input, no-tillage agriculture. *Agric. Ecosyst. Environ.* **27**, 331–345.

House, G.J. and Parmelee, R.W. (1985). Comparison of soil arthropods and earthworms from conventional and non-tilled agroecosystems. *Soil Till. Res.* **5**, 351–360.

House, G.J. and Stinner, B.R. (1983). Arthropods in no-tillage soybean agro-ecosystems: community composition and earthworm interactions. *Environ. Manage.* **7**, 23–28.

House, G.J., Stinner, B.R., Crossley, D.A., Odum, E.P. and Langdale, G.W. (1984). Nitrogen cycling in conventional and no-tillage agroecosystems in the southern piedmont. *J. Soil Water Conserv.* **39**, 194–200.

House, G.J., Worsham, A.D., Sheets, T.J. and Stinner, R.E. (1987). Herbicide effects on soil arthropod dynamics and wheat straw decomposition in a North Carolina no-tillage agro-ecosystem. *Biol. Fertil. Soils* **4**, 109–114.

Houseworth, L.D. and Tweedy, B.G. (1973). Effect of atrazine in combination with captan or thiram upon fungal and bacterial populations in the soil. *Plant Soil* **38**, 493–500.

Huhta, V. and Raatikainen, M. (1974). Spider communities of leys in winter cereal fields in Finland. *Ann. Zool. Fenn.* **11**, 97–104.

Hunt, H.W., Coleman, D.C., Ingham, E.R., Ingham, R.E., Elliott, E.T., Moore, J.C., Rose, S.L., Reid, C.P.P. and Morley, C.R. (1987). The detrital food web in the short-grass prairie. *Biol. Fertil. Soils* **3**, 57–68.

Hyvönen, R., Andersson, S., Clarholm, M. and Persson, T. (1994). Effects of lumbricids and enchytraeids on nematodes in limed and unlimed coniferous mor humus. *Biol. Fertil. Soils* **17**, 201–205.

Ilnicki, R.D. and Enache, A.J. (1992). Subterranean clover mulch: an alternative method of weed control. *Agric. Ecosyst. Environ.* **40**, 249–264.

Ineson, P. and Anderson, J.M. (1985). Aerobically isolated bacteria associated with the gut of the feeding animal macroarthropods *Oniscus asellus* and *Glomeris marginata*. *Soil Biol. Biochem.* **17**, 843–849.

Ingham, E.R., Coleman, D.C. and Moore, J.C. (1989). An analysis of food web structure and function in a short-grass prairie, a mountain meadow, and a lodgepole pine forest. *Biol. Fertil. Soils* **8**, 29–37.

Ingham, E.R., Trofymow, J.A., Ames, R.N., Hunt, H.W., Morley, C.R., Moore, J.C. and Coleman, D.C. (1986). Trophic interactions and nitrogen cycling in a semi-arid grassland soil. 1. Seasonal dynamics of the natural populations, their interactions and effects on nitrogen cycling. *J. Appl. Ecol.* **23**, 597–614.

Ingham, R.E. (1988). Interactions between nematodes and vesicular-arbuscular fungi. *Agric. Ecosyst. Environ.* **24**, 169–182.

Ingham, R.E., Trofymow, J.A., Ingham, E.R. and Coleman, D.C. (1985). Interactions of bacteria, fungi, and their nematode grazers on nutrient cycling and plant growth. *Ecol. Monogr.* **55**, 119–140.

Insam, H. and Domsch, K.H. (1988). Relationship between soil organic carbon and microbial biomass on chronosequences of reclamation sites. *Microb. Ecol.* **15**, 177–188.

Isakeit, T. and Lockwood, J.L. (1990). Evaluation of the soil microbiota in producing the deleterious effect of atrazine on ungerminated conidia of *Cochliobolus sativus* in soil. *Soil Biol. Biochem.* **22**, 413–417.

Ishibashi, N., Muraoko, M., Kondo, E., Yamasaki, H., Kai, H., Iwakiri, T. and Nakahara, M. (1978). Effect of annual applications of herbicides on nematodes, soil mites and springtails in satsuma mandarin orchards. *Agric. Bull. Saga Univ.* **44**, 43–55.

James, T.K. and Rahman, A. (1992). Impact of some factors on the performance of the fatty acid herbicide "Greenscape". *Proc. 45th N.Z. Plant Protection Conf.* 235–238.

Jaworska, T. (1981). Wplyw herbicydow triazyrowych na biegacczowate (Carabidae, Coleoptera) u uprawach kapusty glawiostej. *Pol. Pismo Entomol.* **51**, 323–353.

Jenkinson, D.S. and Rayner, J.S. (1977). The turnover of organic matter in some of the Rothamsted classical experiments. *Soil Sci.* **123**, 298–305.

Juo, A.S.R. and Lal, R. (1979). Nutrient profile in a typical alfisol under conventional and no-till systems. *Soil Sci.* **127**, 168–173.

Kaiser, P. and Reber, H. (1970). Interactions entre la simazine-herbicide et les microorganismes de la rhizosphere du mais. *Meded. Fac. Landbouw. Rijksuniv. Gent.* **35**, 689–705.

Kajak, A., Chmielewski, K., Kaczmarek, M. and Rembialkowska, E. (1993). Experimental studies on the effect of epigeic predators on matter decomposition process in managed peat grasslands. *Pol. Ecol. Stud.* **17**, 289–310.

Kaluz, S. (1984). Aktivia vybranych skupin podnych clankonzcov (Arthropoda) vo vinohrade ostrenom herbicidmi. *Biologia (Bratislava)* **39**, 853–890.

Kareiva, P. and Sahakian, R. (1990). Tritrophic effects of a simple architectural mutation in pea plants. *Nature* **345**, 433–434.

Karg, W. (1964). Untersuchungen uber die wirkung von dinitroorthokestrol (DNOC) auf die microarthropoden des bodens unter berucksichtigung der beziehungen zwischen mikroflora und mesofauna. *Pedobiol.* **4**, 138–157.

Karki, A.B. and Kaiser, P. (1974). Effects du chlorate de sodium sur les microorganismes du sol, leur respiration et leur activite enzymatique. II. Etude d'incubation au laboratorie. *Rev. Ecol. Biol. Sol* **11** 477–498.

Kasprzak, K. (1982). Review of enchytraeid (Oligochaeta, Enychtraeidae) community structure and function in agricultural ecosystems. *Pedobiol.* **23**, 217–232.

Kerbes, R.H., Kotanen, P.M. and Jeffries, R.C. (1990). Destruction of wetland habitats by lesser snow geese: a keystone species on the west coast of Hudson Bay. *J. Appl. Ecol.* **27**, 242–258.

Klinger, K. von (1987). Auswirkungen eingesater randstreifen an einem winterweizen-field auf raubarthropodenfauna und getreideblattlausbefall. *J. Appl. Entomol.* **104**, 47–58.

Klopatek, C.C., O'Neill, E.G., Freckman, D.W., Bledsoe, C.S., Coleman, D.C., Crossley, D.A., Ingham, E.R., Parkinson, D. and Klopatek, J.M. (1992). The sustainable biosphere initiative: a commentary from the U.S. Soil Ecology Society. *Bull. Ecol. Soc. Amer.* **73**, 223–228.

Klyuchnikov, L.Y., Petrova, A.N. and Polesko, Y.A. (1964). Influence of simazin and atrazin on the microflora of sandy soils. *Mikrobiol.* **33**, 879–882.

Kreutzer, W.A. (1963). Selective toxicity of chemicals to microorganisms. *Ann. Rev. Phytopathol.* **1**, 101–126.

172   D.A. WARDLE

Kromp, B. and Steinberger, K-H. (1992). Grassy field margins and arthropod biodiversity: a case study on ground beetles and spiders in eastern Austria (Coleoptera: Carabidae; Arachnida: Aranei, Opiliones). *Agric. Ecosyst. Environ.* **40**, 71–93.

Kruckelman, H.W. (1975). Effects of fertilisers, soils, soil tillage and plant species on the frequency of Endogone chlamydospores and mycorrhizal infection in arable soils. In: *Endomycorrhizas* (Ed. by F.E. Sanders, B. Mosse and P.B. Tinker), pp. 511–525. Academic Press, New York.

Lagerlöf, J. and Andrén, O. (1988). Abundance and activity of soil mites (Acari) in four cropping systems. *Pedobiol.* **32**, 129–145.

Lagerlöf, J. and Andrén, O. (1991). Abundance and activity of Collembola, Protura and Diplura (Insecta, Apterygota) in four cropping systems. *Pedobiol.* **35**, 337–350.

Lagerlöf, J., Andrén, O. and Paustian, K. (1989). Dynamics and contribution to carbon flows of Enchytraeidae (Oligochaeta) under four cropping systems. *J. Appl. Ecol.* **26**, 183–199.

Lagerlöf, J. and Wallin, H. (1993). The abundance of arthropods along two field margins with different types of vegetation composition: an experimental study. *Agric. Ecosyst. Environ.* **43**, 141–154.

Lal, R. (1974). No-tillage effects on soil properties and maize (*Zea mays* L.) production in western Nigeria. *Plant Soil* **140**, 321–331.

Lal, R. and Doran, D.M. van (1990). Influence of 25 years of continuous corn production by three tillage methods on water infiltration for two soils in Ohio. *Soil Till. Res.* **16**, 71–84.

Lal, R., Eckert, D.J., Fausey, N.R. and Edwards, W.M. (1990). Conservation tillage in sustainable agriculture. In: *Sustainable Agricultural Systems* (Ed. by C.A. Edwards, R. Lal, P. Madden, R.H. Miller and G.J. House), pp. 203–225. Soil and Water Conservation Society, Ankeny.

Lal, R., Vleeschauwer, D. de and Nganje, R.M. (1980). Changes in properties of a newly cleared tropical alfisol as affected by mulching. *Soil Sci. Soc. Am. J.* **44**, 827–833.

Lamb, J.A., Peterson, G.A. and Fenster, C.R. (1985). Wheat fallow tillage systems effect on a newly cultivated soil's nitrogen budget. *Soil Sci. Soc. Am. J.* **49**, 353–356.

Lavelle, P. (1995). Mutualism and biodiversity in soils. *Plant and Soil*: in press.

Lee, K.E. (1985). *Earthworms: Their Ecology and Relationships with Soil and Land Use.* Academic Press, Sydney.

Leonard, M.A. (1983). Influence of nitrogen on spatio-temporal interactions between a collembola species and a soil fungus. In: *Nitrogen as an Ecological Factor* (Ed. by J.A. Lee, S. McNeill and I.H. Rorison), p. 441. Blackwell Scientific Publications, Oxford.

Linn, D.M. and Doran, J.W. (1984). Aerobic and anaerobic microbial populations in no-tilled and ploughed soils. *Soil Sci. Soc. Am. J.* **48**, 794–799.

Lofs-Holmin, A. (1980). Measuring growth of earthworms as a method of testing sublethal toxicity of pesticides. *Swed. J. Agric. Res.* **10**, 25–33.

Loots, G.C. and Ryke, P.A. (1967). The ratio Oribatei: Trombidiformis with reference to organic matter content in soils. *Pedobiol.* **7**, 121–124.

Loquet, M., Bhatnagar, T., Bouche, M.B. and Rouelle, J. (1977). Essai d'estimation de l'influence ecologique des lombriciens sur les microorganismes. *Pedobiol.* **17**, 400–417.

IMPACTS OF DISTURBANCE ON DETRITUS FOOD WEBS 173

Loring, S.J., Snider, R.J. and Robertson, L.S. (1981). The effects of three tillage practices on Collembola and Acarina populations. *Pedobiol.* **22**, 172–184.

Lund, V. and Goksoyr, J. (1980). Effects of water fluctuations on microbial mass and activity in soils. *Microb. Ecol.* **6**, 115–123.

Lussenhop, J. (1992). Mechanisms of microarthropod-microbial interactions in soil. *Adv. Ecol. Res.* **23**, 1–33.

Lyons, C., Milsom, N., Morgan, N.G. and Stringer, A. (1972). The effects of repeated applications of the grass suppressant maleic hydroxide on an orchard sward and on the soil fauna. *Proc. 11th British Weed Control Conf.*, 356–359.

Lynch, J.M. and Panting, L.M. (1980a). Cultivation and the soil biomass. *Soil Biol. Biochem.* **12**, 29–33.

Lynch, J.M. and Panting, L.M. (1980b). Variations in the size of the soil biomass. *Soil Biol. Biochem.* **12**, 547–550.

Lynch, J.M. and Panting, L.M. (1982). Effects of season, cultivation and nitrogen fertiliser on the size of the soil microbial biomass. *J. Sci. Food Agric.* **33**, 249–252.

Lys, J.-A. and Nentwig, W. (1994). Improvement of overwintering sites for Carabidae, Staphylinidae and Araneae by strip-management in a cereal field. *Pedobiol.* **38**, 238–242.

Mackay, A.D. and Klavidko, E.J. (1985). Earthworms and the rate of breakdown of soybean and maize residues in soil. *Soil Biol. Biochem.* **17**, 851–857.

Mahn, E-G. and Kastner, A. (1985). Effects of herbicide stress on weed communities and soil nematodes in agro-ecosystems. *Oikos* **44**, 185–190.

Mallow, D., Snider, R.J. and Robertson, L.S. (1985). Effects of different management practices on Collembola and Acarina in corn production systems. II. The effects of moldboard ploughing and atrazine. *Pedobiol.* **28**, 115–131.

Marinissen, J.C.Y. and Bok, J. (1988). Earthworm-amended soil structure: its influence on Collembola populations in grassland. *Pedobiol.* **32**, 243–252.

Martin, N.A. (1982). The effects of herbicides used on asparagus on the growth of the earthworm *Allobophora canliginosa*. *Proc. 35th N.Z. Weed and Pest Control Conf.*, 328–331.

Martinez, N.D. (1992). Constant connectance in community food webs. *Am. Nat.* **139**, 1208–1218.

May, R.M. (1973). *Stability and Complexity in Model Ecosystems*. Princeton University Press, Princeton.

McClaugherty, C.A. (1983). Soluble polyphenols and carbohydrates in throughfall and leaf litter decomposition. *Acta Oecol./Oecol. General* **4**, 375–385.

McGonigle, T.P. and Miller, M.H. (1993). Mycorrhizal development and phosphorous absorption in maize under conventional and reduced tillage. *Soil Sci. Soc. Am. J.* **57**, 1002–1006.

McNaughton, S.J. (1992). The propagation of disturbance in savannas through food webs. *J. Veg. Sci.* **3**, 301–314.

Meikle, L.N., Doran, J.W. and Richards, K.A. (1986). Physical environment near the surface of ploughed and no-till soil. *Soil Till. Res.* **7**, 355–366.

Mekwantanakarn, P. and Sivasithamparam, K. (1987). Effect of certain herbicides on soil microbial populations and their influence on saprophytic growth in soil and pathogencity of take-all fungus. *Biol. Fertil. Soils* **5**, 175–180.

Menge, B.A. and Sutherland, J.P. (1976). Species diversity gradients: synthesis of the roles of predator, competition and spatial heterogeneity. *Am. Nat.* **110**, 351–369.

Menge, B.A., Lubchenco, J., Gaines, S.D. and Ashkenas, L.R. (1986). A test of

174   D.A. WARDLE

the Menge-Sutherland model of community organisation in a rocky tropical intertidal food web. *Oecologia* **71**, 75–89.

Mitchell, D.J., Schenck, N.C., Dickson, D.W. and Gallagher, R.N. (1980). The influence of minimum tillage on populations of soilborne fungi, endomycorrhizal fungi and nematodes in oats and vetch. *Proc. 3rd Annual No Tillage Systems Conf., Gainesville*, pp. 115–119.

Mola, L., Sabatini, M.A., Fratello, B. and Bertolani, R. (1987). Effects of atrazine on two species of Collembola (Onychiuridae) in laboratory tests. *Pedobiol.* **30**, 145–149.

Moore, J.C. and Hunt, H.W. (1988). Resource compartmentation and the stability of real ecosystems. *Nature* **333**, 261–263.

Moore, J.C., Snider, R.W. and Robertson, L.S. (1984). Effects of different management practices on Collembola and Acarina in corn production systems. 1. The effects of no-tillage and atrazine. *Pedobiol.* **26**, 143–152.

Moore, J.C., Walter, D.E. and Hunt, H.W. (1988). Arthropod regulation of micro- and meso-biota in below-ground detrital food webs. *Ann. Rev. Entomol.* **33**, 419–439.

Moorhouse, L.A. (1909). Some soil problems in Oklahoma. *Proc. Am. Soc. Agron.* **1**, 234–238.

Moorman, T.B. and Dowler, C.C. (1991). Herbicide and rotation effects on soil and rhizosphere microorganisms and crop yield. *Soil Biol. Biochem.* **35**, 311–325.

Morgan, W.C. (1989). Alternatives to herbicides. *Plant Prot. Quart.* **4**, 33–37.

Morgan, W.C. (1992). Strategies to reduce dependence on herbicides. *Proc. 1st Intern. Weed Control Congr. Vol. 1*, 289–294.

Müller, G. (1971). Laboruntersuchungen zur wirkung von heriziden auf Carabiden. *Arch. Pflanzen*, **7**, 351–364.

Mulognoy, K. (1986). Microbial biomass and maize nitrogen uptake under a *Psophocarpus palustris* live mulch grass on a tropical alfisol. *Soil Biol. Biochem.* **18**, 395–398.

Newell, K. (1984). Interaction between two decomposer bosidiomycetes and a collembolan under Sitka spruce: distribution, abundance and selective grazing. *Soil Biol. Biochem.* **16**, 227–233.

Newman, E.I. (1985). The rhizosphere—carbon sources and microbial populations. In: *Ecological Interactions in Soil* (Ed. by A.H. Fitter), pp. 107–121. Blackwell Scientific Publications, Oxford.

Nicholson, K.S. and Wardle, D.A. (1994). Decomposition of weed tissues in a pasture and maize crop situation: implications for nutrient cycling. *Proc. 47th N.Z. Plant Protection Conf.* 29–33.

Niemalä, J. (1993). Interspecific competition in ground beetle assemblages (Carabidae): what have we learned? *Oikos,* **66**, 325–335.

Norstadt, F.A. and McCalla, T.M. (1969). Microbial populations in stubble mulched soil. *Soil Sci.* **107**, 188–193.

Nuutinen, V. (1992). Earthworm community response to tillage and residue management on different soil types in southern Finland. *Soil Till. Res.* **23**, 221–239.

O'Connor, F.B. (1967). The enchytraeidae. In: *Soil Biology* (Ed. by A. Burges and F. Raw), pp. 213–257. Academic Press, London.

Odeyemi, O., Salami, A. and Ugoji, E.O. (1988). Effect of common pesticides used in Nigeria on soil microbial populations. *Ind. J. Agric. Sci.* **58**, 624–628.

Odum, E.P. (1969). The strategy of ecosystem development. *Science* **164**, 262–270.

Olson, B.M. and Lindwall, C.W. (1991). Soil microbial activity under chemical fallow conditions: effects of 2,4–D and glyphosate. *Soil Biol. Biochem.* **23**, 1071–1075.

Ou, L.T., Davidson, J.M. and Rothwell, D.F. (1978). Response of soil microflora to high 2,4–D applications. *Soil Biol. Biochem.* **10**, 443–445.

Paine, R.T. (1980). Food webs: linkage, interaction strength and community infrastructure. *J. Anim. Ecol.* **49**, 667–686.

Paine, R.T. (1988). Food webs: road maps of interactions or grist for theoretical development? *Ecology* **69**, 1648–1654.

Paine, R.T. (1992). Food web analysis through field measurement of per capita interaction strength. *Nature* **355**, 73–75.

Paoletti, M.G., Favretto, M.R., Stinner, B.R., Purrington, F.F. and Bater, J.E. (1991). Invertebrates as bioindicators of soil use. *Agric. Ecosyst. Environ.* **34**, 341–362.

Parkinson, D., Visser, S. and Whittaker, J.B. (1979). Effects of collembolan grazing on fungal colonization of leaf litter. *Soil Biol. Biochem.* **11**, 529–535.

Parle, J.N. (1963). A microbiological study of earthworm casts. *J. Gen. Microbiol.* **31**, 13–23.

Parmelee, R.W. and Alston, D.G. (1986). Nematode trophic structure in conventional and no-tillage agroecosystems. *J. Nematol.* **18**, 403–407.

Parmelee, R.W., Beare, M.H., Cheng, W., Hendrix, P.F., Rider, S.J., Crossley, D.A. and Coleman, D.C. (1990). Earthworms and enchytraeids in conventional and no-tillage agroecosystems: a biocide approach to assess their role in organic matter breakdown. *Biol. Fertil. Soils* **10**, 1–10.

Paustian, K., Andrén, O., Clarholm, H., Hansson, A-C., Johansson, G., Lagerlöf, J., Lindberg, T., Pettersson, R. and Sohlenius, B. (1990). Carbon and nitrogen budgets in four agro-ecosystems with annual and perennial crops, with and without N-fertilization. *J. Appl. Ecol.* **27**, 60–84.

Perdue, J.C. and Crossley, D.A. (1990). Vertical distribution of soil mites (Acari) in conventional and no-tillage agricultural systems. *Biol. Fertil. Soils* **9**, 135–138.

Perucci, P. (1990). Effects of the addition of municipal solid-waste compost on microbial biomass and enzyme activities in soil. *Biol. Fertil. Soils* **10**, 221–226.

Peters, R.H. (1991). *A Critique of Ecology.* Cambridge University Press, Cambridge.

Petersen, H. and Luxton, M. (1982). A comparative analysis of soil fauna populations and their role in decomposition processes. *Oikos* **39**, 287–388.

Phillips, R.E., Blevins, R.L., Thomas, G.W., Frye, W.W. and Phillips, S.H. (1980). No-tillage agriculture. *Science* **208**, 1108–1113.

Phillips, S.H. (1984). Introduction. In: *No Tillage Agriculture: Principles and Practices* (Ed. by R.E. Phillips and S.H. Phillips), pp. 1–10. van Nostrand Reinhold Company, New York.

Pimentel, D., Andow, D., Dyson-Hudson, R., Gallaghan, D., Jacobson, S., Irish, M., Kroop, S., Moss, A., Schreiner, I., Shepherd, M., Thompson, T. and Vincent, B. (1980). Environmental and social costs of pesticides—a preliminary assessment. *Oikos* **34**, 126–140.

Pimentel, D., Berard, G.M. and Fast, S. (1983). Energy efficiency of farming systems—organic and conventional agriculture. *Agric. Ecosyst. Environ.* **9**, 359–372.

Pimm, S.L. (1982). *Food webs.* Chapman and Hall, London.

Pimm, S.L. (1991). *The Balance of Nature. Ecological Issues in the Conservation of Species and Communities.* University of Chicago Press, Chicago.

Pimm, S.L. and Lawton, J.H. (1977). The number of trophic levels in ecological communities. *Nature* **268**, 329–331.

Pimm, S.L. and Lawton, J.H. (1980). Are food webs divided into compartments? *J. Anim. Ecol.* **49**, 879–898.

Pimm, S.L., Lawton, J.H. and Cohen, J.E. (1991). Food web patterns and their consequences. *Nature* **350**, 669–674.

Polis, G.A. (1991). Complex trophic interactions in deserts: an empirical critique of food-web theory. *Am. Nat.* **138**, 123–155.

Polis, G.A. (1994). Food webs, trophic cascades and community structure. *Aust. J. Ecol.* **19**, 121–136.

Popovici, I., Stan, G., Stefan, V., Tomescu, R., Dumea, A., Tarta, A. and Dan, F. (1977). The influence of atrazine on the soil fauna. *Pedobiol.* **17**, 209–215.

Powell, W., Dean, G.J. and Dewar, A. (1985). The influence of weeds on polyphagous arthropods in winter wheat. *Crop Prot.* **4**, 298–312.

Power, M.E. (1990). Effects of fish in river food webs. *Science* **250**, 411–415.

Power, M.E. (1992). Top-down and bottom-up forces in food webs: do plants have primacy? *Ecology* **73**, 733–746.

Powlson, D.S. and Jenkinson, D.S. (1981). A direct comparison of the organic matter, biomass adenosine triphosphate and mineralisable nitrogen contents of ploughed and direct-drilled soils. *J. Agric. Sci.* **97**, 713–721.

Prasse, J. (1975). The effect of the herbicides 2,4–D and simazin on the coenosis of Collembola and Acari in arable soil. In: *Progress in Soil Zoology* (Ed. by J. Vanek) pp. 469–480. Academia, Prague.

Prasse, J. (1978). Die struktur von mikroarthropodenönosen in Agro-Ökosystemen und ihrebeeinflussung durch herbizide. *Pedobiol.* **18**, 381–383.

Purvis, G. and Curry, J.P. (1984). The influence of weeds and farmyard manure on the activity of Carabidae and other ground-dwelling arthropods in a sugar beet crop. *J. Appl. Ecol.* **21**, 271–284.

Racke, K.D. (1990). Pesticides in the soil microbial ecosystem. In: *Enhanced Degradation of Pesticides in the Environment* (Ed. by K.D. Racke and J.R. Coats), pp. 1–12. American Chemical Society, Washington D.C.

Raffaelli, D. and Hall, S.J. (1992). Compartments and predation in an estuarine food web. *J. Anim. Ecol.* **61**, 551–560.

Rai, J.P.N. (1992). Effects of long-term 2,4–D application on microbial populations and biochemical processes in cultivated soil. *Biol. Fertil. Soils* **13**, 187–191.

Rapoport, E.H. and Cangioli, G. (1963). Herbicides and the soil fauna. *Pedobiol.* **2**, 235–238.

Read, D.J. (1991). Mycorrhizas in ecosystems. *Experientia* **47**, 376–391.

Reganold, J.P. (1994). Statistical analysis of soil quality (reply to Wardle). *Science* **264**, 282–283.

Reganold, J.P., Palmer, A.S., Lockhart, J.C., Macgregor, A.N. (1993). Soil quality and financial performance of biodynamic and conventional farms in New Zealand. *Science* **260**, 344–349.

Reinecke, A.J. and Nash, R.G. (1984). Toxicity of 2,3,7,8–TCDD and short-term bioaccumulation by earthworms. *Soil Biol. Biochem.* **16**, 45–49.

Rejmanek, M. and Stary, P. (1979). Connectance in real biotic communities and critical values of stability for model ecosystems. *Nature* **280**, 311–313.

Riechert, S.E. and Bishop, L. (1990). Prey control by an assemblage of generalist predators: spiders in test garden systems. *Ecology* **71**, 1441–1450.

Roberts, B.L. and Dorough, H.W. (1984). Relative toxicities of chemicals to the earthworm *Eisenia foetida*. *Environ. Toxicol. Chem.* **3**, 67–78.

Roslycky, E.B. (1977). Response of soil microbiota to selected herbicide treatments. *Can. J. Micro.* **23**, 426–433.

Roslycky, E.B. (1982). Glyphosate and the response of the soil microbiota. *Soil Biol. Biochem.* **14**, 87–92.

Rovira, A.D., Smetten, K.R.J. and Lee, K.E. (1987). Effect of rotation and conservation tillage on earthworms in a red-brown earth under wheat. *Aust. J. Agric. Res.* **38**, 829–834.

Sabatini, M.A., Pederzoli, A., Fratello, B. and Bertolani, R. (1979). Microarthropod communities in soil treated with atrazine. *Boll. Zool.* **46**, 333–341.

Saffigna, P.G., Powlson, D.S., Brookes, P.C. and Thomas, G.A. (1989). Influence of sorghum residues and tillage on soil organic matter and soil microbial biomass in an Australian vertisol. *Soil Biol. Biochem.* **21**, 759–765.

Sály, A. and Staňová, A. (1983). Vplyv herbicidov na spolocenstva volne zijucich nematov v pode kultuny silazneja kukurice. *Biologia (Bratislava)* **38**, 155–165.

Sanocka-Woloszyn, E. and Woloszyn, B.W. (1970). The influence of herbicides on the mesofauna of the soil. *Meded. Fac. Landbouw. Rijksuniv. Gent* **35**, 731–738.

Santos, P.F., Phillips, J. and Whitford, W.G. (1981). The role of mites and nematodes in early stages of litter decomposition in a desert. *Ecology* **62**, 664–669.

Santos, P.F. and Whitford, W.G. (1981). The effects of microarthropods on litter decomposition in a Chihuahuan desert ecosystem. *Ecology* **62**, 654–663.

Santrúcková, H. and Vraný, J. (1990). Microbial biomass in various agro-ecosystems. *Agrokem. Talaj.* **39**, 476–480.

Satchell, J.E. (1983). Earthworm microbiology. In: *Earthworm Ecology* (Ed. by J.E. Satchell), pp. 351–364. Chapman and Hall, New York.

Schnürer, J., Clarholm and Rosswall, T. (1986a). Fungi, bacteria and protozoa in soil from four arable cropping systems. *Biol. Fertil. Soils* **2**, 119–126.

Schnürer, J., Clarholm, M., Bostrom, S. and Rosswall, T. (1986b). Effects of moisture on soil microorganisms and nematodes: a field experiment. *Microb. Ecol.* **12**, 217–230.

Schoener, T.W. (1989). Food webs from the small to the large. *Ecology* **70**, 1559–1589.

Schreven, D.A. van, Lindenbergh, D.J. and Koridon, A. (1970). Effects of several herbicides on bacterial populations and activity and persistence of three herbicides in soil. *Plant Soil* **33**, 513–532.

Schreiber, B. and Brink, N. (1989). Pesticide toxicity using protozoans as test organisms. *Biol. Fertil. Soils* **7**, 289–296.

Schuster, E. and Schröder, D. (1990a). Side-effects of sequentially applied pesticides on non-target soil microorganisms: field experiments. *Soil Biol. Biochem.* **22**, 367–373.

Schuster, E. and Schröder, D. (1990b). Side-effects of sequentially- and simultaneously- applied pesticides on non-target soil organisms: laboratory experiments. *Soil Biol. Biochem.* **22**, 375–383.

Seastadt, T.R. (1984). The role of microarthropods in decomposition and mineralisation processes. *Ann. Rev. Entomol.* **29**, 25–46.

Shams, M.N., Snider, R.J. and Robertson, L.S. (1981). Preliminary investigations on the effects of no-till corn production methods on specific soil mesofaunal populations. *Comm. Soil Sci. Plant Analys.* **12**, 179–188.

Sharma, L.N. and Saxena, S.N. (1974). Influence of 2,4–D on soil microorganisms with special reference to *Azotobacter*. *J. Indian Soc. Soil Sci.* **22**, 168–171.

Shields, J.A., Paul, E.A. and Lowe, W.E. (1974). Factors influencing the stability of labelled microbial materials in soils. *Soil Biol. Biochem.* **6**, 31–37.

Sieverding, E. and Leihner, D.E. (1984). Effects of herbicides on population dynamics of V-A mycorrhizae with Cassava. *Agnew. Botanik.* **58**, 283–294.

Small, R.W. (1987). A review of the prey of predatory soil nematodes. *Pedobiol.* **30**, 179–206.

Small, R.W. and Grootaert, P. (1983). Observations on the predation abilities of some soil dwelling predatory nematodes. *Nematol.* **29**, 109–118.

Smith, J.L. and Paul, E.A. (1990). The significance of soil biomass estimates. In: *Soil Biochemistry, Volume 6.* (Ed. by J.M. Bollag and G. Stotzky), pp. 357–396. Marcel-Dekker, New York.

Sohlenius, B. and Bostrom, S. (1984). Colonization, population development and metabolic activity of nematodes in buried barley straw. *Pedobiol.* **27**, 67–78.

Sohlenius, B., Bostrom, S. and Sandor, A. (1987). Long-term dynamics of nematode communities in arable soil under four cropping systems. *J. Appl. Ecol.* **24**, 131–144.

Sohlenius, B., Bostrom, S. and Sandor, A. (1988). Carbon and nitrogen budgets of nematodes in arable soil. *Biol. Fertil. Soils* **6**, 1–8.

Soulas, G., Chaussod, R. and Verguet, A. (1984). Chloroform fumigation technique as a means of determining the size of specialized microbial populations: application to pesticide-degrading microorganisms. *Soil Biol. Biochem.* **16**, 497–501.

Sparling, G.P., Shephard, T.G. and Kettles, H.A. (1992). Changes in soil organic C, microbial C and aggregate stability under continuous maize and cereal cropping, and after restoration to pasture in soils from the Manawatu region, New Zealand. *Soil Till. Res.* **24**, 225–241.

Speight, M.R. and Lawton, J.H. (1976). The influence of weed cover on the mortality imposed on artificial prey by predatory ground beetles in cereal fields. *Oecologia* **23**, 211–223.

Sprague, M.A. (1986). Overview. In: *No Tillage and Surface Tillage Agriculture* (Ed. by M.A. Sprague and G.B. Triplett), pp. 1–18. John Wiley and Sons, New York.

Sprules, W.G. and Bouwman, J.E. (1988). Omnivory and food chain length in zooplankton food webs. *Ecology* **69**, 418–426.

Stassart, P. and Gregorie-Wibo, C. (1983). Influence du travail du sol sur des populations de Carabides en grande culture, resultats preliminaires. *Meded. Fac. Landbouw. Rijksuniv. Gent* **48**, 465–474.

Stearman, G.K. and Matocha, J.E. (1992). Soil microbial population and nitrogen-transforming bacteria from 11-year sorghum tillage plots. *Soil Till. Res.* **22**, 299–310.

Steen, E. (1983). Soil animals in relation to agricultural practices and soil productivity. *Swed. J. Agric. Res.* **13**, 157–165.

Steinberger, Y., Freckman, D.W., Parker, L.W. and Whitford, W.G. (1984). Effects of simulated rainfall and litter quantities on desert soil biota: nematodes and microarthropods. *Pedobiol.* **26**, 267–274.

Steinbrenner, K., Naglitsch, Fr. and Schlicht, I. (1960). Der einfluß der herbicide simizin und W6658 auf die bodenmikroorganismen und die bodenfauna. *Albrecht Thaer Archiv.* **4**, 611–631.

Stinner, B.R. and Crossley, D.A. (1982). Nematodes in no-tillage agroecosystems.

In: *Nematodes in Soil Ecosystems* (Ed. by D.W. Freckman), pp. 14–28. University of Texas Press, Austin.

Stinner, B.R. and House, G.J. (1990). Arthropods and other invertebrates in conservation-tillage agriculture. *Ann. Rev. Entomol.* **35**, 299–318.

Stinner, B.R., Crossley, D.A., Odum, E.P. and Todd, R.L. (1984). Nutrient budgets and the internal cycling of N, P, K, Ca and Mg in conventional tillage, no tillage and old-field systems on the Georgia piedmont. *Ecology* **65**, 354–369.

Stinner, B.R., Hoyt, G.D. and Todd, R.L. (1983). Changes in soil properties following a 12-year fallow: a two year comparison of conventional-tillage and no-tillage agroecosystems. *Soil Till. Res.* **3**, 277–290.

Stinner, B.R., Kreuger, H.R. and McCartney, D.A. (1986). Insecticide and tillage effects on pest and non-pest arthropods in corn agroecosystems. *Agric. Ecosyst. Environ.* **15**, 11–21.

Stinner, B.R., McCartney, D.A. and Doran, D.M. van (1988). Soil and foliage arthropod communities in conventional and no-tillage corn (maize, *Zea mays* L.) systems: a comparison after 20 years of continual cropping. *Soil Till. Res.* **11**, 147–158.

Stout, J.D. (1980). The role of protozoa in nutrient cycling and energy flow. *Adv. Microb. Ecol.* **4**, 1–50.

Strong, D.R. (1992). Are trophic cascades all wet? Differentiation and donor-control in species ecosystems. *Ecology* **73**, 747–754.

Swift, M.J., Heal, O.W. and Anderson, J.M. (1979). *Decomposition in Terrestrial Ecosystems*. University of California Press, Berkeley.

Tajovský, K., Santruckova, H., Hanel, L., Balik, V. and Lukesova, A. (1992). Decomposition of faecal pellets of the millipede *Glomeris hexasticha* (Diplopoda) in forest soil. *Pedobiol.* **36**, 146–158.

Tan, Y.K. and Chua, L.H. (1987). The response of three soil fungi to dalapon and diuron treatments. *Pestic. Sci.* **20**, 233–240.

Teuteberg, A. von (1968). Der einfluß einer herbizidbehandlung auf antibiotisch wirksame mikroorganismen des bodens. *Zeitschr. Pflanzenkrank. Pflanzenschutz.* **75**, 72–86.

Thomas, M.B., Wratten, S.D. and Sotherton, N.W. (1992). Creation of "island" habitats in farmland to manipulate populations of beneficial arthropods: predator densities and species composition. *J. Appl. Ecol.* **29**, 524–531.

Thomas, S.J. (1978). Population densities of nematodes under seven tillage regimes. *J. Nematol.* **10**, 24–27.

Thompson, L., Thomas, C.D., Radley, J.M.A., Williamson, S. and Lawton, J.H. (1993). The effects of earthworms and snails in a simple plant community. *Oecologia* **95**, 171–178.

Tilman, D. and Downing, J.A. (1994). Biodiversity and stability in grasslands. *Nature* **367**, 363–365.

Torstensson, L. (1988). Testing pesticide fate and side-effects in the terrestrial environment. *Toxic. Assess.* **3**, 407–414.

Trappe, J.M., Molina, R. and Castellano, M. (1984). Reactions of mycorrhizal fungi and mycorrhiza formation to pesticides. *Ann. Rev. Phytopath.* **22**, 331–359.

Triplett, G.B. and Worsham, A.D. (1986). Principles of weed management with surface-tillage systems. In: *No-Tillage and Surface-Tillage Agriculture* (Ed. by M.A. Sprague and G.B. Triplett), pp. 319–346. John Wiley and Sons, New York.

Trujillo-Arriaga, J. and Altieri, M. (1990). A comparison of aphidophagous arth-

180   D.A. WARDLE

ropods on maize polycultures and monocultures in central Mexico. *Agric. Eco-syst. Environ.* **31**, 337–349.

Tscharntke, T. (1992). Cascade effects among four trophic levels: bird predation on galls affects density-dependent parasitism. *Ecology* **73**, 1689–1698.

Tu, C.M. and Bollen, W.B. (1968). Effect of paraquat on microbial activities in soils. *Weed Res.* **8**, 28–37.

Tucker, G.M. (1992). Effects of agricultural practices on field use by invertebrate-feeding birds in winter. *J. Appl. Ecol.* **29**, 779–790.

Unger, P.W. and Cassell, D.K. (1991). Tillage implement disturbance effects on soil properties related to soil and water conservation: a literature review. *Soil Till. Res.* **19**, 363–382.

Visser, S. (1985). Role of soil invertebrates in determining the composition of soil microbial communities. In: *Ecological Interactions in Soil* (Ed. by A.H. Fitter, D. Atkinson, D.J. Read and M.B. Usher), pp. 297–317. Blackwell Scientific Publications, Oxford.

Vitousek, P.M. and Hooper, D.U. (1993). Biological diversity and terrestrial eco-system biogeochemistry. In: *Biodiversity and Ecosystem Function* (Ed. by E.-D. Schulze and H.A. Mooney), pp. 3–14. Springer-Verlag, Berlin.

Vythilingham, I. and Chua, T.H. (1982). The effect of herbicides on soil Collembola in oil palm estates. *Trop. Ecol.* **23**, 165–172.

Wacha, A.G. and Tiffany, K.H. (1979). Soil fungi isolated from fields under different tillage and weed control regimes. *Mycologia* **71**, 1215–1226.

Wallin, H. (1985). Spatial and temporal distribution of some abundant carabid beetles (Coleoptera: Carabidae) in cereal fields and adjacent habitats. *Pedobiol.* **28**, 19–34.

Wallwork, J.A. (1982). *Desert Soil Fauna.* Praeger Publishers, New York.

Walter, D.E. and Kaplan, D.T. (1990). Feeding observations on two astigmatid mites, *Schwiebiea rocketii* (Acaridae) and *Histiostoma bakeri* (Histiostomati-dae) associated with *Citrus* feeder roots. *Pedobiol.* **34**, 281–286.

Walter, D.E., Hugens, P.A. and Freckman, D.W. (1986). Consumption of nematodes by fungivorous mites, *Tyrophagus* spp. (Acarina: Astigmata: Acaridae). *Oecologia* **70**, 357–361.

Walter, D.E., Kaplan, D.T. and Permer, T.A. (1991). Missing links: a review of methods used to estimate trophic links in food webs. *Agric. Ecosyst. Environ.* **34**, 399–405.

Wardle, D.A. (1991). Free lunch? *Nature* **352**, 482.

Wardle, D.A. (1992). A comparative assessment of factors which influence microbial biomass carbon and nitrogen levels in soils. *Biol. Rev.* **67**, 321–358.

Wardle, D.A. (1993). Changes in the microbial biomass and metabolic quotient during leaf litter succession in some New Zealand forest and scrubland ecosystems. *Funct. Ecol.* **7**, 346–355.

Wardle, D.A. (1994). Statistical analyses of soil quality. *Science* **264**, 281–282.

Wardle, D.A. and Parkinson, D. (1990a). Effects of three herbicides on soil microbial biomass and activity. *Plant and Soil* **122**, 21–28.

Wardle, D.A. and Parkinson, D. (1990b). Influence of the herbicide glyphosate on soil microbial community structure. *Plant and Soil* **122**, 29–37.

Wardle, D.A. and Parkinson, D. (1991a). Relative importance of the effect of 2,4-D, glyphosate and environmental variables on the soil microbial biomass. *Plant and Soil* **134**, 209–219.

Wardle, D.A. and Parkinson, D. (1991b). Analysis of co-occurrence in a fungal community. *Mycol. Res.* **95**, 504–507.

Wardle, D.A. and Parkinson, D. (1992). Influence of the herbicides 2,4–D and glyphosate on soil microbial biomass and activity: a field experiment. *Soil Biol. Biochem.* **24**, 185–186.

Wardle, D.A. and Yeates, G.W. (1993). The dual importance of competition and predation as regulating forces in terrestrial ecosystems: evidence from decomposer food-webs. *Oecologia* **93**, 303–306.

Wardle, D.A., Nicholson, K.S. and Rahman, A. (1994). The influence of herbicide application on the decomposition, microbial biomass and microbial activity of pasture shoot and root litter. *N.Z. J. Agric. Res.* **37**, 29–39.

Wardle, D.A., Nicholson, K.S. and Yeates, G.W. (1993a). Effect of weed management strategies on some soil associated arthropods in maize and asparagus ecosystems. *Pedobiol.,* **37**, 257–269.

Wardle, D.A., Yeates, G.W., Watson, R.N. and Nicholson, K.S. (1993b). Response of soil microbial biomass and plant litter decomposition to weed management strategies in maize and asparagus cropping systems. *Soil Biol. Biochem.* **25**, 857–868.

Wardle, D.A., Yeates, G.W., Watson, R.N. and Nicholson, K.S. (1995a). The detritus food-web and diversity of soil fauna as indicators of disturbance regimes in agro-ecosystems. *Plant and Soil*: in press.

Wardle, D.A., Yeates, G.W., Watson, R.N. and Nicholson, K.S. (1995b). Development of the decomposer food web, trophic relationships and ecosystem properties during a three-year primary succession of sawdust. *Oikos*, in press.

Warren, P.H. (1990). Variations in food web structure: the determinants of connectance. *Am. Nat.* **136**, 689–700.

Warren, P.H. (1994). Making connections in food webs. *Trends Ecol. Evol.* **9**, 136–143.

Wasilewska, L. (1979). The structure and function of soil nematode communities in natural ecosystems and agro-coenoses. *Pol. Ecol. Stud.* **5**, 97–145.

Wasilewska, L., Jakubczyk, H. and Paplinska, E. (1975). Production of *Aphelenchus avenae* Bastian (Nematoda) and reduction of saprophytic fungi by them. *Pol. Ecol. Stud.* **1**, 61–73.

Werner, M.R. and Dindal, D.L. (1990). Effects of conversion to organic agricultural practices on the soil biota. *Am. J. Altern. Agric.* **5**, 24–32.

Whitford, W.G., Freckman, D.W., Elkins, N.Z., Parker, L.W., Parmelee, R.W., Phillips, J. and Tucker, S. (1981). Diurnal migration and responses to simulated rainfall in desert soil microarthropods and nematodes. *Soil Biol. Biochem.* **13**, 417–425.

Whitford, W.G., Freckman, D.W., Santos, P.F., Elkins, N.Z. and Parker, L.R. (1982). The role of nematodes in decomposition in desert ecosystems. In: *Nematodes in Soil Ecosystems* (Ed. by D.W. Freckman), pp. 98–116. University of Texas Press, Austin.

Whittaker, R.H. (1975). *Communities and Ecosystems. Second Edition*. Macmillan, New York.

Wilson, E.O. (Ed.) (1988). *Biodiversity*. National Academy of Sciences, Washington D.C.

Winemiller, K.O. (1989). Must connectance decline with species richness? *Am. Nat.* **134**, 960–968.

Wingfield, G.I., Davies, H.A. and Greaves, M.P. (1977). The effect of soil treatment on the response of the soil microflora to the herbicide dalapon. *J. Appl. Bacteriol.* **43**, 39–46.

Winter, J.P., Voroney, E.P. and Ainsworth, D.A. (1990). Soil microarthropods in long-term no tillage and conventional tillage corn production. *Can. J. Soil Sci.* **70**, 641–653.

Witkowski, T. (1973). Znaczenie chemizacji srodowiska dla nicieni (Nematoda) glebowych zesc II. Obserwacje nad przezywaniem nicieni glebowych w wodnych roztworach preparatow chwastobojczych. *Rocz. Glebozn. T.* **24**, 367–387.

Witkowski, T. (1974). Znarzenie chemizacji srodowlska dla nicieni (Nematoda) glebowych. cz. III. Obserwacje nad wplywem wodnych roztworow preparatow chwastobojczych o dzialainiu kontakawym na dua saprobiotyczne gatuski nicieni. *Zesz. Problem. Poste. Nauk. Roln.* **154**, 443–456.

Wolters, V. (1989). The influence of omnivorous elaterid larvae on the microbial carbon cycle in different forest soils. *Oecologia* **80**, 405–413.

Wolters, V. and Joergenson, R.G. (1992). Microbial carbon turnover in beech forest soils by *Aporrectodea caliginosa* (Savigny) (Oligochaeta: Lumbricidae). *Soil Biol. Biochem.* **24**, 171–177.

Wood, C.W. and Edwards, J.H. (1992). Agroecosystem management effects on soil carbon and nitrogen. *Agric. Ecosyst. Environ.* **39**, 123–138.

Yeates, G.W. (1979). Soil nematodes in terrestrial ecosystems. *J. Nematol.* **11**, 213–229.

Yeates, G.W. (1981). Soil nematode populations depressed in the presence of earthworms. *Pedobiol.* **22**, 191–195.

Yeates, G.W. (1987). How plants affect nematodes. *Adv. Ecol. Res.* **17**, 61–113.

Yeates, G.W. (1994). Modification and qualification of the nematode maturity index. *Pedobiol.* **38**, 97–101.

Yeates, G.W. and Bird, A.F. (1994). Some observations on the influence of agricultural practices on the nematode fauna of some South Australian soils. *Fundament. Appl. Nematol.* **17**, 133–145.

Yeates, G.W. and Foissner, W. (1995). Testate amoebae as predators of nematodes. *Biol. Fertil. Soils* **20**, 1–7.

Yeates, G.W. and Hughes, K.A. (1990). The effects of three tillage regimes on plant and soil nematodes in an oats/maize rotation. *Pedobiol.* **34**, 379–387.

Yeates, G.W., Bongers, T., de Goede, R.G.M., Freckman, D.W. and Georgeiva, S.S. (1993a). Feeding habits in nematode families—an outline for soil ecologists. *J. Nematol.* **25**, 315–331.

Yeates, G.W., Wardle, D.A. and Watson, R.N. (1993b). Relationships between nematodes, soil microbial biomass and weed management strategies in maize and asparagus cropping systems. *Soil Biol. Biochem.* **25**, 869–876.

Yodzis, P. (1981). The stability of real ecosystems. *Nature* **284**, 544–545.

APPENDIX 1

The following studies were used for compiling Figure 2. Abbreviations for food web components are: C (% soil organic carbon); N (% total soil nitrogen); B (bacteria); F (fungi); MT (total microflora); TN (total nematodes); BF (bacterial-feeding nematodes); FF (fungal-feeding nematodes); P (predatory/omnivorous nematodes); CO (collembola); TM (total mites); PR (prostigmatid mites); ME (mesostigmatid mites); CR (cryptostigmatid mites); AS (astigmatid mites); EN (enchytraeids); EA (earthworms); S (spiders); CA (carabid beetles).

Altieri *et al.* (1985): S, CA; Andersen (1987): EA; Andrén and Lagerlöf

(1983): BF, TN, EN, CO, CR, PR, AS, TM; Andrén *et al.* (1988): F, B, TN, EN, CO, CR, TM, EA; Angers *et al.* (1992): C, MT; Arshad *et al.* (1990): C, N; Barnes and Ellis (1979): EA; Bauer and Black (1981): C, N; Blevins *et al.* (1977, 1983): C, N; Blumberg and Crossley (1983): S; Boles and Oseto (1987): CO, CR, ME, PR, AS, TM; Bostrom and Sohlenius (1986): F, FF, BF, P, TN; Broder *et al.* (1984): C, N; Brust and House (1990): ME, CA; Buchanan and King (1992): MT; Carter (1986): MT; Carter (1991): C, N, MT; Carter (1992): C, N; Carter and Mele (1992): C, N, MT; Carter and Rennie (1982): C, N, MT; Crossley *et al.* (1992): CO, CR, ME, PR, AS, TM; Curry (1970): CO, CR, ME, PR, AS, TM; Dalal *et al.* (1991): C, N; Desender and Alderweireldt (1988): CA; Desender *et al.* (1989): S, CA; De St. Remy and Daynard (1982): EA; Dick (1983): C; Doane and Dondale (1979): S; Doran (1980a): F, B; Doran (1980b): C; Doran (1987): MT; Edwards (1975a): CO, CR, ME, PR, TM; Edwards (1975b): CO, TM, EA; Edwards (1984): CO, CR, ME, PR, TM, EA; Edwards and Lofty (1975): CO, CR, ME, PR, TM, EA; Edwards and Lofty (1982): EA; Ehlers (1975): EA; Elliott *et al.* (1984): MT, CO, TM; Emmanuel *et al.* (1985): CR, ME, PR, AS, TM; Fleige and Baumer (1974): C, N; Follett and Schimel (1989): N, MT; Freckman and Ettema (1993): FR, BF, P, TN; Gerard and Hay (1979): EA; Ghilarov (1975): CO, TM, EA; Granatstein *et al.* (1987): C, N, MT; Gupta and Germida (1988): C, N, F, B, MT; Haines and Uren (1990): C, MT, EA; Hargrove *et al.* (1982): C, N; Hassall *et al.* (1992): S, CA; Haukka (1988): EA; Havlin *et al.* (1990): C, N; Haynes and Knight (1989): C, N; Hendrix *et al.* (1986): C, N, FI, BF, TN, EN, CO, TM, S, CA, EA; Hendrix *et al.* (1990): CO, CR, ME, PR, AS, TM, EA; Holland and Coleman (1987): F, MT; House (1989): S, CA; House and All (1981): CA; House and Alzugaray (1989): CA; House and Parmelee (1985): EN, CO, CR, MG, PR, AS, TM, S, CA, EA; House and Stinner (1983): S, CA; Juo & Lal (1979): C, N; Kromp and Steinberger (1992): S, CA; Lagerlöf and Andrén (1988): CR, ME, PR, AS, TM; Lagerlöf & Andrén (1991): CO; Lagerlöf *et al.* (1989): EN; Lal (1974): C, N; Lamb *et al.* (1985): C, N; Linn and Doran (1984): C, F, B; Lynch and Panting (1980a, b; 1982): C, MT; Mackay and Kladivko (1985): EA; Mallow *et al.* (1985): CO, CR, ME, PR, TM; Meikle *et al.* (1986): C; Moore *et al.* (1984): CO, CR, ME, PR; Norstadt and McCalla (1969): F, B; Nuutinen (1992): EA; Parmelee and Alston (1986): FF, BF, P, TN; Paustian *et al.* (1990): C, N, F, B, MT, BF, P, EN, EA; Perdue and Crossley (1990): EN, CR, ME, PR, AS, TM, EA; Powlson and Jenkinson (1981): C, N, MT; Rovira *et al.* (1987): EA; Saffigna *et al.* (1989): C, N, MT; Schnürer *et al.* (1986a): F, B; Shams *et al.* (1981): C, TM; Sohlenius *et al.* (1987): FF, BF, P, TN; Sohlenius *et al.* (1988): FF, BF, P, TN; Sparling *et al.* (1992): C, MT; Stassart and Gregorie-Wibo (1983): CA; Stearman and Matocha (1992): B; Steen (1983): EN, CO, PR, TM; Stinner and

184 D.A. WARDLE

Crossley (1982): TN; Stinner *et al.* (1983, 1984): N; Stinner *et al.* (1986): CO, CR, TM, S, CA; Stinner *et al.* (1988): CO, CR, ME, TM, S, CA; Thomas (1978): TN; van de Bund (1970): CO, TM; von Klinger (1987): CA; Wacha and Tiffany (1979): F; Wardle *et al.* (1993a): S, CA; Wardle *et al.* (1993b): C, N, F, B, MT; Wasilewska (1979): FF, BF, TN; Winter *et al.* (1990): C, CO, CR, ME, PR, AS, TM; Yeates and Hughes (1990): FF, BF, P, TN; Yeates *et al.* (1993b): FF, BF, P, TN.

APPENDIX 2

The following studies were used for compiling Tables 3 and 4. Abbreviations as described in Appendix 1 except that PR = protozoa and MA = macroarthropods.

Abdel-Fattah *et al.* (1983): F; Abdel-Mallek and Moharram (1986): F; Andariese and Vitousek (1988): MT; Andrén and Lagerlöf (1983): EN, TM, CO; Arthur and Frea (1988): B, F; Atlavinyte *et al.* (1977): EA; Baarschers *et al.* (1988): B, F; Baudissin (1952): CO; Bhattacharya and Joy (1977): CO, TM; Boiteau (1984): MA; Breazeale and Camper (1970): B, F; Brust (1990): MA; Camper *et al.* (1973): B, F; Caseley and Eno (1966): EA; Chakravarty and Chatarpaul (1990): B, F; Chakravarty and Sidhu (1987): F; Chandra (1964): B, F; Chiverton and Sotherton (1991): CO, MA; Cooper *et al.* (1978): B, F; Curry (1970): TM, CO; Curry and Purvis (1981): TM, CO; Czulakow and Zarasow (1975): B, F; Davis (1965): TM, CO; Duah-Yentumi and Johnson (1986): B, F, MT; Dumontet and Perucci (1992): MT; Edwards (1970): EN, TM, CO; Edwards and Stafford (1979): TM, CO, EA; Eijsackers (1975): CO; Eijsackers (1978a): MA; Eijsackers (1978b,c): CO; Eijsackers (1985): CO, MA; Everts *et al.* (1989): MA; Fayolle (1979): EA; Fedorko *et al.* (1977): TN; Finlayson *et al.* (1975): EA, MA; Fischer (1989): EA; Fox (1964): TM, CO, EA, MA; Fratello *et al.* (1985): TM, CO, MA; Frey (1976): TN; Gel'cer and Geptner (1976): PR; Ghabbour and Imam (1967): EA; Gomez and Sagardoy (1985): B, F, TM, CO; Haque and Ebing (1983): EA; Heimbach (1984): EA; Heinonen-Tanski *et al.* (1986): F; Helmecke *et al.* (1977): B, F; Hickisch *et al.* (1986): B, F, TM, CO; Honczarenko (1969): EA, MA; House *et al.* (1987): MA; Houseworth and Tweedy (1973): B, F; Isakeit and Lockwood (1990): MT; Ishibashi *et al.* (1978): TN, TM, CO; Jaworska (1981): MA; Kaluz (1984): TM, CO; Karg (1964): TM, CO; Karki and Kaiser (1974): MT; Klyuchnikov *et al.* (1964): B, F; Lofs-Holmin (1980): EA; Lyons *et al.* (1972): TN, EN, TM, CO; Mahn and Kastner (1985): TN; Mallow *et al.* (1985): TM, CO; Martin (1982): EA; Mekwantanakarn and Sivasithamparam (1987): B, F; Mola *et al.* (1987): CO; Moore *et al.* (1984): TM, CO; Müller (1971): MA; Odeyemi *et al.* (1988): B, F, PR; Olson and Lindwall (1991): MT; Ou *et al.* (1978): B; Perucci (1990): MT; Popovici *et al.* (1977): PR, TN, EN, TM, CO; Powell *et al.* (1985): MA; Prasse (1975, 1978): TM, CO; Purvis and Curry (1984): MA; Rai (1992): B, F, MT; Rapoport and Cangioli (1963): TM, CO; Reinecke and Nash (1984): EA; Roberts and Dorough (1984): EA; Roslycky (1977, 1982): B, F; Sabatini *et al.* (1979): TM, CO; Sály and Staňová (1983): TN; Sanocka-Woloszyn and Woloszyn (1970): TM, CO; Santrúcková and Vraný (1990): MT; Schrieber and Brink (1989): PR; Schuster and Schröder (1990a,b): MT; Sharma and Saxena (1974): B, F; Soulas *et al.* (1984): MT; Steinbrenner *et al.* (1960): B, F, TN, TM, CO; Tan and Chua (1987): F; Teuteberg (1968): B; Tu and Bollen (1968): B, F; van Gestel *et al.* (1992): EA; van Schreven *et al.* (1970): B; Vythilingham and Chua (1982): CO; Wacha and Tiffany (1979): F; Wardle and Parkinson (1990a, 1991a): B, F, MT; Wardle and Parkinson (1990b): B, F; Wardle and Parkinson (1992): MT; Wardle *et al.* (1993a): TM, CO, MA; Wardle *et al.* (1993b): B, F, MT; Wardle *et al.* in press: MT, TN, TM, CO, EA; Wingfield *et al.* (1977): B; Witkowski (1973, 1974): TN; Yeates *et al.* (1993b): TN.

The Cichlid Fish Assemblages of Lake Tanganyika: Ecology, Behaviour and Evolution of its Species Flocks

A. ROSSITER

I. Summary 187
II. Introduction 189
   A. Biodiversity in Lake Tanganyika 189
   B. Fishes in the Lake 189
   C. The Cichlid Species Flocks of Lake Tanganyika 190
III. Speciation in Lake Tanganyika 192
   A. The Lake Environment 192
   B. Evolution of the Cichlid Fishes 199
IV. Breeding Habits 211
   A. Behaviour, Strategies and Tactics 211
   B. Substrate Spawners 212
   C. Mouthbrooders 215
   D. Commensalism and Parasitism 218
   E. Effects of Breeding Mode on Speciation 219
V. Species Diversity and Coexistence within the Rocky Shore
   Communities 220
   A. Mechanisms Promoting High Species Diversity and Coexistence 220
   B. Foods and Feeding 223
   C. Partitioning of Spatial Resources 229
   D. Partitioning within Communities and Effects of Relatedness . 231
VI. Community Composition and Species Diversity 234
   A. Importance of Size Differences for Species Diversity and
     Speciation 234
   B. Species Packing and Species Invasions 235
   C. Stability and Persistence 237
   D. Is Community Composition Predictable? 237
VII. Concluding Remarks, Caveats and Perspectives 240
References 243

I. SUMMARY

The occurrence and coexistence of large numbers of species within tropical

ADVANCES IN ECOLOGICAL RESEARCH VOL. 26
ISBN 0–12–013926–X

communities has long been of particular interest to evolutionists and ecologists alike. Noteworthy in this context are the tropical fish assemblages of the African Great Lakes—Malawi, Victoria and Tanganyika—which are the most speciose lacustrine fish faunas known (Fryer, 1991a,b,c). Among these lakes, the Tanganyikan fish fauna is taxonomically the most diverse, having more families and more endemic genera.

The fish faunas of all three lakes are characterized by their preponderance of species of the family Cichlidae (Teleostei: Perciformes), which have undergone an explosive speciation and have diversified to an extent unmatched in any other vertebrate family, and have done so independently in each lake. These species flocks are outstanding examples of evolution in isolation, and endemicity is so prevalent as to be almost the rule. Radiation has been both quantitative and qualitative, and such is the morphological diversity of these fishes that they occupy niches that one might expect to be occupied by fishes of different families. They display supralimital specializations (Myers, 1960), the endemic species transcending the familial limits of other cichlid faunas combined.

This chapter will focus on the cichlid fauna of Lake Tanganyika, the least studied of the African Great Lakes and about whose cichlid species surprisingly little is known. It will place particular emphasis on the faunas of the rocky shore habitats, and will explore how this great species diversity may have arisen. Alternative, but not mutually exclusive, evolutionary scenarios within the lake are described and related to the biology of these fishes. Several of the more parsimonious explanations as to how the extant species diversity is maintained are considered, and examples of resource partitioning among coexisting species along a variety of resources axes, the most likely mechanism of coexistence, are given. Several examples of the diversity of feeding and breeding habits seen both within groups and within species are described.

The cichlids of Lake Tanganyika offer splendid, perhaps unique, opportunities for studies of ecology, behaviour and evolution, yet there is a paucity of quantitative information concerning the ecology and evolution of its species flocks, such that our present level of knowledge is insufficient on which to form comprehensive theories regarding factors structuring these communities. The time is ripe for rigorous testing of ecological predictions. Several avenues of potential research interest are suggested. However, findings to date suggest that any rules that may exist concerning maintenance of species coexistence may be community- or even population-specific. Anthropogenic threats lend a sense of urgency to further research efforts on this fascinating ecosystem.

II. INTRODUCTION

A. Biodiversity in Lake Tanganyika

Lake Tanganyika has perhaps the highest diversity of species of any lake, harbouring almost 1300 species of vertebrate, invertebrate and plant (Coulter, 1991). The paucity of comprehensive studies precludes accurate comparison, but it seems that only Lake Baikal approaches this number. Much of the lake remains unexplored scientifically, and many groups of organisms are poorly known, so this number seems certain to increase with further study. The fauna of the lake shows enormous adaptive radiation, displayed by organisms of various groups—sponges, isopods, shrimps, crabs, gastropod molluscs, fishes and others. Many of these include endemic genera. The richness of these faunas is not simply a reflection of the richness of the faunas from which they are derived, and in most cases results from intralacustine speciation.

Many of the faunal groups show superficial similarities to marine forms: some of the cichlid fishes resemble the marine damselfishes both in gross morphology and breeding behaviour, two species of freshwater clupeid (*Limnothrissa* and *Stolothrissa*) are present, as are freshwater crabs and shrimps, a diverse fauna of gastropods (some of which show a remarkable "thalassoid" aspect), isopod crustaceans (which occur in no other African lake), endemic ostracods with close relatives in marine or brackish waters, and a freshwater jellyfish (*Limnocnida*). These similarities, noted among the few earliest collections made during the mid- to late 1800s, prompted speculation that Lake Tanganyika was once linked to the sea, and that the extant fauna was a relic of a Jurassic marine incursion. It was this possibility that provided the stimulus for the earliest scientific expeditions to the lake, in 1895 and 1899, the findings of which supported this theory (Moore, 1903). But subsequent geological and faunal studies of this and adjacent lakes all provided strong contradictory evidence against it (Cunnington, 1920). Indeed, excepting Brooks (1950), who briefly exhumed the marine origin hypothesis, later studies have all concluded that these faunas have originated largely by intralucustrine speciation, and not by multiple invasions, and firmly established their documented resemblances to be the result of convergent evolution, and not of any marine colonization.

B. Fishes in the Lake

The bulk of our knowledge of the composition of the fish fauna of the lake derives from the Belgian expeditions of 1945–1948 (Poll, 1946, 1948, 1949, 1952, 1956a,b, 1986), but it is only recently (1977–present) that any meaningful ecological surveys have been made, by a Japanese team working together with scientists from some of the riparian African countries (Kawanabe *et al.* 1992, 1994). It is the findings of this group, based mainly on SCUBA-assisted observations of the fishes in their natural surroundings, which provide the bulk of our knowledge of the behaviour and ecology of the lake's cichlids. The lack of previous ecological studies meant that much basic biological information had to be collected, and it is therefore only even more recently that these studies have graduated beyond the descriptive stage. Anecdotal examples are therefore numerous, but it should soon be possible to apply rigorous tests to predictions of ecological factors structuring these communities.

Lake Tanganyika possesses one of the most diverse lacustrine fish faunas known. Over 290 species and subspecies have been recorded in the lake (Coulter, 1991), a number which approaches the *c.* 350 species of fishes, both marine and freshwater, recorded for the whole of northern Europe (Wheeler, 1978). The lake contains small species flocks of bagrid and mochokid catfishes, mastacembelid eels, and centropomids (*Lates*: Nile perch). But it is the cichlid fauna that predominates. As in the other African Great Lakes, in Lake Tanganyika the cichlids have undergone explosive speciations, and a cichlid assemblage of almost 180 species is known today (Table 1). Only *c.* 10% of the shoreline has been explored for fishes (Brichard, 1989), so this number must be subject to change as knowledge increases.

C. The Cichlid Species Flocks of Lake Tanganyika

Almost 97% of the cichlid species inhabiting the lake proper are endemic (Poll, 1986). They have evolved from several ancestral lineages (Fryer & Iles, 1972), with at least eleven founder stocks in this species assemblage (Nishida, 1991), which includes mouthbrooders, as in other African lakes, and substrate spawners. The Tanganyikan cichlid assemblages differ from those of lakes Malawi and Victoria in greater differentiation of its cichlids, into tribes, twelve tribes recognized by Poll (1986), reflecting both the greater number of founder lineages and the great age of the lake. However, despite its greater age (Sturmbauer and Meyer, 1992, 1993) and the polyphyletic origins of its flocks, Tanganyika seems to have fewer cichlid species (*c.* 180), compared with 200–300 for Lake Victoria (Witte *et al.*, 1992) and 500–1000 species for Lake Malawi (Lewis *et. al.*, 1986). In all these cases, this remarkable diversity at the generic and species levels is the result of intralacustrine evolution; a phenomenon partly attributable to the large sizes, tropical location, and great ages of these lakes (Bootsma and Hecky, 1993).

Table 1
Taxonomic composition of the Tanganyikan cichlid fauna

Tribe BATHYBATINI	
Genus *Bathybates*	7 species
Genus *Hemibates*	1 species (+?)
Tribe CYPRICHROMINI	
Genus *Cyprichromis*	2 species (+?)
Genus *Paracyprichromis*	2 species (+?)
Tribe ECTODINI	
Genus *Asprotilapia*	1 species (+?)
Genus *Aulonocranus*	1 species (+?)
Genus *Callochromis*	4 species (+?)
Genus *Cardiopharynx*	1 species
Genus *Cunningtonia*	1 species
Genus *Cyathopharynx*	1 species (+?)
Genus *Ectodus*	1 species
Genus *Enantiopus*	1 species
Genus *Lestradea*	2 species
Genus *Microdontochromis*	1 species (+)
Genus *Opthalmotilapia*	4 species (+?)
Genus *Xenotilapia*	15 species (+)
Tribe ERETMODINI	
Genus *Eretmodus*	1 species (+?)
Genus *Spathodus*	2 species (+?)
Genus *Tanganicodus*	1 species
Tribe HAPLOCHROMINI	
Genus *Astatotilapia*	1 species (+?)
Genus *Astatoreochromis*	2 species
Genus *Ctenochromis*	1 species
Genus *Orthochromis*	1 species
Tribe LAMPROLOGINI	
Genus *Altolamprologus*	2 species (+?)
Genus *Chalinochromis*	2 species (+?)
Genus *Julidochromis*	5 species (+)
Genus *Lamprologus*	8 species
Genus *Lepidiolamprologus*	6 species (+)
Genus *Neolamprologus*	36 species (+)
Genus *Telmatochromis*	5 species (+?)
Tribe LIMNOCHROMINI	
Genus *Baileychromis*	1 species
Genus *Benthochromis*	2 species
Genus *Gnathochromis*	2 species
Genus *Greenwoodochromis*	2 species
Genus *Limnochromis*	3 species
Genus *Reganochromis*	1 species
Genus *Tangachromis*	1 species
Genus *Triglachromis*	1 species

Table 1—*cont.*

Tribe PERISSODINI	
Genus *Perissodus*	2 species
Genus *Plecodus*	4 species
Genus *Haplotaxodon*	1 species
Genus *Xenochromis*	2 species
Tribe TILAPINII	
Genus *Boulengerochromis*	1 species
Genus *Oreochromis*	4 species
Genus *Tilapia*	1 species
Tribe TROPHEINI	
Genus *Cyphotilapia*	1 species
Genus *Lobochilotes*	1 species
Genus *Limnotilapia*	2 species
Genus *Petrochromis*	6 species (+)
Genus *Pseudosimochromis*	1 species
Genus *Simochromis*	6 species (+?)
Genus *Tropheus*	6 species (+?)
Tribe TREMATOCARINI	
Genus *Telotrematocara*	1 species
Genus *Trematocara*	7 species (+?)
Tribe TYLOCHROMINI	
Genus *Tylochromis*	1 species

Key:
(+) indicates new species are currently under description.
(+?) indicates further species are believed to exist, but have yet
to be described.

III. SPECIATION IN LAKE TANGANYIKA

In exploring the origin and maintenance of biodiversity, it is unwise to
divorce ecology from evolutionary considerations, and an understanding of
the origin and maintenance of the speciose communities of endemic cichlids of Lake Tanganyika demands consideration of the mechanisms of
speciation. The generic diversity of the cichlid fishes in Lake Tanganyika is
partly a reflection of the diversity of the founding stocks, but the geology
and history of the lake basin, and the diversity of habitats this presents,
have also played a prominent role in their speciation.

A. The Lake Environment

1. Geological History

Lake Tanganyika is a Graben-type lake formed by rifting, and has a long and
complex geological history. The present-day lake consists of a series of three

major basins, separated at intervals in its history but now joined. The lake margins consist of steep rocky bottoms alternating with more gently sloping shores, usually of sand or mud substrates. Typical of a rift lake structure, the lake is long (c. 650 km), narrow (c. 90 km at its widest point), and extremely deep, reaching 1470 m, making it the second deepest lake in the world.

The lake is extremely old by geological standards, in terms of which most lakes are ephemeral. Although water levels showed great fluctuations during the pluvial and interpluvial periods that corresponded to the inter-glacial and glacial periods in the north, the great depth of the lake enabled the basin to retain water, and an aquatic fauna, during the arid interpluvial periods of the Pleistocene, which resulted in desiccation of shallower lakes elsewhere on the continent (Worthington, 1937). Also, its tropical location (3°–9°S) meant that it was not affected by glaciation during this epoch, although undoubtedly subject to changes in temperature. The antiquity of the lake is confirmed by the presence of lacustrine sediment layers up to 7 km deep, and its age is estimated as 9–12 m.y. (Tiercelin et al., 1988; Tiercelin and Mondeguer, 1991).

2. Hydrology and Limnology

As a water body, Lake Tanganyika presents an extremely stable hydrologi-cal environment. Fed by a catchment area of 220 000 km^2, the lake has a surface area of 32 600 km^2 and a volume of 18 900 km^3 (Gonfiantini et al., 1979), yet the outflow, via a single river, is a mere 2·7 km^3.y^{-1}. The lake is essentially a closed system, and the major loss of water takes place through evaporation (44 km^3.y^{-1}; Coulter and Spigel, 1991). This water budget and its large volume result in the lake having a flushing time of 7000 y and a water residence time of 440 y. The major roles of precipitation and evaporation mean water levels in the lake are very sensitive to climatic changes (Haber-yan and Hecky, 1987; Bootsma and Hecky, 1993), and large fluctuations in lake level have occurred even during the immediate past (Coulter, 1991). In closed lake systems, evaporation and outflow exceeding water input will result in a decrease in surface area. However, the steep sides of Lake Tanganyika mean that even a large drop in water level will result in only a small reduction in surface area: to achieve even a 10% decrease in surface area, Lake Tanganyika must drop 56 m (Bootsma and Hecky, 1993). These changes in water level are of immense evolutionary significance (Section III.B.1.c).

Deep, tropical lakes are particularly prone to oxygen depletion (Lewis, 1987), and it is not generally appreciated that not all of Lake Tanganyika is habitable. The lack of marked seasonality results in a permanent stratifica-tion, and the lake has a conspicuous thermocline and oxycline. The surface temperatures range from 23·3° to 29·5 °C; the hypolimnion remains at around 23 °C throughout the year. Although some mixing, due to wind-induced upwelling, does occur at the southern (upwind) end (Coulter and Spigel, 1991), for most of the year oxygen extends down to only 100 m in the northern basin, and to 200 m in the southern basin. Below these depths and into the profundal zone the waters are permanently deoxygenated, and therefore azoic. There is thus a relatively thin oxygenated layer overlying a vast, permanently oxygenless region. The steeply sloping rocky shores and this limit on habitable depth have important implications for faunal distribution. While life is essentially restricted to a thin layer at the surface, most cichlid species occur only over sandy or rocky substrates. For these, the habitable zone is a narrow ribbon running around the lake shoreline; the littoral (0–10 m), the sublittoral (10–14 m), and the benthic zone inhabited by fishes from 40 m to the oxygen limit (c. 70–240 m in the northern and southern basins and varying seasonally) (Coulter, 1991).

3. Habitat and Faunal Diversity

Lake Tanganyika has three major ecosystems: the pelagic, the deepwater and the littoral. The cichlids occur in all three, but predominate in the littoral/sublittoral zone. Little is known of the interactions between fishes of different zones. In other lake fish faunas, a variety of orders have diversified to utilize different niches, particularly feeding niches. Tanganyika also has a number of endemic non-cichlids (listed in Coulter, 1991), which interact with the cichlids, but it is the cichlids that have shown explosive radiations, such that they now utilize almost every available niche, as they have in the other African Great Lakes Malawi and Victoria.

(a) The pelagic zone. The extensive pelagic zone is dominated by the two endemic clupeids, *Limnothrissa miodon* (Boulenger) and *Stolothrissa tanganicae* Regan, and their centropomid predators, the four *Lates* species (*L. angustifrons* Boulenger, *L. mariae* Steindacher, *L. microlepis* Boulenger and *L. stappersi* (Boulenger)), but specialized pelagic cichlids are present, among which *Boulengerochromis microlepis* (Boulenger) and members of the genera *Hemibates* and *Bathybates* are predominant. Several members of the latter two genera also prey upon the clupeids, but are also themselves food for the larger *Lates* species.

(b) The deepwater zone. The lack of seasonal surface cooling and turnover means that the deeper water in Lake Tanganyika also remains warm; even water at around 200 m has a mean temperature of over 23 °C (Fryer, 1991a). Warmer water is less able to retain dissolved oxygen, and oxygen concentration decreases rapidly with depth—from 85% at 60 m to 10% at 100 m, 4% at 140 m and 2% at 170 m in the north basin. Together with pressure-imposed difficulties, these features make the deepwater zone the most specialized habitat in the lake. Tanganyika has an ecologically distinct and unique benthic fish community of phylogentically diverse species, nearly all endemic and including cichlids. They include shelf-dwelling and bathypelagic species, others characteristic of rocky bottoms, some remaining in deep water, others making diurnal or periodic migrations inshore (Coulter, 1991), where they may interact with the littoral fishes.

Since they live below the limit for SCUBA observations, very little is known of the biology of these species; often the sole information consists of absence/presence data from solitary gill net samplings. Below 100 m over muddy bottoms, the cichlid fauna is dominated by several species of *Xenotilapia*, *Limnochromis* and *Trematocara*. Other deepwater species, which appear to be less sensitive to substrate type and may be bathypelagic, include the predators *Haplotaxodon microlepis* Boulenger, *Bathybates ferox* Boulenger and *B. fasciatus* Boulenger, and the scale-eater *Xenochromis hecqui* Boulenger. These species, and *Gnathochromis permaxillaris* (David), are physiologically noteworthy in having been caught at depths (>120 m) where there was no recordable oxygen (Coulter 1966, 1968).

(c) The littoral zone. Nearly all cichlid species are restricted to the littoral/sublittoral zone. This zone presents an environment that is structurally complex on a wide variety of scales. On the largest scale, of hundreds of metres to kilometres, the major littoral habitats are of two main types: patches of rocky shoreline, which array themselves as discrete areas, separated from neighbouring rocky stretches by expanses of sandy shoreline. Some of the sandy shores are reed-fringed; most are bare. Other littoral habitats are few; estuarine regions, swampy areas, and shingle beaches cover all the main types. Only 3% of the lake bottom lies within the euphotic zone (Hecky and Kling, 1981), and littoral macrophyte vegetation, usually represented by *Vallisneria* or *Ceratophyllum*, is scarce, and discontinuously distributed. Macrophytes are absent from rocky shores. Below 20 m the bottom is generally of sand or mud and mollusc shells and, excepting underwater cliff faces, rocky areas are rare (Lowe-McConnell, 1987; Coulter, 1991). Both rocky and sandy shores show a certain degree of heterogeneity, presenting a fine-scale mosaic of habitats in what might superficially appear uniform regions.

Sandy bottoms may extend into muddy areas, to where species such as *Triglachromis otostigma* (Regan) and *G. permaxillaris* are restricted, or they may contain discrete, but often large, beds of empty gastropod mollusc shells. Except for differences in fine sediment loading, much of the sandy area seems to be homogeneous, although I know of no quantitative examination of this.

The rocky shore presents greater structural complexity, and there is often considerable vertical relief. Almost 65% of the Tanganyikan cichlid species are petrophilous and restricted to this habitat, and it is these diverse communities that have received the most attention. The varied geological formations provide a wide variety of crevices (used as spawning sites by many lamprologine species (Gashagaza, 1991), or as retreats), and of boulder and substrate sites (used as feeding (Hori *et al.*, 1983; Kuwamura, 1992; Kohda and Yanagisawa, 1992) or breeding sites (Kuwamura, 1992, Rossiter and Yamagishi, in press; Rossiter, 1994)). At the micro-scale there are marked structural differences between crevices and smooth bedrock, or between boulders and gravel or sand interfaces. Fine-scale partitioning of these features, for feeding sites or, more especially, breeding sites may be a major factor in permitting coexistence among the littoral cichlids.

(d) The littoral cichlids. Disjunct communities characterize different biotopes, and the change in faunas between rock and sandy zones is abrupt and striking (Table 2). The petrophilous cichlids form by far the largest group, comprising *c.* 130 of the almost 180 species known. These are strongly stenotopic, intimately associated with rocky shores and outcrops, and show a strong site fidelity. Both mouthbrooders and substrate-spawners (see below) are represented. It is among the petrophilous cichlids that the most spectacular examples of geographic variation in colour morph are known (Khoda *et al.*, in press).

There is an indistinct depth-related zonation of species. *Eretmodus*, *Spathodus* and *Tanganicodus* are goby-like cichlids, found only in the surf zone and at depths less than 2 m. Their stocky body and reduced swim bladder are adaptations to withstand buffeting by the surf. Overlapping with these species, and extending down to depths of *c.* 15 m is a large community of algivores, often with highly specialized mouthparts for grazing or browsing, and dominated by *Tropheus* and *Petrochromis*. The vertical distribution of some *Tropheus* species is mediated through competitive interactions (Kohda and Yanagisawa, 1992). From the surface to depths of 40 m or more, where rocks extend that deep, lives an 'infauna' of carnivorous lamprologine cichlids (*Neolamprologus*, *Altolamprologus*, *Lamprologus*, *Julidochromis*), all of which are substrate spawners, and most of which are solitary when not breeding. Within these depths another group of cichlids are "hoverers", living above the substrate surface. These may be either solitary (e.g. *Cunningtonia*, *Opthalmotilapia*, *Cyathopharynx*, *Cyphotilapia*) or gregarious, living in schools (e.g. *Neolamprologus brichardi* (Poll), and the two *Cyprichromis* species).

The cichlids of the sand zone form another distinct group. They tend to be ubiquitous around the lake shoreline, occurring in schools. They are generally pale coloured, and feed on particulate matter, sifted from the sand with spatulate teeth. Typical examples are *Xenotilapia sima*

Table 2

Species composition of cichlids at four quadrat areas in Lake Tanganyika

Species recorded	Location of quadrat (dominant substrate)			
	Wonzye (Sand)	Wonzye (Rock & rubble)	Kasenga (Boulder)	Mahali (Boulder & sand)
Tribe CYPRICHROMINI				
Cyprichromis leptosoma	0	0	417	0
Tribe ECTODINI				
Aulonocranus dewindti	0	157	99	1
Callochromis macrops	0	19	0	0
Cunningtonia longiventralis	0	2	0	0
Cyathopharynx furcifer	0	3	1	241
Enantiopus melanogenys	212	0	0	0
Grammatotria lemairei	14	0	0	2
Opthalmotilapia ventralis	0	5	48	0
Xenotilapia boulengeri	0	0	0	22
X. flavipinnis	11	8	4	0
X. longispinis	0	0	0	157
X. ochrogenys	14	0	0	65
X. sima	158	39	1	10
X. spilopterus	0	41	15	0
Tribe ERETMODINI				
Eretmodus cyanostictus	0	9	69	8
Tribe HAPLOCHROMINI				
Ctenochromis horei	7	4	4	1
Tribe LAMPROLOGINI				
Altolamprologus compressiceps	0	7	18	15
Chalinochromis brichardi	0	0	0	19
Julidochromis ornatus	0	34	87	0
J. "dark"	0	0	17	0
J. regani	0	0	0	83
Lamprologus callipterus	40	58	46	102
L. lemairii	0	31	21	23
L. ocellatus	25	0	0	0
L. ornatipinnis	11	0	0	0
Lepidiolamprologus attenuatus	0	0	21	58
L. cunningtoni	27	9	0	0
L. elongatus	0	9	15	69
L. nkambae	0	0	6	0
L. profundicola	1	11	5	9
Neolamprologus brevis	44	0	0	0
N. brichardi	0	0	0	2724
N. buescheri	0	0	4	0
N. caudopunctatus	0	6	13	196
N. cylindricus	0	4	8	0
N. fasciatus	0	27	134	23

Table 2—*cont.*

Species recorded	Location of quadrat (dominant substrate)			
	Wonzye (Sand)	Wonzye (Rock & rubble)	Kasenga (Boulder)	Mahali (Boulder & sand)
N. furcifer	0	5	7	12
N. meeli	53	0	0	0
N. modestus	0	0	0	36
N. mondabu	0	9	2	0
N. moorii	0	35	398	0
N. mustax	0	1	5	0
N. niger	0	0	0	20
N. prochilus	0	23	6	0
N. pulcher	0	21	116	0
N. savoryi	0	33	124	28
N. sexfasciatus	0	7	15	0
N. tetracanthus	0	9	2	0
N. toae	0	0	0	57
N. tretocephalus	0	0	0	28
Telmatochromis bifrenatus	0	0	0	106
T. dhonti	0	6	0	15
T. temporalis	0	9	12	909
T. vittatus	0	78	389	0
Tribe LIMNOCHROMINI				
Gnathochromis pfefferi	0	3	18	7
Tribe PERISSODINI				
Perissodus microlepis	7	18	37	0
Plecodus straeleni	0	2	6	12
P. paradoxus	0	0	0	27
Tribe TILAPINII				
Boulengerochromis microlepis	30	4	0	0
Oreochromis tanganicae	0	53	0	0
Tribe TROPHEINI				
Lobochilotes labiatus	0	11	17	59
Limnotilapia dardennii	0	3	4	0
Petrochromis famula	0	11	19	59
P. fasciolatus	0	60	10	10
P. macrognathus	0	9	14	0
P. orthognathus	0	7	0	98
P. polyodon	0	5	36	6
P. trewavasae	0	37	93	30
Pseudosimochromis curvifrons	0	3	5	0
Simochromis diagramma	0	22	95	4
S. marginatus	0	0	0	1
S. loocki	0	2	7	0
Tropheus moorii	0	82	173	91

Table 2—*cont.*

Species recorded	Location of quadrat (dominant substrate)			
	Wonzye (Sand)	Wonzye (Rock & rubble)	Kasenga (Boulder)	Mahali (Boulder & sand)
Tribe TYLOCHROMINI				
Tylochromis polylepis	0	2	0	0
Number of species recorded	15	51	47	41
Shannon diversity index	0·935	1·452	1·259	0·906
Evenness	0·795	0·851	0·753	0·562
Jaccard's similarity coefficient (upper triangle)				
Wonzye sand	*	0·138	0·107	0·143
Wonzye rubble	15·398	*	0·750	0·394
Kasenga boulder	3·999	43·348	*	0·397
Mahali boulder & sand	3·950	12·166	9·944	*
Percent similarity (lower triangle)				

Wonzye sand quadrat, 20 m × 10 m, Zambian shoreline (Southern depositional basin) (Rossiter, unpublished); Wonzye rubble quadrat, 20 × 15 m, Zambian shoreline (Rossiter, unpublished); Kasenga Rock quadrat, 10 m × 40 m, Zambian shoreline (Hori, Rossiter and Sato, unpublished); Mahali quadrat, 20 m × 20 m, Tanzanian shoreline (Kigoma depositional basin) (Hori, 1987). Data were standardized prior to calculation of similarity values. The two quadrats at Wonzye were located only 50 m apart, on a line perpendicular to the shoreline. The Kasenga quadrat was located 3 km away on the same, continuous shoreline. Differences in species composition between these three quadrats may be related to differences in habitat structure within each quadrat. The topographically complex rubble substrate shows the highest diversity, and the uniform sand substrate shows the lowest. The low similarity coefficient and percentage similarity values of the sand substrate community and the rocky substrate quadrats clearly show the marked differences in the cichlid communities of these two habitat types. The Mahali quadrat was located over 300 km away, in the Kigoma depositional basin, and differences between this and the other quadrats therefore also reflect differences in the geographical distribution patterns of species.

Boulenger and *Grammatotria lemairei* Boulenger. Little is known of the ecology and behaviour of this group, which is dominated by mouth-brooders.

Lastly, there is a distinct group of cichlids that are ostracophilous, occurring on the sand/mud floors but restricted to the often large beds of abandoned shells of the gastropods *Neothauma*, *Paramelania* and *Lavigeria*. These fishes are all substrate spawners, and use the empty shells for shelter and for breeding and rearing their young (see below).

B. Evolution of the Cichlid Fishes

The precise mechanisms of genetic isolations and subsequent speciation in lacustrine species flocks remain matters of controversy (Echelle and

Kornfield, 1984; Meyer, 1993). However, studies of intralacustrine speciation within the cichlids of the African Great Lakes have supported allopatric divergence hypotheses (Greenwood, 1974; Fryer and Iles, 1972; McKaye and Gray, 1984; Beadle, 1981), with most authorities invoking relatively small and spatially confined populations as a prerequisite for speciation (Fryer, 1977; Greenwood, 1984, 1991; Lowe-McConnell, 1969).

1. Evolutionary Mechanisms

(a) Macroallopatric evolution. A scenario of macroallopatric speciation, entailing separation of lake basins during periods of low lake levels, has been suggested by Greenwood (1974) and Beadle (1981), and is supported by gross, geographical differences in species distributions within the lake basin. Within the Tanganyikan cichlids several cases are known where distribution is restricted to either the north or the south end of the lake basin. Some separate at the species level and seem to occupy identical niches in their respective communities (e.g. *Lestradea perspicax* Poll, *Neolamprologus tretocephalus* (Boulenger) and *Callochromis pleurospilus* (Boulenger), found only in the north, and *Lestradea stappersi* (Poll), *Neolamprologus sexfasciatus* (Trewavas and Poll) and *Callochromis macrops* (Boulenger), found at the south), but some have no niche equivalents (e.g. the southern herbivorous and diurnal *Neolamprologus moorii* (Boulenger), and the carnivorous and noctural *N. toae* (Poll) of the north). Below the species level, evidence of macroallopatric evolutionary events is present in that many of those littoral species ubiquitous throughout the lake basin show distinct northern or southern colour morphs, for example, the northern form of *Cyphotilapia frontosa* (Boulenger) has six vertical bands, whereas the form inhabiting the middle and southern shoreline has seven bands. Similarly, the northern form of *Lobochilotes labiatus* (Boulenger) has distinct vertical banding, whereas in the southern form the banding is indistinct or absent.

Such broad-scale species and morph differences between the northern and southern basins are presumably a reflection of both the geological history of the lake and a lack of fish movements. Geological surveys have shown that the lake consists of three basins, once separate. Seismic data indicate that at about 200 000 to 75 000 y ago (Tiercelin and Mondeguer, 1991), the level of the lake dropped to 600 m below its present level, splitting the lake into three sublakes for several tens of thousands of years (Fig. 1). Within these refugia, allopatric evolution could have occurred within subpopulations. As water levels rose, the lakes rejoined, and presumably separately-evolved faunas were able to mix. However, the new species seem to have remained genetically isolated from their now-sympatric sister species, perhaps through evolution of sexually selected behavioural and colour differences (Dominey, 1984; West-Eberhard, 1983). With a decrease in water level, the basins were again separated, and evolution proceeded separately in each basin. Fluctuations in water level meant that the ancestral lakes, and their faunas, were united and separated repeatedly during the early history of the lake.

This complex interaction between physical and biotic factors produced a template for potential secondary, tertiary and further, evolutionary radiations, and ultimately contributed to the diversity of cichlid species present today. Although absolute ages of formation remain unknown, the most reliable estimate is that the three ancestral palaeolakes have remained connected as a single lake since the middle-upper Pleistocene (Scholtz and Rosendahl, 1988). But just why certain species, some of which are today stenotopic, were successful in invading and colonizing one or both of the other palaeolakes, and thus the entire present lake basin, while other species show a greatly restricted distribution, possibly even having remained within the limits of their ancestral palaeolake, remains a matter for conjecture. Presumably this is related to a complex of features, including a species' colonizing ability, its ecology, and the evolutionary pressures under which it arose.

(b) Microallopatric evolution. The differences between the species diversity and abundance of the complex cichlid communities of the rocky shores, and the spartan communities of the sandy zone (Table 2) suggest that habitat diversity and patchiness may have exerted a major role in shaping their composition and species diversity. Separation of basins may therefore be unnecessary to explain certain (perhaps many) speciations (Fryer, 1959a,b, 1969, 1972; Fryer and Iles, 1972).

Populations of fishes may have become isolated and morphologically divergent as once-continuous habitats became fragmented. This interpretation is supported by present-day distribution patterns. Most of the littoral cichlids, particularly the petrophilous species, are highly stenotopic, and the alternation of sand and rock patches around the shoreline of the lake results in a fragmentation of occupiable habitat. Together, these features act to cause a reduction in the rate of migration and gene flow between adjacent populations, and can potentially result in speciation, as evidenced by the disjunct distribution of species around the shoreline. But evolutionary radiation need not always reflect habitat diversity; *Xenotilapia* of the uniform sandy habitat, and *Bathybates* of the bathypelagic zone are both endemic genera showing extensive radiations.

Colour morphs. The strongest support for a microallopatric scenario comes from variations below the species level. There exist numerous races or varieties, often morphologically identical, and distinguishable only as colour morphs of certain species. Colour variation is often great, and may be

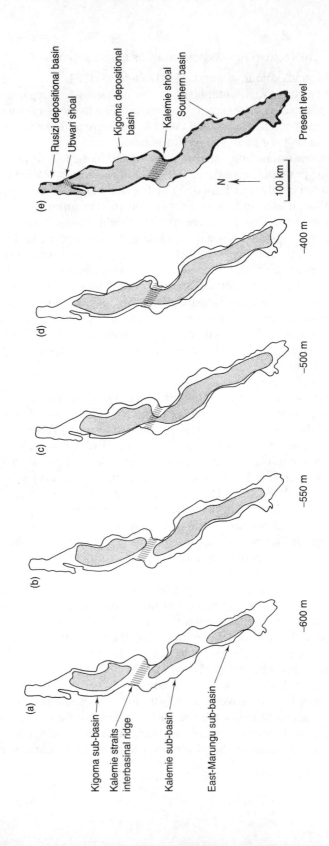

(a)

Kigoma sub-basin

Kalemie straits
interbasinal ridge

Kalemie sub-basin

East-Marungu sub-basin

−600 m

(b)

−550 m

(c)

−500 m

(d)

−400 m

(e)

Rusizi depositional basin

Ubwari shoal

Kigoma depositional
basin

Kalemie shoal

Southern basin

N

100 km

Present level

population-specific. Colour morphs are known within species of the mouth-brooding genera *Asprotilapia*, *Aulonocranus*, *Callochromis*, *Cyathophar-ynx*, *Cyphotilapia*, *Cyprichromis*, *Eretmodus*, *Lobochilotes*, *Opthalmotila-pia*, *Petrochromis*, *Spathodus*, *Tanganicodus*, *Tropheus* and *Xenotilapia*, and in the substrate spawners *Altolamprologus*, *Chalinochromis*, *Julidoch-romis*, *Lamprologus*, *Lepidiolamprologus*, *Neolamprologus* and *Telma-tochromis*. The best known of these is *Tropheus moorii* (Sturmbauer and Meyer, 1992), with over 50 distinctly coloured "races", some of which have overlapping distributions. Species-status has been assigned to certain morph variants in some genera (e.g. *Chalinochromis*, *Lepidiolamprologus*), but for the majority of the genera listed above, the taxonomic status of morphs is unexplored, although studies are in progress (Snoeks and Verheyen, 1994).

Male coloration is basically a sexually selected trait. The situation among these fishes supports the contention that such traits are more likely to undergo rapid divergence in allopatry than are non-sexually selected traits (West-Eberhard, 1983). In the substrate-spawners any population-specific colour variation is usually displayed by both sexes, whereas in the mouthbrooding genera it is usually the males only that show any colour differences, suggesting the importance of sexual selection in the mainly polygamous mating systems of this group (Rossiter, submitted; Rossiter and Yamagishi, in press). The wide range of colours that has evolved in

---

Fig. 1. Role of changes in water level in effecting macroallopatric speciation during the formation of Lake Tanganyika. Figure shows key stages in water levels during the formation of the present-day Lake Tanganyika. At (a), water level is 600 m below present level, and the lake consists of three separate sub-basins, within which faunas can evolve separately. At (b), water level has risen 50 m, and the Kalemie and East-Marungu sub-basins have fused, allowing the mixing of separately-evolved faunas. Evolution proceeds independently in the Kigoma sub-basin. At (c), water level has risen a further 50 m, and all three original basins, and their faunas, are now linked. The fauna of the Kigoma sub-basin can now mix with the fauna of the previously fused Kalemie and East-Marungu sub-basins. Geological records indicate several large increases and decreases in water levels to have occurred during the lake's history. Thus, at (b), differences of only ± 50 m, will result in faunas of two, or three sub-basins either being mixed or allowed to evolve separately. This series of isolations and reintroductions must have had major effects upon community composition and in promoting speciation. Faunas at the Ubwari and Kalemie shoals seem worthy of detailed investigation. From (c) to (e) major effects of water level changes are likely to have been influential in providing conditions conducive to microallopatric speciation. The distribution of major rocky (dark areas) and sand/swamp habitats around the present-day lake shoreline (e), indicate the presence of major geographical barriers to dispersal of stenotopic cichlid species. Investigation of cichlid faunas either side of such barriers also warrants attention. Lesser, yet effective, sand-beach barriers are also evident within many of the rocky stretches illustrated.

essentially identical environments is also in agreement with the observation that sexually selected traits are often arbitrary with respect to the physical environment (West-Eberhard, 1983). It remains unknown how females come to accept, and perhaps select for, the changes.

The distribution of colour morphs of a species are usually disjunct, but may overlap. In contrast to most other fish groups, in the cichlids range overlaps of such morphs or species are not generally accompanied by hybridization (Fryer, 1991b). Just how specific integrity is maintained remains to be explored, but subtle ecological and behavioural differences between overlapping colour morphs have been noted, for example the depth occupied and size of boulder used as a mating site by males of *Opthalmotilapia ventralis* (Boulenger) (Rossiter, unpublished). Females within this population were superficially identical, yet each individual selected only males of one morph with which to mate. Presumably, female selection is related to the complex courtship, involving male coloration and behavioural displays, but the potential role of pheromonal cues also deserves attention.

More challenging is the occurrence of a variety of male morphs within schools of the planktivorous *Cyprichromis leptosoma* (Boulenger). Males may be electric blue all over, blue with yellow fins, blue with yellow fins and tail, or tail alone, or bright banana yellow all over. Other morphs show a dark vertical stripe pattern underlying these colour variations. Assortative mating and phenotype of progeny seem obvious subjects for future investigation.

While research has been inadequate to establish conclusively whether any colour morph or sibling species combinations coexist without interbreeding, the spartan evidence available suggests that, as in Lake Malawi (Dominey, 1894; McKaye *et al.*, 1984), assortative mating and sexual selection may have acted to maintain divergent populations that were in sympatry.

Evidence that limitation in dispersal ability can translate into subsequent morphological divergence between populations due to a reduction in gene flow can only come from an analysis of interpopulation variability of species over similar geographic ranges and in similar habitats. However, for the petrophilous cichlids, their strong stenotopy and the discontinuous habitat distribution act to confound such analysis, and examination of the sand-dwelling and pelagic cichlids is more illuminating. At depths exceeding 20 m, the sand substrate extends uninterrupted for large sections of shoreline, and one would therefore expect less variation within sand-dwelling than rock-dwelling species, and similarly for species inhabiting the continuous pelagic zone. This is clearly the case. No colour morphs are known for any of the true pelagic species, and the only known instance where different colour morphs occur within a sand-dwelling species is in two adjacent populations of *Xenotilapia ochrogenys* (Boulenger), separated by an almost vertical rocky shoreline that extends down well into the anoxic zone (Rossiter, pers. obs.). There are no sand corridors linking these two populations, and the rocky area presents a barrier to their mixing. An examination of similarly physically separated populations of other sand-dwelling species around the lake shoreline will be revealing. Although a simple interpretation, and confounded by effects of breeding system and sexual selection, it seems clear that limitations to mixing of gene pools of adjacent populations, mediated through strong stenotopy and presence of other barriers to mixing, are of crucial importance in the evolution of colour morphs and, ultimately, of species.

(c) Importance of changes in lake level. Substrate type *per se* has played a major role in the evolution of the lake benthos (as clearly shown by Boss (1978) for molluscs). Availability of certain substrate types, in combination with physical processes, notably changes in lake level, has also facilitated intralacustrine allopatric speciation among the cichlids (Fryer, 1977). Migration rates across barriers are low, and through fragmentation of major habitats, changes in lake level offer abundant opportunities for a species to be broken up into many isolated populations. Historically, fluctuations in lake level must have united formerly isolated populations and fragmented formerly continuous ones. Such events must have had trenchant effects on the biogeographical distribution of many petrophilous cichlid species, and probably led to repeated series of widespread extinctions and fusions of formerly isolated populations. In the shallow areas of the lake, petrophilous species might well have disappeared, whereas at steeply shelving rocky shores, populations probably survived, but may have become divided into a series of population patches. Additionally, fluctuations in the level of the anoxic layer may have destroyed intervening populations and further disrupted genetic exchange between population patches. Repetition of this evolutionary process through repeated oscillations in lake level led to a "species pump", and hence to the present high diversity. Fluctuating lake levels may thus have promoted both micro- and macro-allopatric evolutionary events.

2. Phylogeny and Systematics

A sound taxonomic base and clear understanding of phylogenetic relationships is essential for any meaningful study of evolution and ecology. Hitherto, the phylogeny and systematics of these fishes have been approached from an exclusively morphological perspective, but this has recently been complemented by electrophoretic studies (Nishida, 1991) and the analysis of mitochondrial DNA (Sturmbauer and Meyer, 1992, 1993). These studies have produced phylogenetic branching patterns which differ from those derived exclusively from morphological criteria (Liem, 1981; Greenwood, 1983), and also differ slightly between each other. They are, however, unanimous in demonstrating the Tanganyikan assemblage to be polyphyletic, and in showing many Tanganyikan cichlids to be exceptionally closely related, despite having undergone major morphological diversifications. An example of the extreme morphological variation seen within one tribe, the Ectodini, is shown in Figure 2. Equally great intertribal variation in body shape is evident throughout the Tanganyikan cichlid assemblage. Differences in anatomy of the jaw apparatus are related to specialized feeding adaptations (Section V.B.1), but in the ectodine tribe the gross differences in body proportions can be categorized in terms of habitat occupied and mating system (Rossiter and Newlands, in prep.).

Although the Tanganyikan cichlids provide prime examples of evolutionary diversity within a family, and thus appear ideal material for studies of evolutionary mechanisms and speciation, in terms of phyletic interpretation they pose several challenges.

(i) The series of evolutionary events during the long history of the lake has acted to obscure taxonomic affinities within the lake cichlids, such that only one monophyletic genus is currently recognized—the seven *Bathybates* species (Poll, 1986)—although the genus *Perissodus* (Liem and Stewart, 1976, refuted by Poll 1986) and, at a higher level, the *Opthalmotilapia* assemblage (Liem, 1991) have also been suggested as showing monophyly.

(ii) There is an absence of intermediate morphotypes among the morphologically diverse Tanganyikan cichlid lineages and congruent synapomorphies are few, complicating investigation of evolutionary trends and relationships within and between groups. This absence of intermediate morphotypes has been interpreted as evidence of morphological stasis. The main lineages may have been formed shortly after colonization of the lake by several ancestors filling available niches, thereafter remaining virtually unchanged (Greenwood, 1984), perhaps through stabilizing selection as a consequence of species being a part of the tightly packed communities (Sturmbauer and Meyer, 1993).

(iii) Many of the groups present have no close relatives outside the lake basin. This, and the high degree of endemism, compounds the difficulty in recognizing affinities with faunas in other aquatic systems, notably those of adjacent rift lakes and those of the River Zaire watershed, and thus of constructing representative colonization hypotheses.

(iv) The long geological history and the likelihood of many series of invasions, speciations and extinctions make it probable that various

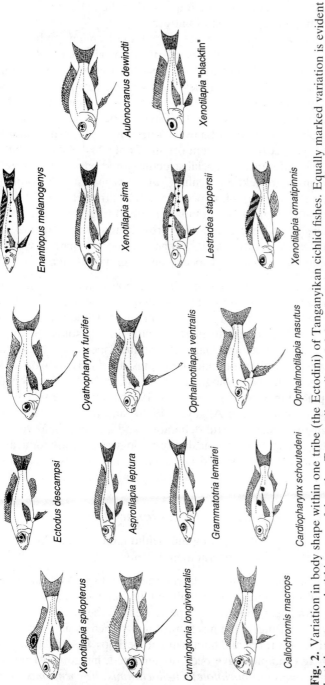

Fig. 2. Variation in body shape within one tribe (the Ectodini) of Tanganyikan cichlid fishes. Equally marked variation is evident both between and within many of the other Tanganyikan tribes. Adult males are shown, and species are not drawn to scale. Names in inverted commas indicate manuscript names for undescribed species. In this tribe, body shape differences are attributable to both mating system and habitat occupied (Rossiter and Newlands, in prep.)

Enantiopus melanogenys

Aulonocranus dewindti

Xenotilapia sima

Xenotilapia "blackfin"

Lestradea stappersii

Xenotilapia ornatipinnis

Cyathopharynx furcifer

Opthalmotilapia ventralis

Opthalmotilapia nasutus

Ectodus descampsi

Asprotilapia leptura

Grammatotria lemairei

Cardiopharynx schoutedeni

Xenotilapia spilopterus

Cunningtonia longiventralis

Callochromis macrops

degrees of relatedness will be found within the major groups present, complicating interpretation of patterns.

(v) Speciation events may have been rapid and frequent. While the permanency and great age of this environment allowed populations to become established, and to persist, such an extended time-frame is not crucial for speciation in these fishes. Some cichlids of Lake Nabugabo, separated from Lake Victoria by a sand spit, have achieved specific distinctness within 4000 y (Greenwood, 1965), and even more remarkably, some of the Malawi colour morphs, "species" *sensu* Ribbink, have an extremely recent origin of only 200–300 y (Owen *et al.*, 1990). The differences in social structure and body sizes between two populations of the lekking cichlid *Cyathopharynx furcifer* (Boulenger) (Rossiter and Yamagishi, in press) and the extremely rapid changes in these features in one population following predator removal (Rossiter, submitted) (Table 3), provide tentative evidence of the potential rapidity with which changes can occur; changes which may be but the first step toward speciation of this population.

Much remains to be learned of the taxonomy and evolution of the Tanganyikan cichlids. Given the large number of what are currently tentatively described as colour morphs, the number of cichlid species recorded from the lake seems certain to increase with further study. Sampling to date has been inadequate and many undescribed species probably remain to be discovered, and taxonomic anomalies are likely to exist. This is especially true in the lamprologines, the most speciose fish taxon in the lake. Long placed in a single genus (Poll, 1956b), they have recently been split into five (Colombe and Allgayer, 1985), and then four (Poll, 1986), genera. Their exact status is still debatable, and a comprehensive analysis of this group, using both classical morphological and anatomical techniques, and modern isozyme and DNA methodologies, is now underway (M.L.J. Stiassny, pers. comm.; A. Meyer, pers. comm).

3. Ecological Pressures and Evolutionary Responses

Differences and similarities among the cichlid faunas of the African Great Lakes can be attributed to several main factors.

(a) Physical similarities in habitat. Given their great ages and similarities of their rocky littoral zones (almost absent in Lake Victoria), initial and subsequent ecological conditions experienced by the founding species in lakes Malawi and Tanganyika have presumably been similar, and similarities among their littoral cichlid faunas might, perhaps, be expected. To date, several striking examples have been documented, including (Tanganyika/ Malawi, respectively) *Lobochilotes/Cheilochromis, Petrochromis/Petrotila-*

Table 3
Differences in the social structure of males of *Cyathopharynx furcifer* in relation to absence or presence of predators

Northern site	Southern site (1988)	Southern site (1991)
Two types of mound owner a) mound built on large rocks b) mound built on sand	One type of mound owner a) mound built on large rocks No mounds built on sand	Two types of mound owner a) mound built on large rocks b) mound built on sand
Floating male population: Sub-adults only	Sub-adults, smaller, and spent males	Sub-adults only
Mound occupied for: a) *c.* 21 days, 24 h/day b) unknown, but less, 12 h/day	a) <13 days, 24 h/day b) Absent	a) 12–17 days, 24 h/day b) 7–15 days
Other features: Kapenta* absent Predators absent	Kapenta abundant Large resident population of predatory *Boulengerochromis* and *Hydrocynus*	Kapenta absent Few *Boulengerochromis* No *Hydrocynus*
	1988 ——————→ 1991	
	Courtship and mating often interrupted by: a) predators in the vicinity b) intrusions by conspecific males	Mass and height of nest mound increased Decrease in frequency of intrusions Increase in time spent in courtship Increase in no. of female visits/h Increase in no. eggs laid/ mound/h Increase in frequency and duration of feeding bouts Increase in male : female size ratio

* Kapenta is the local Lungu name for the sardines *Limnothrissa* and *Stolothrissa*.
Data from Rossiter & Yamagishi (in press) and Rossiter (submitted).

pia and *Tropheus*/*Labeotropheus*. But similarities are not restricted to the fishes of the rocky littoral. Close physical and morphological similarities exist between *Grammatotria*/*Lethrinops* of the sandy bottom, *Bathybates*/*Rhamphochromis* of the pelagic zone, and *Trematocara*/*Trematocranus* of the deepwater zone. In almost all cases, it has not been established whether these similarities have resulted from convergent evolution or are indicative of a possible common phyletic ancestry. While undoubtedly not closely related, the cichlid fishes of these two lakes are likely to have some constituent of the genome in common (Fryer, 1991a). Parallel evolution in each of the lakes may then have acted to produce remarkably similar solutions to similar ecological problems, as it has for *Bathybates* and *Rhamphochromis* (Stiassny, 1981), the only pair of genera so far investigated.

(b) Differences in fauna. Differences between the African Great Lakes in terms of major faunal components, notably feeding groups, seem also to have played a role in adaptive radiation and speciation of the cichlids. The openwater survey currently underway in Lake Malawi is pointing to the probable evolutionary importance of such faunal differences, for example, in Lake Malawi the absence of endemic planktivorous clupeids has allowed for the evolution of an extensive openwater planktivorous cichlid fauna, Utaka; a guild poorly represented in Lake Tanganyika.

(c) The role of predation. Although a topic of some contention, the role of predation pressure in evolution within the cichlids of the African Great Lakes should also be considered here. Predation has been proposed as either retarding (Worthington, 1937, 1940; Jackson, 1961) or promoting speciation (Fryer, 1959b, 1965). In Lake Tanganyika, the presence of large visually dependent predators such as *Boulengerochromis*, *Hydrocynus*, and four species of *Lates*, none of which are naturally present in Lake Malawi or Victoria, would appear to be acting in both directions (Lowe-McConnell, 1975, 1987). In the pelagic and openwater community "cover" is obtained by schooling, and predation pressure appears to lead to uniformity in appearance and behaviour, thereby reducing species diversity. Conversely, amongst the rock-dwelling cichlids, predation promotes speciation by increasing isolation between adjacent populations. Through selective reduction of the population sizes of the most abundant species, predators at the littoral/sublittoral zone may also increase species diversity by reducing competition and permitting coexistence (Glasser, 1979).

The presence of large predators and a rich cichlid diversity in Lake Tanganyika contrasts markedly with the drastic and rapid reduction of the Victorian haplochromine cichlids following the introduction of a single predator, *Lates niloticus*, into the lake in about 1960 (Barel *et al.*, 1985; Witte *et al.*, 1992). In Lake Tanganyika, the cichlid species coevolved with the predator species and have developed elaborate anti-predator reactions, whereas in lakes Malawi and Victoria, where large predatory fishes were absent, the anti-predator reactions are less well developed (Coulter, 1991). Beyond this, the absence of detailed comparative studies necessitates some speculation. The absence of an extensive rocky shoreline in Lake Victoria means the degree of substrate complexity is extremely low, offering little natural cover to the cichlid fauna. Lake Victoria thus seems populated by a relatively naive prey population of cichlids, with no natural refuge from a novel predator. Lastly, the endemic clupeids of Lake Tanganyika may have provided an alternative, abundant food source, helping to reduce direct pressure, and eventual feeding specialization on the cichlids by the Tanganyikan large predatory species. Clearly, further research is required to clarify these points (Lowe-McConnell, 1994a; Worthington and Lowe-McConnell, 1994).

IV. BREEDING HABITS

A. Behaviour, Strategies and Tactics

The breeding behaviour of cichlids involves territoriality, the development of breeding colours, and various permutations of mating system (Kuwamura, 1986a; Rossiter, 1994). It also involves the use of colour in species recognition, and in courtship behaviour, which is often complex and entails species-specific stereotyped movements. There is increasing evidence of the importance of behavioural mechanisms, especially sexually- and socially-selected traits (West-Eberhard, 1983; Dominey, 1984). Rapidly evolving differences in these might arise as reproductive barriers during periods of isolation, facilitating the explosive pace of speciation among these fishes.

Although the cichlids show great variation in form, feeding habits and habitat preference, they are characterized by one feature that separates them from most other fish families—the parental care of eggs and young. Two breeding strategies are prevalent: mouthbrooding, where the eggs and young are incubated and protected in the buccal cavity; and substrate spawning, where eggs are laid on a solid substrate and guarded. Tanganyika is unique among the African Great Lakes in possessing indigenous representatives of both groups; the endemic cichlids of Lakes Malawi and Victoria are all maternal mouthbrooders (Fryer and Iles, 1972; Greenwood, 1974, 1984; Ribbink et al., 1983). These two groups show fundamental differences in their breeding mode and ecology; substrate spawning usually involves strong territoriality, both parents, a firm pair bonding, and guarding of eggs and young (Nagoshi, 1987; Gashagaza and Nagoshi, 1988; Nakano and Nagoshi, 1990), whereas mouthbrooding is usually the prerogative of the female only (Yanagisawa, 1991; Rossiter, submitted; Rossiter and Yamagishi, in press) with the male showing no parental care (but see Kuwamura, 1986a, 1988; Kuwamura et al., 1989; Yanagisawa, 1986; Yanagisawa et al., submitted).

B. Substrate Spawners

One third of Tanganyika's cichlid fauna are substrate spawners. The majority of these are species in the various rock-dwelling lamprologine genera, including several species that spawn in empty gastropod shells on muddy or sand bottoms. The ubiquitous *Boulengerochromis* is also a substrate spawner.

1. Diversity within the Group

Substrate spawners are generally monogamous, and are usually territorial and sedentary; egg laying and raising of young occur at the spawning site. However, variations in mating system and parental care are known; monogamy, polygyny, and even a helper system of brood care have been described (Taborsky and Limberger, 1981; Nagoshi, 1983, 1985; Yanagisawa, 1987). Some species guard the brood for several months (Nagoshi, 1985, 1987; Rossiter, 1991; Nakai, 1993); others, for less than 20 days, until the end of the larval phase, e.g. *Lamprologus lemairii* Boulenger and *Altolamprologus compressiceps* (Boulenger) (Nakai, *et al.*, 1990; Gashagaza, 1991). During their development, *Boulengerochromis* moves its young between pits dug in the sand (Kuwamura, 1986b). Although regarded as a more primitive mode of breeding than mouthbrooding, the abundance of substrate spawning species on the rocky shores attests to its effectiveness and belies this simplistic interpretation of the respective adaptive advantages of these two breeding strategies.

2. Shell Brooders

The lamprologines are exclusively substrate spawners, and depend on a solid substratum for breeding. A consequence of the greater representation of molluscs in Lake Tanganyika is the presence of extensive beds of gastropod shells at from 10–35 m depth on gently sloping bottoms. Several lamprologine species have extended their range to include the open, sandy bottom by occupying these gastropod shells. The 17 or so species to have done so utilize these shells as spawning sites and/or refuges in which they hold territories and raise their offspring.

The abundance of abandoned shells, and the small size necessary to reside in them, has had profound evolutionary effects, and the smallest cichlids known are in this group e.g. *Neolamprologus brevis* (Boulenger), total length (TL) 50 mm max., and *N. multifasciatus* (Boulenger), TL 25 mm max. Where the shells accumulate in piles, *N. multifasciatus* sometimes occurs at high densities, and displays little intraspecific aggression. When threatened, adults retreat not into the empty shells, but to the interstices beneath the shell bed, perhaps indicating the evolutionary origin of shell-brooding behaviour. In the majority of species, adults are too small to transport shells, but they can manipulate them by undermining. *N. multifasciatus* does this such that each family lives in a small (*c.* 15 cm diameter) circular depression within the shell bed. Several of the species living over sand bury their shells, leaving just the openings exposed. The smaller species are monogamous, and both individuals may occupy a shell together. Medium-sized species are polygynous, and the male and each of its mates has its own separate shell within the male's territory. The larger species are also polygynous, but only females occupy shells, the males being too large. The small size of individuals and the diversity of breeding systems seen within this group has provided suitable material for laboratory investigations of breeding behaviour and of the adaptive advantages of a given breeding tactic (Haussknecht and Kuenzer, 1991; Walter, 1991; Walter and Trillmich, 1994).

In *Lamprologus callipterus* (Boulenger) the male is three to five times larger than the female, and holds a territory of *c.* 50 cm diameter, containing a number of empty shells of the whelk-like genus, *Neothauma*. Shells are carried to the territory by the owner male, and deposited in a flat, circular cluster. Within a cluster, many of the shells are occupied by individual females, which are the sole caretakers of their young. Neighbouring territory owners steal shells from one another, and by sequestering resources necessary for reproduction (shells), certain males can achieve a disproportionately high percentage of matings. In this harem system, larger males have an adaptive advantage, both in terms of collecting/ stealing shells for their own nest, and in defending their nests against other males (Sato, 1994). Females have had to retain their small size to permit entry into the shell. Interestingly, the distribution of gastropod genera around the lake shore is not uniform, and this has implications for female size (see later).

3. Spawning Synchroneity

Substrate spawning species breed several times during the year, with no clear seasonality. Although *Boulengerochromis* is suggested as having three spawning peaks in a year (Matthes, 1961), it shows no such pattern along the Zambian shoreline (pers. obs.), and is most definitely not semelparous (cf. Konings, 1988). Recent theoretical studies have suggested that predation may select for or against reproductive synchrony, depending on the ecological setting. In some lamprologines, avoidance of nocturnal predation, likely from catfishes, has been accomplished through a lunar spawning synchroneity (Nakai *et al.*, 1990; Rossiter, 1991). Lunar spawning species show clear benefits in terms of brood survivorship, particularly during the vulnerable,

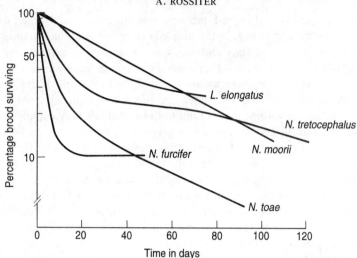

Fig. 3. Decrease in brood size during period of parental care in five lamprologine cichlid fishes. *Neolamprologus moorii*[1] and *Lepidiolamprologus elongatus*[2] are monogamous species with a lunar spawning synchroneity. Broods of these two species show a relatively constant mortality rate, with little mortality during the early, vulnerable, developmental stages. *N. tretocephalus*[2] and *N. toae*[2] are monogamous species but do not have a lunar spawning cycle, and both show a high mortality during the early brood stages. *N. furcifer*[3] is a polygynous species with no lunar spawning cycle. The extremely high mortality seen during the early brood stages of this species reflects both the lack of lunar-spawning cycle and consequential vulnerability to nocturnal predators, and suggests the effectiveness, in the other species, of constant biparental care in reducing brood losses through predation.

[1] Rossiter, 1991; [2] Nagoshi, 1985, 1987; [3] Yanagisawa, 1987.

early brood stages, over those showing no synchroneity (Rossiter, 1991) (Fig. 3). Species with a lunar-spawning synchroneity show Type 2 (Pearl, 1928) mortality curves, with a relatively uniform rate of brood reduction. Species with no lunar synchroneity show Type 3 curves, with significant mortality during the early brood stages, when eggs and larvae are clumped and relatively immobile and particularly susceptible to predation. Nakai *et al.* (1990) proposed that only species with more open, and hence less defensible, spawning sites, show a lunar spawning synchroneity, but their inclusion of the shell-brooder *L. callipterus* as a lunar spawner suggests this hypothesis to be in need of revision.

4. Helpers

Among the smaller substrate-spawners, six species, of two closely-related genera, have been found to have non-breeding helpers (*Neolamprologus brichardi, N. pulcher* (Trewavas and Poll), *N. savoryi* (Poll), *Julidochromis marlieri* Poll, *J. ornatus* Boulenger, *J. regani* Poll), where conspecifics other than the parents participate in various aspects of brood care (Taborsky and Limberger, 1981). The tasks may be shared unequally between family members, according to size and status. These family groups form closely integrated units, where breeders recognize their helpers individually (Hert, 1985; Taborsky, 1985, 1987). In *N. brichardi*, families generally consist of the parents and several broods of their offspring, all defending a common territory. The helpers' defence is mainly (83%) against competitors for spawning sites (crevices), rather than against predators on the eggs and young (Taborsky, 1984), suggesting the importance of species-specific spatial partitioning of the rocky substratum.

C. Mouthbrooders

1. Diversity within the Group

Mouthbrooding cichlids have a *k*-selected life history, geared towards the production of precocial young and characterized by the production of relatively few, large-yolked eggs, a reduced larval stage, larger offspring size at independence, and a high parental investment in a long period of intensive brood care in the buccal cavity. In Lake Tanganyika, most mouthbrooders are polygynous and perform maternal mouthbrooding, where the female takes up the eggs in the mouth and broods them with no parental assistance from the male (Kuwamura, 1986a; Yanagisawa & Nishida, 1991; Ochi, 1993; Rossiter, 1994; Rossiter & Yamagishi, in press). However, a wide variety of alternative parental care patterns are also present in the Tanganyikan mouthbrooders. Representatives of several genera show monogamy and share the mouthbrooding role by shifting parental duties during the brooding cycle (*Xenotilapia flavipinnis* Poll: Yanagisawa, 1986; *Haplotaxodon microlepis*: Kuwamura, 1988; *Tanganicodus irsacae* Poll: Kuwamura *et al.*, 1989; *Microdontochromis* sp., Yanagisawa *et al.*, submitted; *X. spilopterus* Poll and Stewart: Rossiter, in prep.; *X.* "Chituta Bay yellow": Rossiter, unpublished). This contrasts sharply with the cichlids of Lakes Victoria and Malawi, which show exclusively maternal mouthbrooding and no pair formation.

Females of several species are known to browse during buccal incubation, to feed themselves and/or their brood; *Tropheus duboisi* Marlier and *T. moorii* Boulenger (Yanagisawa and Sato, 1990; Yanagisawa, 1993), *Cyphotilapia frontosa* (Yanagisawa and Ochi, 1991) and *Microdontochromis* sp. (Yanagisawa *et al.*, submitted). Mortality of eggs and young while under parental care can be almost negligible (Rossiter and Yamagishi, in press).

2. The Special Case of the Perissodini

Members of the scale-eating tribe Perissodini show a breeding system that is considered intermediate between substrate brooding and the more-advanced types of mouthbrooding. The best studied species is *Perissodus microlepis*. Adults are monogamous. Semi-adhesive eggs are deposited on the substrate and, following fertilization, are then picked up by the female and brooded in the mouth. When they reach a size too large to be brooded in the mouth of the female parent, both parents guard the free-swimming juveniles (Yanagisawa and Nshombo, 1983; Nshombo *et al.*, 1985; Kuwamura, 1986a).

The pair formation and biparental guarding seen in several Tanganyikan mouthbrooders (Yanagisawa, 1986; Kuwamura *et al.*, 1989) appears to be intermediate between the supposedly ancestral breeding behaviour expressed in the Perissodini and several other derived patterns found in the other lineages of Lake Tanganyika cichlids. But while an accurate estimate of the phylogenetic position of the Perissodini would seem fundamental to an interpretation of the evolution of the different patterns of mouthbrooding (Sturmbauer and Meyer, 1993), the wide variety of parental patterns present today may actually have evolved in some of the ancestral stocks before they invaded the lake: representatives of most types of parental care seen in the lake are present in the cichlids of African rivers. This dilemma clearly needs to be resolved.

3. Midwater Spawning

In the Cyprichromini, a transition from spawning in a territory held over a hard substrate (*Paracyprichromis*) to spawning in a three-dimensional, mobile territory held in midwater (*Cyprichromis*) has occurred. In both these cases females release eggs while in a head-down position, and catch the eggs in the mouth as they fall. *P. brienni* (Poll) occasionally allows eggs to reach the substrate before taking them into the mouth. In *Cyprichromis*, fertilization occurs directly in the mouth of the female, but in *Paracyprichromis* fertilization can also occur in the water column or on the rock surface. These tactics seem to represent different stages of an evolutionary trend within this tribe toward emancipation from the substrate.

4. Lekking Behaviour

Several species of mouthbrooder show lekking behaviour, where males congregate in arenas, which females visit for mating purposes. Most of these species utilize the substate to construct structures within or upon which mating takes place.

Rossiter (1994) examined the social structure and the mating success of males at a lek of the maternal mouthbrooder *Aulonocranus dewindti* Boulenger. Males build elaborate mounds (mating craters), each located within a territory, whose sole function is to attract females for mating. Distinct inter-population differences in social structure were evident. In one, geographically isolated, population, social structure was very complex, with a variety of alternative mating tactics (Fig. 4). In addition to 'normal'

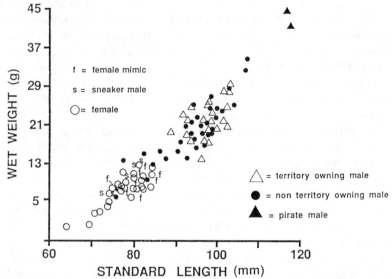

Fig. 4. The complex social structure within one population of the lek-forming cichlid *Aulonocranus dewindti*, where severe resource limitation has resulted in the evolution of a variety of alternative male mating tactics. Several of these tactics involve both morphological and behavioural adaptations. Males in 'normal' populations show territory-owning or non territory-owning mating tactics only (Rossiter, 1994, submitted).

territory-owning, males within this population showed female mimicry, sneaking behaviour, satellite behaviour, and piracy (*sensu* van den Berghe, 1988; Taborsky, 1994). These tactics entailed both behavioural and morphological variation being shown by non crater-owning males (Rossiter, 1994). In other populations examined, social structure was much simpler. In the isolated population, severe resource limitation, *viz* availability of sites and materials suitable for crater construction, and commensurately greater pressure for the non crater-owning male population to achieve fitness seems to have resulted in the evolution of the variety of male mating tactics seen (Rossiter, submitted).

Inter-population differences in social structure are also evident in another

218 A. ROSSITER

lekking species, *Cyathopharynx furcifer* (Section VI.D.2). Typical of lek mating systems, in both these species the distribution of male mating success showed a strong skew.

D. Commensalism and Parasitism

As in Lake Malawi (Ribbink *et al.*, 1980, 1981), provision of parental care toward foreign young, both conspecific and heterospecific, has been observed. Experimental removal of one individual of a brooding pair of *Perissodus microlepis* Boulenger (unusually, a biparental mouthbrooder), resulted in the remaining parent transporting its young and leaving them in the care of another pair of *P. microlepis*. The species can thus behave as a "cuckoo" (Yanagisawa, 1985).

Parental care shown toward the young of other cichlid species is also known. Again, a similar phenomenon occurs in Lake Malawi (Ribbink *et al.*, 1980), but as substrate spawners are absent from this lake, both host and foreign species are always mouthbrooders. In Lake Tanganyika, host species are nearly always substrate spawners, and foreign young may be either substrate spawners or mouthbrooders. Examples include juveniles of the mouthbrooders *Microdontochromis* sp. (Yanagisawa *et al.*, submitted), *X. spilopterus* Poll and Stewart, *P. microlepis* and *Cyprichromis* mingling with guarded broods of the piscivorous substrate spawner *Lepidiolamprologus elongatus* (Boulenger) (Rossiter, unpublished). Although not a true mouthbrooder, in addition to hosting conspecifics, *P. microlepis* sometimes hosts foreign young, e.g. *Haplotaxodon microlepis* Boulenger, but does so only when buccal incubation is over and the young are being guarded in typical substrate spawner fashion (Ochi *et al.*, 1995).

The planktonic food source of the juveniles of these species is superabundant, and the presence of foreign juveniles would seem unlikely to result in resource depression to the host's brood. This exploitation of the parental behaviour of another conspecific or species is therefore perhaps best interpreted as a combination of a facultative brood commensalism and the selfish herd effect (Hamilton, 1971); foreign young gain parental protection, and the increased numbers in the brood reduce the likelihood of predation for individuals of both foreign and host progeny.

A similar interpretation may be valid for the case of parasitism of the nest of *L. callipterus*. The number of shells within a nest exceeds the number of conspecific females at that nest, suggesting shell number, or nest size, to be a signal of a male's status. Many of these vacant shells are occupied by other shell brooding species; *N. brevis*, *N. fasciatus*, *N. caudopunctatus*, *T. vittatus* and *T. dhonti* have all been observed breeding in *L. callipterus* nests. Females of *N. brevis* and *N. fasciatus* are often numerous within the nest. *N. caudopunctatus* occurs less frequently. These species utilize both the shell resource, and the guarding ability of the much larger *L. callipterus* male. In these three species males are also present, but their primary purpose would seem to be to guard their mates against mating with conspecifics. The two *Telmatochromis* species occur infrequently, but generally comprise brooding females only. In this situation they are totally dependent on the parental protection of the *L. callipterus* male. These females also feed on the eggs and young of other females breeding in the same nest, and their presence can be described as brood parasitic.

Although mouthbrooding is the most advanced form of parental care seen in these fishes, several mouthbrooding species are parasitized by the mochokid catfish, *Synodontis multipunctatus* (Sands, 1983), which introduces its eggs into the buccal cavity of brooding females, where the catfish young hatch and feed on the eggs and larvae of the mouthbrooder, eventually destroying the entire brood. This phenomenon has been recorded at one location in the lake (Sato, 1986), where females of six species were found to be parasitized, in one species at an incidence of 15%. However, the phenomenon seems to occur only infrequently, at best in other areas. Brichard (1989) reported it as being extremely rare along the Burundi shoreline, and along the Zambia shoreline, only one individual (*Enantiopus melanogenys*) (Boulenger) among several thousand brooding females of a wide variety of species examined was found to be parasitized (Rossiter, pers. obs.; T. Veall, pers. comm.; R. Bills, pers. comm.), despite the large numbers of *S. multipunctatus* present. The catfish would appear not to be an obligate parasite, and the phenomenon deserves detailed examination.

E. Effects of Breeding Mode on Speciation

Small population and clutch size and spatially confined populations with limited dispersal abilities will increase the probability of genetic isolation and, consequently, of speciation. The fundamental differences between mouthbrooders and substrate spawners may influence the respective dispersal abilities of mouthbrooding or substrate spawning species, and result in different degrees of isolation and, consequently, of divergence, but the contention that brooding capability is important in effecting speciation in lacustrine organisms (Cohen and Johnson, 1987) is not supported by a comparison of the degree of intraspecific morphological or colour variation within members of these two groups.

For the substrate spawners, new populations cannot be established by a single founder female. Conversely, for the mouthbrooders, chance dispersal of a single brooding female which enables her to overcome a dispersal barrier, can lead to the founding of a new population. As such, the potential for gene pool dilution between populations of mouthbrooding species would seem greater, and that for variation within species, smaller, than that of the substratum spawning (mainly lamprologine) species. Yet, in Lake Tanganyika, the converse appears true, suggesting alternative factors to be important. Evidence from other groups (the species flocks of Hawaiian drosophilid flies (Carson & Kaneshiro, 1976) and the endemic cyprinid fish flock of Lake Lanao (Kornfield and Carpenter, 1984) indicates that brood care plays an equivocal role in effecting speciation.

V. SPECIES DIVERSITY AND COEXISTENCE WITHIN THE ROCKY SHORE COMMUNITIES

Whereas the offshore fish communities are characterized by having a rather simple structure dominated by the two clupeids and several centropomids, the rocky shore communities show a complex structure and great species diversity (Lowe-McConnell, 1987; Coulter, 1991). Cichlids dominate, and may be present at densities exceeding 5000 individuals per 20×20 m (Hori *et al.*, 1983). This great species diversity challenges one of ecology's basic assumptions—that competition is more severe between closely related species than between distantly or unrelated species, and thus, under stable conditions, closely related species cannot coexist indefinitely.

A. Mechanisms Promoting High Species Diversity and Coexistence

In the following section, I explore several of the ways in which coexistence may have been achieved in the cichlid assemblages found around the lake shore, mainly in terms of alternative hypotheses forwarded to explain the high species diversity often found in other tropical ecosystems.

1. Community Instability

Where a community is unstable or is subjected to repeated physical disturbances, the resultant suppression of competitive interactions may mean that niche differentiation is unnecessary (e.g. Levin and Paine, 1974; Abele, 1976; Sousa, 1984; Dethier, 1984).

The physical environment of Lake Tanganyika shows a remarkable stability. Lake Tanganyika has existed for several million years, and parallels the oceans in the predictability of habitat. The great depth and volume of the lake buffers it against short-term environmental perturbations (Section III.A). Seasonal variations are minimal, and major perturbations resulting from violent storms, and consequent damage and change to community and habitat structure (so important in shaping marine and riverine assemblages), are absent. Similarly, studies of species composition within communities at several parts of the lake have suggested a remarkable degree of community stability (Hori, 1991, see below). The community instability hypothesis is inapplicable to the Tanganyikan cichlid fauna. Indeed, the converse seems more appropriate, in that the stable environment has provided a foundation upon which specialization has been selected for and can persist.

There are five major ways in which coexisting cichlid species can subdivide resources; they can differ in where they live, in what, where and how they eat, and when they are active. Ecological differences along each of these niche axes should reduce competition and facilitate coexistence.

2. Within-patch Habitat Structure

Theoretical considerations have shown that the coexistence of similar species is possible in a patchy environment (Slatkin, 1974; Chesson, 1981; Hanski, 1987), and that patch quality, rather than the number of different kinds of patch in the habitat, is the key factor to permitting coexistence (Levins, 1979).

Because they are strongly stenotopic, of small size, have life spans of several years, and may occupy the same territory for extended periods (Yanagisawa and Nishida, 1991; Rossiter, 1994), the littoral cichlids will be most influenced by small-scale spatial differences. The complex topo-graphical structure of the rocky shore makes it spatially a highly hetero-geneous habitat. This is a feature which may lead to the finer partitioning of resources—be they spatial or trophic—and hence to a subtle diversification of various facets of ecology and behaviour, facilitating microgeographical isolation and promoting speciation and coexistence. As many species live within the rock interstices, habitat diversity is also important in considering community or assemblage composition, both on a size and species basis. Smaller individuals may take refuge in small holes in which they are inac-cessible to larger predators or conspecifics. If the cost of this relationship is low, it may be uneconomical for the territory owner to exclude the smaller individual(s) from its territory. This may partly explain the high densities of fishes seen over rock rubble habitats. Species may also show some degree of selection in terms of crevice size used as breeding holes (Gashagaza, 1991, see below). Differences in patch structure probably play an important role in the maintenance of diversity in the rocky shore communities.

3. Diel Separation

Diel separation is expected to be most evident in time of feeding, especially in habitats supporting structure-oriented fishes which tend to forage in restricted areas, primarily on plankters. The rapid renewal of this resource makes diel separation in food use a viable approach to resource partitioning, and the only known examples of diel partitioning within members of the rocky littoral community do indeed occur within the planktivorous cichlids.

Preliminary studies of four coexisting species of the tribe Cyprichromini suggest a diel partitioning of feeding times; *Paracyprichromis* feeds more at dawn and dusk, whereas *Cyprichromis* feeds through the day (Rossiter, pers. obs.). Gashagaza (1988) identified a diel feeding pattern in the planktivorous *N. brichardi*, which he related to predictable differences in plankter density mediated through biotic (nocturnal migration of plankters) and physical (afternoon winds causing an onshore current) processes. This diel pattern in feeding seems to be functional purely in terms of optimization of food intake, and is not related to competitive related processes.

The impact of predation in restricting diel differences in feeding activity of fishes may also be important in the rocky littoral environment; evidence to date suggests that, among the cichlids of the rocky littoral, only *Neolamprologus toae* (Poll) and *Lamprologus lemairii* Boulenger are nocturnal feeders. All other species appear inactive at nighttime (pers. obs.; T. Veall, pers. comm.). The bathybenthic *Trematocara* species undertake nocturnal migrations to shallow water, following the slope of the bottom (Poll, 1956b). These movements are probably for the purpose of feeding in the sublittoral and littoral zones during darkness, when the resident cichlid community is inactive and when darkness provides a measure of protection against predators.

Approached from a theoretical perspective, unlike trophic or habitat partitioning, diel separation would be expected to be significantly more important in less related species, and may thus reflect historical effects, rather than coevolution within a particular community. This argument may well be relevant to the case of *Trematocara* and their nocturnal invasion of the habitats of the lamprologines and other groups, but it clearly does not apply to the coexisting, closely-related cyprichromines.

4. Partitioning along Trophic and Spatial Axes

The dominant view has been that tropical assemblages are in equilibrium, and that tropical species, both aquatic and terrestrial, are more specialized ecologically than those in sub-tropical and temperate climes. This view emphasizes ways of efficient partitioning of resources, facilitated through a reduction of niche breadth, which permits less resource overlap among species. In tightly packed, diverse tropical communities there will be a trend for resource partitioning, detectable as small, clear differences in ecological properties, to develop (Pianka, 1973; Schoener, 1974).

Resources used by fish have traditionally been examined along two major resource axes: food and space. However, our knowledge of the manner in which component species within the Tanganyikan assemblages utilize either resource is rather limited. Food utilization is summarized in the next section.

B. Foods and Feeding

Studies of feeding habits have dominated ecological research in Lake Tanganyika. Detailed underwater studies of foraging methods and success have revealed intricate, and often complicated, interactions among heterospecific individuals. These are described below. What general results have emerged (Yamaoka, 1991) have paralleled those found for the cichlids of Lake Malawi (Fryer, 1959a; Ribbink *et al.*, 1983; Reinthal, 1989, 1990). The cichlids of Lake Tanganyika display an apparent paradox in that while they show specialized feeding morphologies and occupy very narrow trophic niches, they have retained the ability to exploit a wide range of food resources, e.g. the algivorous *T. moorii* sometimes feeds in midwater on pelagic larvae of the clupeid *Limnothrissa*, or on eggs and larvae of other cichlid species. As also in the other African Great Lakes, this combination of specialization and flexibility in cichlids may have contributed to their evolutionary and ecological success (Liem, 1980a,b; Ribbink *et al.*, 1983).

1. Feeding Specialization and Trophic Adaptations

Feeding opportunities have affected the pattern of evolutionary development, with the main morphological trends being towards trophic specialization. Food types are very diverse, reflecting the great diversity of algae and invertebrates (and fish for piscivores), and the cichlid fishes show clear segregation patterns in food utilization. Twelve feeding guilds have been identified among the lake's cichlids (Hori, 1987), and are represented in all rocky shore communities thus far examined (Hori, 1991). Reduction of niche breadth is achieved through specialization, which results in splitting of niches to create specialists that are more efficient than a generalist in utilizing a given resource, and will therefore ultimately replace it. These specializations can be both behavioural and morphological (Yamaoka, 1991).

The pronounced morphological specialization of the feeding apparatus of these fishes is a feature attributed to a predisposition toward evolutionary plasticity (Liem, 1974, 1978). The highly plastic pharyngeal arch and tooth structures form the "bauplan" upon which variation, and ultimately, speciation, could be built (Liem and Stewart, 1976). Cichlids have enjoyed a particular freedom in the use of their oral jaws because peculiarities in their pharyngeal evolution have resulted in a pharyngeal mill, which performs the duties of mastication and manipulation of food toward the gut. The sole responsibility of the protrusible jaws is food collection, and they have undergone remarkable adaptive changes, enabling these fishes selectively to exploit almost all imaginable food sources in a wide range of habitats: detritus, algae, plankters, insects, sponges, and other fishes—whether adults, or eggs or juveniles—and even fish scales, are all selectively fed upon by various species. Feeding behaviour and dental morphology of the algal-feeding cichlid guild of Lake Tanganyika is discussed in detail by Yamaoka (1982, 1983, 1985, 1987) and Yamaoka et al. (1986). Younger fishes show less specialization in their diet (Nshombo et al., 1985; Gashagaza and Nagoshi, 1986), which may reduce competition with adults.

However, despite feeding specialization, there is often a considerable overlap in food resources utilized by members of the same guild, for example, the trophic differences observed among several species of benthivore which coexist locally can be extremely small indeed, as are differences within co-occurring groups of lamprologine, ectodine, and shell-dwelling cichlids (Table 4). In general, the similarities in food taken by members of the same feeding guild are such that the differences observed in randomly selected subsets of species of similar size within the assemblage in these localities strongly counter the conventional niche argument. Some co-occurring, closely-related species actually show complete overlap in the foods taken (Rossiter, submitted).

2. Co-occurrence and Food Overlap

Where species feed on the same food, it may be regarded as a sharing of abundant resources rather that as an overlap of diet, inferring competition. Both of the only examples known support this interpretation: the scale-eaters (see V.B.2.e) and the plankter feeders.

(a) Abundant resources—an example from the plankter feeders. As in Lake Malawi (Fryer, 1959a; Fryer and Iles, 1972), plankter feeding has been accompanied by the adoption of a schooling (sensu Pitcher, 1983) habit. The specialist planktivores Benthochromis, Microdontochromis and Cyprichromis all school. Other species that regularly forage, solitarily or in pairs, on benthic materials may show opportunistic plankter feeding. At times of food abundance, territoriality breaks down and schooling occurs while plankter feeding. At such times, species may even form mixed schools, e.g. Neolamprologus caudopunctatus (Poll) and Xenotilapia spilopterus, although foods taken are identical. When plankters are few, territoriality reasserts itself, and schools break up.

A more extreme example, where the species are very closely related, is seen in the Cyprichromini. This tribe is composed of four species, equally divided between two genera, Cyprichromis and Paracyprichromis. All species feed on zooplankters in the water column, and all are restricted to

Examples of food overlap within three groups of closely related, coexisting Tanganyikan cichlids: (a) the Ectodini, (b) the shell-brooding Lamprologini, (c) the epibenthophagous Lamprologini

(a)	A. dewindti	Cy. furcifer	Ca. macrops	O. ventralis	X. spilopterus
A. dewindti	*				
Cy. furcifer	0·397	*			
Ca. macrops	0·421	0·855	*		
O. ventralis	0·549	0·727	0·691	*	
X. spilopterus	0·459	0·662	0·517	0·886	*

(b)	N. multifasciatus	N. meeli	N. ornatipinnis	L. callipterus	N. brevis	N. ocellatus	N. fasciatus
N. multifasciatus	*						
N. meeli	0·322	*					
N. ornatipinnis	0·554	0·697	*				
L. callipterus	0·268	0·916	0·612	*			
N. brevis	0·636	0·438	0·731	0·430	*		
N. ocellatus	0·564	0·452	0·922	0·366	0·844	*	
N. fasciatus	0·018	0·059	0·039	0·437	0·169	0·023	*

(c)	A. compressiceps	L. callipterus	N. fasciatus	N. cylindricus	N. prochilus	N. furcifer	N. mondabu	N. savoryi	N. pulcher	N. sexfasciatus
A. compressiceps	*									
L. callipterus	0·962	*								
N. fasciatus	0·176	0·473	*							
N. cylindricus	0·964	0·906	0·078	*						
N. prochilus	0·906	0·902	0·448	0·893	*					
N. furcifer	0·889	0·875	0·203	0·955	0·952	*				
N. mondabu	0·226	0·257	0·129	0·492	0·611	0·642	*			
N. savoryi	0·272	0·267	0·016	0·451	0·511	0·486	0·000	*		
N. pulcher	0·126	0·167	0·167	0·183	0·242	0·167	0·000	0·672	*	
N. sexfasciatus	0·813	0·783	0·111	0·911	0·903	0·963	0·000	0·531	0·198	*

Overlap indices were calculated using Pianka's overlap index. In all groups, extensive food overlap is evident between several of the component species. Coexistence between these species cannot be explained in terms of simple partitioning along the dietary axis.

vertical rock faces in deeper (>15 m) water. All species are gregarious. At some sites all four species can be found coexisting, yet food types taken are identical (Rossiter, in prep). Feeding methods were also closely similar. Although some partitioning of depth and feeding time was detected, this was insufficient to explain the coexistence of all four species. In this instance, the occurrence of several sympatric species showing nearly identical food specialization would not be expected if the specializations had evolved as a means of partitioning the food resources available. Instead, it suggests that those food items specialized upon are in abundant supply. Studies are now underway to evaluate this possibility.

(b) Differences in the method of resource use. Some coexisting species use the same resource in a stable, non-competitive fashion through differences in the method of resource use. This has been most clearly demonstrated by the study of Hori (1983) on a group of lamprologine species coexisting at a rocky shore. Species belonged to one of three tropic guilds: plankter feeders, piscivores, and zoobenthos feeders. Within each guild, component members showed clear segregation along at least one of several axes. The two plankter feeders, *N. savoryi* and *N. brichardi* were spatially segregated. Among the zoobenthophagous species, segregation occurred along a variety of axes; *N. toae* was the only nocturnal feeder, *N. tretocephalus* fed largely on gastropod molluscs, but the other species in this guild ate basically similar food items. Species within the piscivore guild also ate similar food items. Where great overlap in food items occurred, species-specific differences in characteristic and elaborate hunting techniques were evident in both guilds. Some segregation of microhabitat was also detected. *N. leleupi* (Poll) and *N. furcifer* (Boulenger) fed on the rock substrate, the former feeding at horizontal faces and the latter at vertical faces, *L. callipterus* and *A. compressiceps* fed over rubble, and *L. modestus* (Boulenger) fed over sand. All three members of the diurnal piscivore guild, *Lepidiolamprologus elongatus*, *L. profundicola*, and *N. fasciatus*, prey upon small fishes and all hunt by stealth, stalking the prey item and then darting in from a distance. However, these species show subtle behavioural differences in the technique of approaching the prey: *L. elongatus* searches for prey from a position immediately above the substrate, and darts downward to capture it; *L. profundicola* (Poll) weaves through rocks and boulders and darts horizontally, and *N. fasciatus* (Boulenger) approaches prey from midwater, and darts vertically downward (Hori, 1983, 1987). Hence, species may utilize the same food items, but in a different manner, so that partitioning of the food may take place through different capture or handling techniques.

(c) Facultative commensalism within feeding guilds. The benthivores. Where members of the same guild utilize the same resource, such differences may have led to the evolution of facultative commensalism. A typical example

concerns the benthivores *Lobochilotes labiatus*, which feeds on aquatic insect larvae, *Lamprologus callipterus*, a shrimp and chironomid larvae eater, and *Gnathochromis pfefferi* (Boulenger), a shrimp eater. These species sometimes forage in a mixed school. In this situation the latter two species gain a feeding advantage because *L. labiatus* disturbs substrate and sediment while foraging, exposing otherwise inaccessible prey items (Hori, 1987). This phenomenon has recently been explored in greater depth by Yuma (1993, 1994).

The algivores. Many algivorous species establish feeding territories that are maintained against both conspecifics and other algivorous food competitors. Traditional ecological thought has emphasized how territorial herbivores usually attempt to exclude from their feeding areas any feeding competitors, but relationships where different species of territorial herbivore that share and defend a common food resource are known (Takamura, 1983, 1984).

Individuals of the genera *Petrochromis* and *Tropheus* are numerically dominant among the algivorous cichlid guild of Lake Tanganyika. Members of these two genera are aggressive to all other members of the algivore guild, and together effectively monopolize much of the algal resource. Although highly aggressive towards all other algivorous species, two members of these genera, *P. polyodon* Boulenger and *T. moorii*, share grazing sites and feeding times. The two species have a different dentition, and this is reflected in their feeding on different fractions of the algal resource. *P. polyodon* is adapted for combing unicellular algae from the rock surface, whereas the dentition of *T. moorii* is adapted for scraping filamentous algae. The feeding actions of *P. polyodon* result in the removal of sand and silt particles from the rock surface. These detritus-free conditions appear favourable to *T. moorii* when scraping filamentous algae, presumably in terms of increased feeding efficiency, and this species selectively feeds on algal areas previously browsed by *P. polyodon*. Although this relationship was described as symbiotic (Takamura, 1983), it is difficult to envisage the benefits, if any, accrued to *P. polyodon*.

(d) Benefits to heterospecifics. Although the small differences in food types in the above cases could be invoked as evidence of micro-partitioning, this explanation is not applicable to species that feed exclusively on an identical resource. Two examples are known where foraging success increased in the presence of a nearby heterospecific individual feeding on the same food items; the piscivores and the scale-eaters.

The lamprologine piscivores. As noted above, the three diurnal predators of small fishes coexisting on a rocky shore showed species-specific differences in hunting technique. Each of the three species enjoys a higher hunting success when a heterospecific individual is nearby than when hunting solitarily (Hori, 1987). When the nearby individual is a conspecific, no increase in hunting efficiency is seen. This strongly suggests that prey fishes are both aware of the different hunting techniques of these predators, and able to respond appropriately to the first predator species they detect. This seems to reduce their awareness of other predators, allowing an alternative attack method to be successful.

The perissodine scale-eaters. Members of the tribe Perissodini are lepido-phagous, feeding almost exclusively on fish scales (Nshombo, 1994a,b). Two of these species, Perissodus microlepis Boulenger and Plecodus strae-leni Poll, are common along rocky shores. In both species, target fishes are approached from behind to within a metre or so, and then attacked in a swift lunge. Attacks are always sublethal; the hook-like dentition is used to rip several scales from the body of unwary target fish before the attacker flees (Nshombo et al., 1985). Although foods are identical, species-specific differences exist in the mode of hunting, which may reduce competition. As with the lamprologine piscivores, these slight differences in hunting technique may prove mutually beneficial. For the scale-eaters, the close proximity (within 50 cm) of a heterospecific scale-eater increases hunting success (Hori, 1987). Again it seems the differences in hunting technique serve to confuse the target individual, which directs its attention toward avoiding attack by one scale-eating species at a given time, increasing the likelihood of successful attack by the other species.

(e) Intra-population differences in hunting tactics. Dichromatism in pisci-vores and scale eaters. Many predatory species show evidence of crypsis, as a camouflage for hunting. Remarkably, many of these species also show a dichromatism, where individually distinct and stable pale and dark colour morphs occur within the same population (Yanigisawa et al., 1990; Nshombo, 1994b; Mboko and Kohda, in press). Dichromatism has been described in 19 species of carnivorous cichlid, all of which hunt highly mobile prey (Kohda and Hori, 1993). Detailed studies of one such species, L. profundicola, showed that the dark form targets prey from a shaded position under rocks, whereas the light form does so from well-illuminated locations in midwater (Kohda and Hori, 1993; Kohda, 1994). Dichroma-tism here functions to increase hunting possibilities within a population, thereby optimizing intra-population foraging efficiency.

Individually distinct hunting repertoires. The advantage accrued to co-existing individuals which utilize different hunting techniques has been fully exploited by L. profundicola. In one population, within each morph-type, individuals showed a further specialization in hunting tactics (Kohda, 1994). These feeding specializations were consistent and are learned be-haviours. Through utilization of an individually distinct hunting repertoire, individuals in this population reduce direct intraspecific competition. The differences may also act to confuse prey, resulting in a within-population advantage to conspecifics similar to that described for the piscivorous lamprologines and scale-eaters. Clearly, detailed underwater studies of individual fishes are revealing facets of the ecology of these communities that would otherwise have remained undetected.

Morphologically determined attack direction in a scale-eater. Interactions between predators and prey can occur that are neither strictly successful or unsuccessful for either the predator or its prey. Intra-population variations in the outcomes of such sublethal attacks may provide opportunities for understanding how different combinations of traits affect the outcome of interactions between predator and prey. A classic example of this is seen in the laterality of the jaw structure of one scale-eater, which provides an example of a purely frequency-dependent selection. Individuals of *P. microlepis* show either a right- (dextral) or left- (sinistral) handedness in the direction of mouth-opening. Target fishes are always attacked from behind, and the laterality of the mouth-opening means that sinistral individuals can only attack the right flank of their prey, and *vice versa* (Hori, 1993). The ratio of handedness in populations of *P. microlepis* oscillates around unity over a 5-year period. This fluctuation is maintained through frequency-dependent natural selection, a differential hunting success between the two types of handedness (Takahashi and Hori, 1994). At any given time, target fishes will be more alert to attacks to their, say, left flank by the more abundant dextral individuals, giving the rarer sinistral individuals a selective advantage (Hori, 1993). The period of the oscillation is believed to be due to a time lag between the effects of differential hunting success being expressed as an increased fecundity, and the recruitment of these individuals into the population (Matsuda *et al.*, 1993, 1994).

Interactions such as these, which consistently result in only sublethal damage to the victim but provide some nutrition to the attacker, may be consequences of a diffuse coevolution between attacker and victim. The scale-eaters are one of the oldest lineages in Lake Tanganyika, and a coevolution of increased hunting stealth and specialized dental apparatus on the part of the scale-eater, and an alertness by the prey species, is likely to have been undergone for millions of years. Scales are readily regenerated by preyed upon fish, and constitute a readily renewable resource (Rossiter, 1993). The abundance of the scale-eaters throughout the rocky shores speaks volumes of the efficiency of this mode of resource utilization.

C. Partitioning of Spatial Resources

Among the rock habitat specialists, spatial partitioning appears important in promoting the maintenance of high species diversity. However, there exists a high within-habitat structural diversity. Some species are predominantly specialists, and these finely partition the resources available. It follows that examination of their resource use should reveal that most species will be specialized with respect to either within- or between-habitat differences, and that they should be specialized on different resources and exhibit little overlap in their use of these resources. It is important here to emphasize the importance of integrating several resource axes when considering species coexistence. To date this question has been addressed in only three of the component groups of the eulittoral cichlid assemblage, a group of members of the tribe Ectodini, and in two groups of lamprologines.

1. Mating Sites of the Ectodine Cichlids

Rossiter (submitted) examined the mating habits of several species of Ectodini coexisting at the southern shores of the lake. In all species, during breeding, males establish a territory, from which all other conspecific males were excluded. Within the territory is located a mating crater or a spawning site. Territories are always discrete intraspecifically, but often show extensive, and sometimes complete, overlap interspecifically. Studies of this group revealed limited segregation along a depth gradient, and little or no segregation of food. What was evident was an interspecific partitioning of size and location of rocks used as nest sites, and of nest materials (sand, silt and gravel) used in nest construction. This fine-scale partitioning may facilitate maximum utilization of available space, thereby maximizing the species packing of members of this tribe within a given area.

2. Breeding Sites of the Lamprologine Cichlids

The sedentary habits of the petrophilous cichlids are conducive to the development of narrow habitat requirements, especially if one considers the diverse array of potential habitats provided on the rocky littoral. As habitat specialists the eulittoral cichlids can finely partition the living space available to them as long as each species uses a slightly different type of space. Gashagaza (1991) examined the utilization of microhabitats for breeding purposes among 12 lamprologine species coexisting at a rocky shore, and found clear differences in the characteristics of the breeding site used by each species. For hole-nesting species, the dimensions of the entrance to the hole were different between species, while for species using rock surfaces or concavities, a combination of the incline, shape, and configuration of the spawning site was important. Unlike food resources, partitioning of breeding sites in this group appears absolute: there is no evidence of sharing or utilization of the same type of breeding site by different species. Gashagaza suggests this fine-scale partitioning to have been achieved through interspecific competitive processes, and that partitioning of breeding sites can determine the species diversity within a given area.

3. Partitioning within the Shell-brooders

However, just as feeding habits of coexisting food specialists may coincide, so the habitat requirements of habitat specialists may overlap greatly. As with food resources, those species with highly specialized requirements often coexist with other species showing the same specialization, and show segregation on an alternative axis. The clearest example of this comes from the shell-brooding cichlids. All species utilize abandoned gastropod shells as a spawning substrate, and some live within the shell, emerging only to feed and court. One would therefore expect competition to be particularly intense within this group. As predicted, segregation occurs along other axes, reducing competition and permitting coexistence in these fishes (Rossiter, unpublished) (Fig. 5).

D. Partitioning within Communities and Effects of Relatedness

The occurrence of interspecific competition and competitive exclusion within these communities makes it of theoretical interest to determine the types and importance of different mechanisms that maintain their species composition. The traditional view has emphasized the minimization of interspecific competition through resource partitioning. However, this view has come to be challenged by studies which have found that behaviour indicative of interspecific interference competition commonly occurs, and that groups of species with similar, sometimes identical, resource requirements often co-occur.

Resource partitioning studies generally deal with assemblages of closely-related species or assemblages united by some common resource requirement. The complex evolutionary history has resulted in various degrees of relatedness among the groups present, but taxonomically related species assemblages are common. Findings to date suggest the degree of relatedness of component species within these assemblages to have had an insignificant effect on ecological separation for both congeneric-confamilial and confamilial-conordinal species pairs, although less related pairs show some evidence of greater separation in resource (food) use (Hori, 1987). Instead, interspecific utilization of a common food resource appears to be rather common, and such relationships may contribute significantly to maintaining high species diversity at the within-habitat scale. But before we can establish whether they are important in maintaining diversity on a large scale, it will

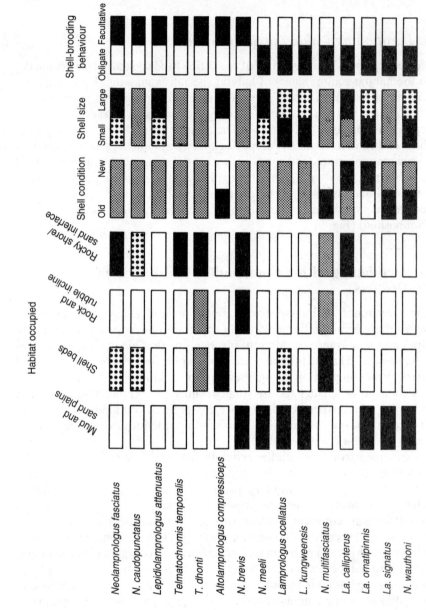

be necessary to determine the degree to which these relationships are obligatory, rather than facultative, for the dependent species.

Substantial niche separation along one or more axes has been demonstrated for the majority of pairs of sympatric cichlid species, but empirical shortcomings make it difficult to argue convincingly that any one particular resource dimension is more important. These increasingly detailed studies do, however, demonstrate how structured these littoral cichlid communities are, and suggest that differences are not exclusively food based, but entail differences in utilization of space for feeding and breeding.

Yet it is not necessary for the equilibrium view that species finely partition the resources available. Theoretical studies indicate that more overlap in resource requirements may be tolerated in a community if coexisting species have leptokurtic resource utilization curves (Roughgarden, 1974), or if predation is more intense (Roughgarden and Feldman, 1975). No field studies in Lake Tanganyika have addressed either of these possibilities.

In drawing conclusions from such studies, it must be acknowledged that, despite evidence of partitioning for space, and other resources, coexistence within these communities may well be explicable solely in terms of factors

---

Fig. 5. Resource and habitat partitioning within the shell-brooding cichlid fishes of Lake Tanganyika. Habitats were divided into four categories, as shown. All shell-brooding species are members of the tribe Lamprologini, and are substrate spawners. Many of the obligate shell-dwellers are restricted to the mud and sand plains, where their small size enables them to utilize shells as shelters and spawning sites, in an otherwise unoccupiable habitat. This move away from the ancestral rocky shore habitat has been accompanied by the evolution of obligate shell-brooding behaviour. Excepting *N. multifasciatus*, all species found at the rocky shore/sand interface occur within the nest of *L. callipterus*, either as facultative commensals or as brood parasites. Shell condition was categorized according to the degree of damage to the shell and the amount of secondary calcification present. Shell size was based on the size range of *Neothauma* shells found around the Zambian shoreline. Although they all utilize an identical resource as a spawning site, none of the above species show complete overlap in all the various factors examined. Dietary differences are also evident between some species (Table 4), but are insufficient to explain coexistence between many species combinations. Data from Rossiter (in prep.), T. Veall (pers. comm.) and R. Bills (pers. comm.)

Key

■ Common, where present

▨ Moderately abundant, where present

▦ Rare, where present

□ Absent

which are habitat- or even community-specific. It would be instructive to compare assemblages from other parts of the lake, but studies are too few, such that comparative quantitative data are conspicuous by their absence. It should also be acknowledged that comparisons of niche overlap between assemblages of different taxonomic structure will be compromised by effects of phylogenetic inertia. The control (or awareness) of biases due to historical effects or sampling design have yet to be addressed, and together with a rigorous experimental approach, will be important components of future studies of resource partitioning in these assemblages.

VI. COMMUNITY COMPOSITION AND SPECIES DIVERSITY

Although differences among species in their use of space resources have been documented, it remains to be demonstrated that these differences exist because of a competitively directed process of niche diversification. They may well exist simply because the species are different in their ecology as well as their morphology. Were this so, we would expect more complex habitats, which offer more fine-partitionable crevice spaces, to contain greater species diversity than other, topographically simpler habitats. On a broad scale this is true. Differences between sand and rocky substrates are marked, and a comparison between rocky habitats differing in topographical complexity also showed some, but lesser, differences (Rossiter, submitted) (Table 2). Further studies, incorporating several sets of adjacent sites with different topography, are needed to clarify this point. Preliminary analysis of available data in terms of alternative species abundance models supports a broken-stick scenario (MacArthur, 1957; Sugihara, 1980), which could have arisen through either an ecological or evolutionary mechanism (Rossiter, submitted).

A. Importance of Size Differences for Species Diversity and Speciation

Size can be as important an influence as age on the demography of a population and on interactions between species. Although nearly all the littoral cichlids are <18 cm TL adult size, within any cichlid assemblage in the rocky littoral, a great size range is evident among the component species, both in terms of species-specific adult size and in the various life stages of each species present. In this context it is interesting to note that the largest (*Boulengerochromis microlepis*; TL 80 cm) and the smallest (*Neolamprolo gus multifasciatus*; TL 2·5 cm) cichlid fishes known are both endemic to Lake Tanganyika. Tanganyikan cichlid communities clearly have potential for size-structuring.

One effect of size-structuring within these communities would be to reduce the potential for separation, at least on the food-size axis, and thus to increase the importance of habitat separation. That this latter prediction is not borne out by the available literature may be a consequence of no studies having adequately addressed the full age/size range of species in an assemblage. Different sized individuals probably require different sized living spaces. Fish that use only a shelter site may have to move to larger quarters, and territorial individuals may have to increase the size of the area defended. This competition is for space for an individual—space that will become vacant again as soon as that individual dies, deserts or is evicted.

Other species occupy different biotopes according to their size, feeding behaviour, and gonad state. A variety of interactions may potentially occur between different life history stages of species, and between species of different sizes. However, only in one species has the effect of body size on resource use of Tanganyikan fishes been documented. While under parental care the young of *Boulengerochromis* are microphagous bottom feeders (Kuwamura, 1986b), but graduate to omnivory when 1 cm TL. When parental care ceases, at 15 cm, they hunt in schools of 100–5000 individuals, breaking into smaller schools (<200 individuals) as they grow, and switching to an almost exclusively fish diet, hunted in the pelagic zone at depths of 15–50 m. Mature fish live in pairs at the littoral zone, feeding on fishes, crabs, molluscs and insect larvae (Matthes, 1961).

In a discussion of speciation in the African Great Lakes, Cohen and Johnston (1987) proposed large body size as one factor responsible for decreased disperal ability, and hence increased speciation. I suggest the converse to be true, in that a large body size enables an individual to travel further on limited reserves, and to reduce the likelihood of being predated while doing so. As body size increases, the suite of potential predatory fishes decreases while contact between adjacent populations will result in gene exchange, which can then act to reduce variation. Witness *B. microlepis*, *Lepidiolamprologus elongatus* and *L. profundicola*, which are large, vagile fishes, ubiquitous throughout the rocky littoral of lake, and showing no colour morphs. Smaller rock-dwelling species are less well able to disperse over sand areas, and are thus more prone to speciate.

B. Species Packing and Species Invasions

While the spectacular array of colours and high species diversity seen in the rocky shore cichlid assemblages of Lakes Tanganyika and Malawi often draw comparison with the marine fish fauna of the coral reefs, these two ecosystems show a fundamental difference in that most coral reef fishes show no parental care and have pelagic eggs and/or larvae. At the population level, this allows for mixing of gene pools; juvenile individuals newly colonizing a particular portion of reef will probably be unrelated to any conspecifics at that site. This mode of recruitment also means that populations are in a state of flux as different species gain access to the reef (Sale, 1978, 1980; Emery, 1978). At the community level therefore, recruitment is stochastic, and species composition on coral reefs is thus unstable and unpredictable.

In contrast to coral reef ecosystems, recruitment to the rocky shore communities of Lakes Tanganyika and Malawi is not stochastic. The extended brood care of the cichlids, in combination with the often strong territoriality and absence of any dispersive larval phase, means that recruitment and, consequently, community composition, is fundamentally a highly deterministic process, at least on a spatial scale. In this ecosystem, however, it is a topic which has received scant attention to date.

Given that species within a community exhibit a fine-scale partitioning of various resources within that habitat, and given that communities typically consist of many species, it should theoretically be difficult for invading species to establish themselves. While we have no evidence for this in Lake Tanganyika, in Lake Malawi this does not appear to be the case. Aliens "accidentally" introduced to areas where that morph or species did not occur naturally were able to establish themselves (Ribbink et al., 1983). Current theory (e.g. Holdgate, 1986) suggests that these fishes, probably from only a few founders, should have great difficulty in establishing themselves in such complex communities of closely related species, yet they clearly have been able to do so. Some studies have examined just how they have accomplished this (Trendall, 1988a, 1988b; Hert, 1990), and the roles of ecology and behaviour in maintenance of fish diversity and speciation in all three African Great Lakes have recently been reviewed by Lowe-McConnell (1994b).

The small size of the rocky shore cichlids will have helped maintain a high species diversity by permitting dense species packing within this habitat. Small species are better at dividing the habitat mosaic than are larger ones, as evidenced by the use of crevice space by lamprologine species (Gashagaza, 1991). Additionally, recent studies of other groups have shown that territory overlap, in conjunction with territory-related interactions among different sized species, can facilitate coexistence and promote species diversity in these communities (Kuwamura, 1992; Kohda and Mboko, 1994; Kohda, in press, a,b; Kohda and Tanida, in press). The rich multitude of niches afforded in the stable, structurally complex environment colonized by specialists, makes it more likely that additional species will be able to find an empty niche than in generalist-populated temperate habitats (Fryer, 1991c).

C. Stability and Persistence

At a given site, the cichlid fish community seems stable, and characterized by a long-term persistence and resilience of structure. Census results in a quadrat at Luhanga, Zaire shoreline, showed little change in the number of species present throughout the 10-year monitoring period. The species composition of this community was also highly persistent, suggesting stochastic recruitment to be unimportant. Removal experiments at a different site, at the Zambian shore, supported these findings. At this site, the community had almost recovered to its initial state after only 1 year (Hori, 1991).

Several interrelated factors contribute to this persistence. The high degree of stenotopy shown by most of the littoral cichlids, especially members of the numerically dominant lamprologine genera, means population size is limited by the suitability of substrate type, which is associated with reproductive territoriality. Territories are occupied by the same individuals for long periods, sometimes for years. The rocky substrate habitat is effectively saturated at all times, and only when a territory owner dies or is evicted does a newcomer gain a territory. Recolonization may occur by invasion from adjacent regions. Generally, though, the poor dispersal of young, resulting from parental care, and specificity of substrate types and boulder and crevice sizes favoured by different species, mean that the occupier of the vacated territory is generally a conspecific. Rossiter (unpublished) removed all 48 territorial males of one species, *Aulonocranus dewindti*, and their mating craters from one quadrat. Three years later, 42 territorial males of this species were resident in the quadrat, 38 of which had constructed new mating craters and set up territories at exactly the same locations as removed individuals. Pressure to obtain a territory can be intense; experimentally vacated territories were sometimes occupied by a conspecific in as little as 5 s (Rossiter, 1994).

D. Is Community Composition Predictable?

This deterministic recruitment and community stability suggest that the forces structuring these communities are theoretically predictable, and can therefore be generalized. But this viewpoint is premature; too few communities have been examined to permit inference of any general rules. In addition, generalizations ignore the most fascinating aspect revealed by studies of these communities—their diversity and uniqueness. As knowledge accumulates it has become increasingly evident that there is great intraspecific variation in the behaviour, and sometimes the morphology, of these fishes, differences which are often population-specific. Should we not therefore assume equivalent or greater ecological differences to be also present?

In terms of physical topography each rocky patch represents a unique biotope, where differences in habitat structure and complexity will influence the composition of associated fish species. For example, the algal biocover is sensitive to the relative surface area of substrate receiving light, which influences the algivorous fish community, and the degree of structural complexity can influence the composition and density of the cichlid infauna. Marked differences will probably be apparent between communities at the topographically simple, laval-formed geological structures of some northern areas, and those at the more complex boulder and rubble substrates in the south. Even on the local scale, differing species components of communities at different sites can mean that each is subject to a suite of species interactions unique to that community.

The following examples serve to illustrate the ecological and behavioural plasticity of the Tanganyikan cichlids, a feature which has surely contributed to their evolutionary and ecological success within this lake, but one which also underlies the inadvisability of making generalized statements, even within species.

1. Physical Influences on Inter-population Differences

Lepidiolamprologus elongatus is one of the top predators in the littoral cichlid communities. It is a substrate spawner, and parental care is continued for at least three months (Nakai, 1993). Within a given population, the mating system varies according to the depth at which the male owns a territory. Males with territories in shallower areas are monogamous, while those in deeper areas are polygamous (Nakai, unpublished).

Lepidiolamprologus attenuatus is a polygynous substrate spawner, where both parents guard the young for up to 3 months after hatching (Nagoshi and Gashagaza, 1988; Gashagaza, 1991). Populations at different localities utilize different spawning substrates, and this affects the male : female size dimorphism. In populations using crevices in boulder-strewn sites, females are markedly larger (TL) than those in populations in which females use gastropod shells as spawning sites (Y. Yanagisawa, pers. comm.). An identical trend occurs in *Altolamprologus compressiceps* (pers. obs.). In both these examples, male size in these populations was approximately equal, and female size has evolved solely in response to the need to access spawning sites. A different trend is seen in the normally shell-brooding *Lamprologus brevis*. Certain populations do not show shell brooding and instead occupy rubble interstices. Here, both male and female are over twice as large as in shell-brooding populations (pers. obs.).

A different permutation occurs in the obligate shell-brooder, *L. callipterus*. The distribution of gastropod fauna around the lake shore is not uniform. At some locations the more common *Neothauma* may be absent, and replaced by the smaller *Paramelania damoni*, or the even smaller *Lavigeria grandis*. *L. callipterus* females show population-specific size differences corresponding to the relative size of the ambient gastropod fauna, those in the *Neothauma* localities being largest and those in the *L. grandis* localities being smallest (T. Sato, pers. comm.). In all these examples, female size seems determined by a single environmental factor; the dimension of the spawning site entrance. The implications of these differences in terms of fecundity and social system deserve investigation, but for these and all other species, data on brood sizes and survivorship are too sparse to permit many deductions to be made. In the mouthbrooders, population-specific differences in egg and brood size have been documented for *Cyathopharynx furcifer* (Rossiter and Yamagishi, in press), but it is unknown whether these are adaptive.

2. Biotic Influences on Inter-population Differences

Excepting the effects of handedness in scale-eaters, interactions with predators have been little explored. However, results of a study of the lekking species, *C. furcifer*, suggest predation can prove a powerful force in affecting community composition and social organization (Rossiter and Yamagishi, in press; Rossiter, submitted) (Table 3). At one lek, at the south end of the lake, a resident population of large predators was present. Here, *Cyathopharynx* males constructed mating craters exclusively on large boulders, with none found on the sand substrate, unlike those noted in a northern population of this species. Social structure also differed between these two populations. Crater-owning males are highly territorial. At the southern site, non crater-owners joined with sub-adult males to form a population of non territorial floaters. These hovered around the lek in schools, and often tried to displace a crater owner or obtain a sneak mating with a female attracted to the crater. In contrast, at the northern site, two categories of crater owner were present—males that built a crater on boulders, as seen in the southern population, and those that built craters on sand. At this site the floating population consisted almost exclusively of sub-adult individuals. The most parsimonious interpretation of these differences is that, at the northern site, adult males unable to obtain a territory and crater on a boulder, which constituted a limited resource in this context, were able to occupy an alternative but inferior location, the sand floor (males on boulders had a higher mating success than those on sand). The two sites were physically similar, and the major difference between them was in predator density. At the northern site, predators were few in number; *Boulengerochromis* were rare and *Hydrocynus* were absent. The high numbers of these predators at the southern site seemed to have prevented the colonization of the sand zone there.

Three years after the original study, the southern lek was re-examined. In the interim period, severe overfishing has eradicated clupeids from the study area, and large predators are now rare or absent. One consequence of this has been a marked change in the social structure of *C. furcifer*, which has changed to that seen at the northern site—mating craters are now found both on boulders, including exposed ones formerly unoccupied, and also on sand. The floating population now consists almost entirely of sub-adults, with only a few recently dispossessed males among them. The littoral cichlid community has also shown marked changes, notably the large increase in the numbers of territorial males of other ectodine species, particularly at the sand/rock margins.

It seems that display areas or mating craters on boulders have the advantage of being close to a refuge, whereas those on sand expose the owner to predation. However, the effect may not be absolute, as individuals occupying the sand zone may be able, on occasion, to flee to safety. Rather, the effect is one of resource depression—the energy invested in crater construction, mating and territorial display is not commensurate with returns in terms of mating success when the close proximity of predators causes displays to be interrupted and territories to be temporarily abandoned.

VII. CONCLUDING REMARKS, CAVEATS AND PERSPECTIVES

The diverse cichlid assemblages present today are a consequence of an interplay among many series of invasions, speciations and extinctions, partly precipitated by a complex geological history occurring over an extended time-frame. Crucial has been the stenotopic habits of many of the littoral cichlids and the disjunct distribution of suitable habitat around the lake shoreline. Even for related, stenotopic cichlids that are equally subject to environmental barriers to dispersal, simple fragmentation of habitat is sufficient to explain different degrees of morphological divergence within species, and of community composition between rocky patches. The structure and slope of the rock substrate is an important factor here, and can contribute to the diversity and uniqueness of the community at that site.

Success seems to have been achieved partly through becoming specialists, each species occupying a narrow niche. This is particularly apparent within feeding groups, where, even when food type taken is similar, species show slight differences in the method by which that food is obtained. As the diets and feeding mechanisms of these fishes are usually reflected by the structure of their mouthparts, there is scope for comprehensive studies of their morphology, ecology, habits, food and feeding mechanisms. However, specialization does not restrict a particular species to a particular food resource, and switching, albeit opportunistically, does occur (Nshombo, 1991).

Several features of the biology of the cichlids have also contributed toward their apparent predisposition toward speciation. These, and the stable, predictable environment, have facilitated fine-scale niche partitioning and species packing within these communities. The result has been a cichlid assemblage that is morphologically, ecologically and behaviourally the most diverse among the African Great Lakes (Lowe-McConnell, 1993).

A sound taxonomy must be in place before meaningful ecological conclusions can be drawn. The debate on speciation and variation is open, and Tanganyika is an ideal natural laboratory where questions concerning phenotypic and genotypic diversification can be explored. The great length of the lake shoreline has allowed the development of orthodox geographical variation within species. Some of these colour morphs may simply be colour variants of known species, others may prove to be distinct, but closely related, species. It has yet to be established whether any colour morph or sibling species combinations coexist without interbreeding, but preliminary findings suggest that assortative mating and sexual selection may maintain divergent populations that are in sympatry. The exploration of differentiation below the species level, through examination of mechanisms of assortative mating and incipient speciation within isolated or partly isolated stenotopic populations, is a promising area for future research. Studies in this area will simultaneously provide a more rigorous evaluation of species' status, degrees of relationship, and the relative importance of the various mechanisms of speciation.

Longer time-series are required of quantitative measurements of the interactions between the biological components of the lake. Our present level of knowledge suggests that biotic interactions, such as predation, competition and commensalism, are important components of the observed patterns, but are by no means the only causative factors. The optimal approach to understanding the structuring of the Tanganyikan cichlid assemblages and the importance of resource partitioning to community structure lies in manipulative field experiments, and studies incorporating both descriptive field observations, and field and laboratory work. These are still at an embryonic stage.

Understanding the patterns in resource use and the role of resource partitioning in the evolution and structuring of species assemblages has been hampered by emphasis on single factors in isolation. Elucidation of these factors will only proceed once we acknowledge that multicausation of differential resource use is the rule, rather than the exception. Future research should redress the balance from its presents focus on feeding and habitat resources toward those of breeding. Breeding territory and stenotopy, in turn related to substratum composition, seem to be the major forces influencing the community composition and species diversity of the rocky shore cichlid communities.

The cichlids of Lake Tanganyika are good subjects for the study of the variables of theoretically predictable influences on behaviour, social structure and community ecology. They are strongly stenotopic and territorial. There exists a large species pool, varying fractions and combinations of which occur in isolated communities. This situation has permitted different species to occupy similar niches in different areas, and for species to be exposed to community- or population-specific interactions. The clear waters of the lake permit direct observation of fishes, crucial in identifying fine-scale mechanisms of coexistence. Most species have a fast brood succession, relatively short generation time, a diverse behavioural repertoire, and are easily observed and manipulated under experimental and semi-natural conditions.

The great length of the lake shoreline provides a field situation with a large number of disjunct communities occurring in a similar physical environment and habitat structure. This is true for a variety of habitat types, facilitating a rigorous exploration of the effects of habitat complexity on community composition. Although their diversity may well defeat any efforts to over-generalize, the cichlid assemblages of Lake Tanganyika may still be viewed as magnificent models for the investigation of the functional mechanisms regulating complex social and community systems.

However, time may be short. Recent increases in pollution (Rossiter, 1993), and rapid deforestation of the lake shoreline, resulting in rapid erosion of topsoils and increased sedimentation at the littoral zone, are causes for serious concern (Coulter and Mubamba, 1993). The consequences of deforestation are likely to be dramatic. Settling sediment will blanket benthic algae, cutting off light and killing it (Grobbelaar, 1985), thereby undermining the production base of the littoral zone. Filter-feeding zooplankters will be interfered with by suspended sediment particles in the water column, and phytoplankters will be affected by reduced light penetration (Cairns, 1968). The more rapid light attenuation will reduce the depth of the euphotic zone, decreasing the littoral fish habitat, a trend further compounded by a rise in the anoxic layer as a result of increased oxygen demand (Bootsma and Hecky, 1993). Sediments will also reduce habitat complexity and heterogeneity, initially by filling in small cracks and crevices, utilized now by invertebrate fauna—a major food source for many of the littoral cichlids—and eventually filling in larger crevices and holes, which currently provide refuges and species-specific breeding sites for many cichlid fishes (Cohen *et al.*, 1993). The ecology of the lake is as little-known as the fishes inhabiting it. It offers splendid, perhaps unique, opportunities for ecological, evolutionary and behavioural research (Hori *et al.*, 1993; Lowe-McConnell, 1994b; Nakai *et al.*, 1994), and in this, the era of "biodiversity", it would be particularly tragic if the complex cichlid communities were to become depleted, especially before being scientifically understood. There is thus an urgent need for more research to unravel the intricacies of the ecology and evolution of the cichlid fishes of Lake Tanganyika.

REFERENCES

Abele, L.L. (1976). Comparative species richness in fluctuating and constant environments: coral-associated decapod crustaceans. *Science* **192**, 461–463.

Barel, C.D.N., Dorit, R., Greenwood, P.H., Fryer, G., Hughes, N., Jackson, P.B.N., Kawanabe, H., Lowe-McConnell, R.H., Nagoshi, M., Ribbink, A.J., Trewavas, E., Witte, F. and Yamaoka, K. (1985) Destruction of fisheries in Africa's lakes. *Nature (London)* **315**, 19–20.

Beadle, L.C. (1981). *The Inland Waters of Tropical Africa: An Introduction to Tropical Limnology* 2nd edition. Longman, London.

van der Berghe, E.P. (1988). Piracy: A new alternative male reproductive tactic. *Nature (London)* **334**, 697–698.

Bootsma, H.A. and Hecky, R.E. (1993). Conservation of the African Great Lakes: A limnological perspective. *Cons. Biol.* **7**, 644–656.

Boss, K.L. (1978). On the evolution of gastropods in ancient lakes. In: *Pulmonates: v. 2a: Systematics, Evolution and Ecology* (Ed. by V. Fretter and J. Peake) pp. 385–428. Academic Press, London.

Brichard, P. (1989). *Pierre Brichard's Book of Cichlids and all the other Fishes of Lake Tanganyika*. T.F.H. Publications, Inc. Neptune City, N.J., USA.

Brooks, J.L. (1950). Speciation in ancient lakes. *Quart. Rev. Biol.* **25**, 30–60, 131–176.

Cairns, J. (1968). Suspended solids standards for the protection of aquatic organisms. *Purdue Univ. Engl. Bull.* **129**, 16–27.

Carson, H.L. and Kaneshiro, K.Y. (1976) *Drosophila* of Hawaii: systematics and ecological genetics. *Ann. Rev. Ecol. Syst.* **7**, 311–345.

Chesson, P.L. (1981). Models for spatially distributed populations: the effect of within-patch variability. *Theor. Pop. Biol.* **19**, 288–325.

Cohen, A.S. and Johnston, M.R. (1987). Speciation in brooding and poorly dispersing lacustrine organisms. *Palaios* **2**, 426–435.

Cohen, A.S., Bills, R., Cocquyt, C.Z. and Caljon, A.G. (1993). The impact of sediment pollution on biodiversity in Lake Tanganyika. *Cons. Biol.* **7**, 667–677.

Colombe, J. and Allgayer, R. (1985). Description de *Variabilichromis, Neolamprologus*, et *Paleolamprologus* genres nouveaux du Lac Tanganika, avec redescription des genres *Lamprologus* Schilthuis, 1891 et *Lepidiolamprologus* Pellegrin, 1904 (Pisces: Teleostei: Cichlidae). *Revue française des Cichlidophiles* **49**, 9–16. 21–28.

Coulter, G.W. (1966). *Hydrobiological processes and the deepwater fish community in Lake Tanganyika*. PhD thesis. Queens University, Belfast.

244   A. ROSSITER

244   A. ROSSITER

Coulter, G.W. (1968). The deep benthic fishes at the southern end of Lake Tanganyika with special reference to distribution and feeding in *Bathybates* species, *Hemibates stenosoma* and *Chrysichthys* species. *Fish. Res. Bull, Zambia* **4**, 33–38.

Coulter, G.W. (1991). (Ed.) *Lake Tanganyika and its Life.* pp. 354. Oxford University Press, Oxford.

Coulter, G.W. and Mubamba, R. (1993). Conservation in Lake Tanganyika, with special reference to underwater parks. *Cons. Biol.* **7**, 678–685.

Coulter, G.W. and Spigel, R.H. (1991). Hydrodynamics. In: *Lake Tanganyika and its Life* (Ed. by G.W. Coulter), pp 49–75. Oxford University Press. Oxford.

Cunnington, W.A. (1920). The fauna of the African lakes: A study in comparative limnology with special reference to Lake Tanganyika. *Proc. zool. Soc. Lond.* **3**, 507–622.

Dethier, M.N. (1984). Disturbance and recovery in intertidal pools: maintenance of mosaic patterns. *Ecol. Monog*, **54**, 99–118.

Dominey, W.J. (1984). Effects of sexual selection and life history on speciation: Species flocks in African cichlids and Hawaiian *Drosophila*. In: *Evolution of Fish Species Flocks* (Ed. by A.A. Echelle and I. Kornfield) pp. 231–248. University of Maine, Orono Press.

Emery, A.R. (1978). The basis of fish community structure: marine and freshwater comparisons. *Env. Biol. Fishes* **3**, 33–47.

Echelle, A.A. and Kornfield, I. (1984). (Eds.) *Evolution of Fish Species Flocks.* University of Maine, Orono Press.

Fryer, G. (1959a). The trophic interrelationships and ecology of some littoral communities of Lake Nyasa with especial reference to the fishes, and a discussion of the evolution of a group of rock-frequenting Cichlidae. *Proc. zool. Soc. Lond.* **132**, 153–281.

Fryer, G. (1959b). Some aspects of evolution in Lake Nyasa. *Evolution* **13**, 440–451.

Fryer, G. (1965). Predation and its effects on migration and speciation in African fishes: a comment with further comments by P.H. Greenwood, a reply by P.B.N. Jackson and a footnote and postscript by G. Fryer. *Proc. zool. Soc. Lond.* **144**, 301–322.

Fryer, G. (1969). Speciation and adaptive radiation in African lakes. *Verh. Internat. Verein. Limnol.* **17**, 303–322.

Fryer, G. (1972). Conservation of the Great Lakes of East Africa: a lesson and a warning. *Biol. Cons.* **4**, 256–262.

Fryer, G. (1977). Evolution of species flocks of cichlid fishes in African lakes. *Zeit. zool. syst. evol.* **15**, 141–165.

Fryer, G. (1991a). Comparative aspects of adaptive radiation and speciation in Lake Baikal and the great rift lakes of Africa. *Hydrobiologia* **211**, 137–146.

Fryer, G. (1991b). The evolutionary biology of African cichlid fishes. *Ann. mus. Roy. Afr. centr., sc. zool.* 13–22.

Fryer, G. (1991c). Biological invasions in the tropics: Hypotheses versus reality. In: *Ecology of Biological Invasion in the Tropics* (Ed. by P.S. Ramakrishnan), pp. 87–101. International Scientific Publications. New Delhi.

Fryer, G. and Iles, T.D. (1972) *The Cichlid Fishes of the Great Lakes of Africa. Their Biology and Evolution.* Oliver & Boyd. Edinburgh.

Gashagaza, M.M. (1988). Feeding activity of a Tanganyikan cichlid fish *Lamprologus brichardi. African Study Monogr.* **9**, 1–9.

Gashagaza, M.M. (1991). Diversity of breeding habits in lamprologine cichlids in Lake Tanganyika. *Physiol. Ecol. Japan* **28**, 29–65.

Gashagaza, M.M. and Nagoshi, M. (1986). Comparative study on the food habits of six species of *Lamprologus* (Osteichthyes: Cichlidae). *African Study Monogr.*, **6**, 37–44.

Gashagaza, M.M. and Nagoshi, M. (1988). Growth of the larvae of a Tanganyikan cichlid, *Lamprologus attenuatus*, under parental care. *Jpn. J. Ichthyol.* **35**, 392–395.

Gonfiantini, R., Zuppi, G.M., Eccles, D.H. and Ferro, W. (1979). Isotope investigation of Lake Malawi. In: *Isotopes in Lake Studies*. International Atomic Energy Agency. Vienna, Austria.

Glasser, J.W. (1979). The role of predation in shaping and maintaining the structure of communities. *Am. Nat.* **113**, 631–641.

Greenwood, P.H. (1965). The cichlid fishes of Lake Nabugabo, Uganda. *Bull. Brit. Mus. (Nat. Hist.), Zool.*, **12**, 315–317.

Greenwood, P.H. (1974). Cichlid Fishes of Lake Victoria, East Africa: The biology and evolution of a species flock. *Bull. Brit. Mus. (Nat. Hist.), Zool.* Suppl. **6**, 1–134.

Greenwood, P.H. (1983). The *Opthalmotilapia* assemblage of cichlid fishes reconsidered. *Bull. Brit. Mus. (Nat. Hist.), Zool.* **45**, 249–290.

Greenwood, P.H. (1984). African cichlids and evolutionary theories. In: *Evolution of Fish Species Flocks* (Ed. by A.A. Echelle and I. Kornfield), pp. 141–154. University of Maine, Orono Press.

Greenwood, P.H. (1991). Speciation. In: *Cichlid Fishes: Behaviour, Ecology and Evolution* (Ed. by M.H.A. Keenleyside), pp. 86–102. Chapman and Hall, London, England.

Grobbelaar, J. (1985). Phytoplankton productivity in turbid waters. *J. Plankton Res.* **7**, 653–663.

Haberyan, K.A. and Hecky, R.E. (1987). The late pleistocene and holocene stratigraphy and palaeolimnology of Lakes Kivu and Tanganyika. *Palaeogeography, palaeoclimatology, palaeoecology* **51**, 161–197.

Hamilton, W.D. (1971). Geometry for the selfish herd. *J. theor. Biol.* **31**, 295–311.

Hanski, I. (1987). Colonization of ephemeral habitats. In: *Colonization, Succession and Stability* (Ed. by A.J. Gray, M.J. Crawley and P.J. Edwards), pp. 155–185. Blackwell Scientific Publications, Oxford.

Haussknecht, T. and Kuenzer, P. (1991). An experimental study of the building behaviour sequence of a shell-breeding cichlid fish from Lake Tanganyika (*Lamprologus ocellatus*). *Behaviour* **116**, 127–142.

Hecky, R.E. and Kling, H. (1981). The phytoplankton and protozooplankton of the euphotic zone of Lake Tanganyika: Species composition, biomass, chlorophyll content and spatio-temporal distribution. *Limnol. Oceanogr.* **26**, 548–564.

Hert, E. (1985). Individual recognition of helpers by the breeders in the cichlid fish *Lamprologus brichardi* (Poll, 1974). *Zeit. Tierpsychol.* **68**, 313–325.

Hert, E. (1990). Factors in habitat partitioning in *Pseudotropheus aurora* (Pisces: Cichlidae), an introduced species in a species-rich community of Lake Malawi. *J. Fish Biol*, **36**, 853–865.

Holdgate, M.W. (1986). Summary and conclusions: Characteristics and consequences of biological invasions. *Phil. trans. Roy. Soc. B* **314**, 655–674.

Hori, M. (1983). Feeding ecology of thirteen species of *Lamprologus* (Teleostei: Cichlidae) coexisting at a rocky shore of Lake Tanganyika. *Physiol. Ecol. Japan* **20**, 129–149.

Hori, M. (1987). Mutualism and commensalism in a fish community in Lake Tanganyika. In: *Evolution and Coadaptation in Biotic Communities* (Ed. by S. Kawano, J.H. Connell and T. Hidaka), pp. 219–239. University of Tokyo Press, Tokyo.

246   A. ROSSITER

Hori, M. (1991) Feeding relationships among cichlid fishes in Lake Tanganyika: Effects of intra- and interspecific variations of feeding behavior on their coexistence. *Ecol. Internat. Bull.* **19**, 89–101.

Hori, K. (1993). Frequency-dependent natural selection in the handedness of scale-eating cichlid fish. *Science* **260**, 216–219.

Hori, M., Yamaoka, K. and Takamura, K. (1983). Abundance and micro-distribution of cichlid fishes on a rocky shore of Lake Tanganyika. *African Study Monogr.* **3**, 25–38.

Hori, M., Gashagaza, M.M., Nshombo, M. and Kawanabe, H. (1993). Littoral fish communities in Lake Tanganyika: irreplaceable diversity supported by intricate interactions among species. *Cons. Biol.* **7**, 657–666.

Jackson, P.B.N. (1961). The impact of predation, especially by the tiger fish (*Hydrocyon vittatus* Cast.) on African freshwater fishes. *Proc. zool. Soc. Lond.* **136**, 603–622.

Kawanabe, H., Kwetuenda, M.K. and Gashagaza, M.M. (1992). Ecological and limnoloical studies of Lake Tanganyika and its adjacent regions between African and Japanese scientists; An introduction. *Mitt. Internat. Verein. Limnol.* **23**, 79–83.

Kawanabe, H., Gashagaza, M.M. and Hori, M. (1994). A conservation issue of biodiversity in Lake Tanganyika, with special reference to inshore fishes. *Verh. Internat. Verein. Limnol.* **25**, 2182.

Kohda, M. (1994). Individual specialized foraging repertoires in the piscivorous cichlid fish *Lepidiolamprologus profundicola*. *Anim. Behav.* **48**, 1123–1131.

Kohda, M. (in press a). Does male-mating attack in the herbivorous cichlid *Petrochromis polyodon* facilitate the coexistence of herbivorous congeners? *Ecol. Freshw. Fish.*

Kohda, M. (in press b). Territoriality of male cichlid fishes in Lake Tanganyika. *Ecol. Freshw. Fish.*

Kohda, M. and Hori, M. (1993). Dichromatism in relation to the trophic biology of predatory cichlid fishes in Lake Tanganyika, East Africa. *J. Zool.* **229**, 447–455.

Kohda, M. and Mboko, S.K. (1994). Territorial attacks against larger heterospecific intruders in Lake Tanganyika. *Afr. Study Monogr.* **15**, 69–75.

Kohda, M. and Tanida, K. (in press). Overlapping territory of the benthophagous cichlid, *Lobochilotes labiatus*, in Lake Tanganyika. *Env. Biol. Fishes.*

Kohda, M. and Yanagisawa, Y. (1992) Vertical distribution of two herbivorous cichlid fishes of the genus *Tropheus* in Lake Tanganyika, Africa. *Ecol. Freshw. Fish* **1**, 99–103.

Kohda, M., Yanagisawa, Y., Sato, T., Nakaya, K., Niimura, Y., Matsumoto, K. and Ochi, H. (in press). Geographic colour variations of cichlid fishes at the southern end of Lake Tanganyika. *Env. Biol. Fishes.*

Konings, A. (1988). *Tanganyika Cichlids.* Verduijn cichlids, Zevenhuizen, Holland.

Kornfield, I. and Carpenter, K.E. (1984). Cyprinids of Lake Lanao, Philippines: Taxonomic validity, evolutionary rates and speciation scenarios. In: *Evolution of Fish Species Flocks* (Ed. by A.A. Echelle and I. Kornfield), pp. 69–84. University of Maine, Orono Press.

Kuwamura, T. (1986a). Parental care and mating systems of cichlid fishes in Lake Tanganyika: a preliminary field survey. *J. Ethol.* **4**, 129–146.

Kuwamura, T. (1986b). Substratum spawning and biparental guarding of the Tanganyikan cichlid, *Boulengerochromis microlepis*, with notes on its life history. *Physiol. Ecol. Japan* **23**, 31–43.

Kuwamura, T. (1988). Biparental mouthbrooding and guarding in a Tanganyikan cichlid *Haplotaxodon microlepis*. *Jpn. J. Ichthyol.* **35**, 62–68.

Kuwamura, T. (1992). Overlapping territories of *Pseudosimochromis curvifrons* males and other herbivorous fishes in Lake Tanganyika. *Ecol. Res.* **7**, 43–53.

Kuwamura, T., Nagoshi, M. and Sato, T. (1989). Female-to-male shift of mouthbrooding in a cichlid fish, *Tanganicodus irsacae*, with notes on breeding habits of two related species in Lake Tanganyika. *Env. Biol. Fishes* **24**, 187–198.

Levin, S.A. and Paine, R.T. (1974). Disturbance, patch formation and community structure. *Proc. natl. Acad. Sci. USA* **71**, 2711–2747.

Levins, R. (1979). Coexistence in a variable environment. *Am. Nat.* **114**, 765–783.

Lewis, D.S.C., Reinthal, P. and Trendall, J. (1986). *A Guide to the Fishes of Lake Malawi National Park*. WWF World Conservation Center, Gland, Switzerland.

Lewis, W.M. Jr. (1987). Tropical limnology. *Ann. Rev. Ecol. Syst.* **18**, 159–184.

Liem, K.F. (1974). Evolutionary strategies and morphological innovations: Cichlid pharyngeal jaws. *Syst. zool.* **22**, 425–441.

Liem, K.F. (1978). Modulatory multiplicity in the functional repertoire of the feeding mechanism in cichlid fishes. Part 1: Piscivores. *J. Morphol.* **158**, 323–360.

Liem, K.F. (1980a). Adaptive significance of intra and interspecific differences in the feeding repertories of cichlid fishes. *Amer. Zool.* **20**. 25–314.

Liem, K.F. (1980b). Acquisition of energy by teleosts: adaptive mechanisms and evolutionary patterns. In: *Environmental Physiology of Fishes*. (Ed. by M.A. Ali), pp. 299–334. Plenum, New York.

Liem, K.F. (1981). A phyletic study of the Lake Tanganyika cichlid genera *Asprotilapia*, *Ectodus*, *Lestradea*, *Cunningtonia*, *Opthalmochromis* and *Opthalmotilapia*. *Bull. Mus. Comp. Zool.* **149**, 19–214.

Liem, K.F. (1991). Functional morphology. In: *Cichlid Fishes: Behaviour, Ecology and Evolution* (Ed. by M.H.A. Keenleyside), pp. 129–150. Chapman and Hall, London, England.

Liem, K.F. and Stewart, D.J. (1976). Evolution of the scale-eating fishes of Lake Tanganyika: a generic revision with a description of a new species. *Bull. Mus. Comp. Zool.* **147**, 319–350.

Lowe-McConnell, R.H. (1969). Speciation in tropical freshwater fishes. *Biol. J. Linn. Soc.* **1**, 51–75.

Lowe-McConnell, R.H. (1975). *Fish Communities in Tropical Freshwaters*. Longman, London.

Lowe-McConnell, R.H. (1987). *Ecological Studies in Tropical Fish Communities*. Cambridge University Press, Cambridge.

Lowe-McConnell, R.H. (1993). Fish faunas of the African Great Lakes: origins, diversity, and vulnerability. *Cons. Biol.* **7**, 634–643.

Lowe-McConnell, R.H. (1994a). Threats to, and conservation of, tropical freshwater fishes. *Mitt. Internat. Verein. Limnol.* **24**, 47–52.

Lowe-McConnell, R.H. (1994b). The roles of ecological and behaviour studies of cichlids in understanding fish diversity and speciation in the African Great Lakes: a review. *Arch. Hydrobiol.* **44**, 335–345.

MacArthur, R.H. (1957). On the relative abundance of bird species. *Proc. natl Acad. Sci USA*. **45**, 293–295.

Matsuda, H., Abrams, P.A. and Hori, M. (1993). The effect of anti-predator behavior on exploitative competition and mutualism between predators. *Oikos* **68**, 549–559.

Matsuda, H., Hori, M. and Abrams, P.A. (1994). Effect of predator-specific defence on community complexity. *Evol. Ecol.* **8**, 628–638.

248   A. ROSSITER

Matthes, H. (1961). *Boulengerochromis microlepis*, a Lake Tanganyika fish of economical importance. *Bull. Aquat. Biol.* **3**, 1–15.

McKaye, K.R. and Gray, W.N. (1984). Extrinsic barriers to gene flow in rock-dwelling cichlids of Lake Malawi: Macrohabitat heterogeneity and reef coloniz-ation. In: *Evolution of Fish Species Flocks.* (Ed. by A.A. Echelle and I. Kornfield), pp. 169–184. University of Maine. Orono Press.

McKaye, K.R., Kocher, T., Reinthal, P., Harrison, R. and Kornfield, I. (1984). Evidence for allopatric and sympatric differentiation among color morphs of Lake Malawi cichlid fish. *Evolution* **38**, 215–219.

Mboko, S.K. and Kohda, M. (in press). Pale and dark dichromatism related to microhabitat in a herbivorous Tanganyikan cichlid fish, *Telmatochromis tempora-lis. J. Ethol.* **13**.

Meyer, A. (1993). Phylogenetic relationships and evolutionary processes in East African cichlid fishes. *TREE* **8**, 279–284.

Moore, J.E.S. (1903). *The Tanganyika Problem.* Hurst & Blacket, London.

Myers, G.S. (1960). The endemic fish faunas of Lake Lanao, and the evolution of higher taxonomic categories. *Evolution* **14**, 323–333.

Nagoshi, M. (1983). Distribution, abundance and parental care of the genus *Lam-prologus* (Cichlidae) in Lake Tanganyika. *African Study Monogr,* **3**, 39–47.

Nagoshi, M. (1985). Growth and survival in larval stage of the genus *Lamprol-ogus* (Cichlidae) in Lake Tanganyika. *Verh. Internat. Verein. Limnol.* **22**, 2663–2670.

Nagoshi, M. (1987). Survival of broods under parental care and parental roles of the cichlid fish, *Lamprologus toae*, in Lake Tanganyika. *Jpn. J. Ichthyol.* **34**, 71–75.

Nagoshi, M. and Gashagaza, M.M. (1988). Growth of the larvae of a Tanganyikan cichlid, *Lamprologus attenuatus*, under parental care. *Jpn. J. Ichthyol.* **35**, 392–395.

Nakai, K. (1993). Foraging of brood predators restricted by territoriality of sub-strate-brooders in a cichlid fish assemblage. In: *Mutualism and Community Organization: Behavioural, Theoretical and Food-web Approaches* (Ed. by H. Kawanabe, J.E. Cohen and K. Iwasaki), pp. 84–108. Oxford Science Publi-cations. Oxford.

Nakai, K., Yanagisawa, Y., Sato, T., Niimura, Y. and Gashagaza, M.M. (1990). Lunar synchronization of spawning in cichlid fishes of the tribe Lamprologini in Lake Tanganyika. *J. Fish Biol.* **37**, 589–598.

Nakai, K., Kawanabe, H. and Gashagaza, M.M. (1994). Ecological studies on the littoral cichlid communities of Lake Tanganyika: the coexistence of many en-demic species. *Arch. Hydrobiol.* **44**, 373–389.

Nakano, S. and Nagoshi, M. (1990). Brood defence and parental role in a biparental cichlid fish, *Lamprologus toae*, in Lake Tanganyika. *Jpn. J. Ichthyol.* **36**, 468–476.

Nishida, M. (1991). Lake Tanganyika as an evolutionary reservoir of old lineages of east African cichlid fishes: inferences from allozyme data. *Experientia* **47**, 974–979.

Nshombo, M. (1991). Occasional egg-eating by the scale-eater *Plecodus straeleni* (Cichlidae) of Lake Tanganyika. *Env. Biol. Fishes* **31**, 207–212.

Nshombo, M. (1994a). Foraging behaviour of the scale-eater *Plecodus straeleni* (Cichlidae, Teleostei) in Lake Tanganyika, Africa. *Env. Biol. Fishes* **39**, 59–72.

Nshombo, M. (1994b). Polychromatism of the scale-eater *Perissodus microlepis* (Cichlidae, Teleostei) in relation to foraging behaviour. *J. Ethol.* **12**, 141–161.

Nshombo, M., Yanagisawa, Y. and Nagoshi, M. (1985). Scale-eating in *Perissodus microlepis* (Cichlidae) and change of its food habits with growth. *Jpn. J. Ichthyol.* **32**, 66–73.

Ochi, H. (1993) Mate monopolization by a dominant male in a multi-male social group of a mouthbrooding cichlid, *Ctenochromis horei*. *Jpn. J. Ichthyol.* **40**, 209–218.

Ochi, H., Yanagisawa, Y. and Omori, K. (in press). Intraspecific brood-mixing of the cichlid fish *Perissodus microlepis* in Lake Tanganyika. *Env. Biol. Fishes*.

Owen, R.B., Crossley, R., Johnson, T.C., Tweddle, D., Kornfield, I., Davidson, S., Eccles, D.H. and Engstrom, D.E. (1990). Major low levels of Lake Malawi and their implications for speciation rates in cichlid fishes. *Proc. Roy. Soc. Lond. B* **240**, 519–553.

Pearl, R. (1928). *The Rate of Living.* Knopf, New York.

Pianka, E.R. (1973). The structure of desert lizard communities. *Ann. Rev. Ecol. Syst.* **4**, 53–74.

Pitcher, T.J. (1983). Heuristic definitions of fish shoaling behaviour. *Anim. Behav.* **31**, 611–613.

Poll, M. (1946). Revision de la faune ichthyologique du Lac Tanganika. *Annales du Musée r. du Congo belge, Bruxelles, Zoologie* **4**, 141–364.

Poll, M.(1948). Descriptions de Cichlidae nouveaux recueillis par la Mission hydrobiologique belge au Lac Tanganika. (1946–1947). *Bulletin du Musée r. d'histoire naturelle de Belgique* **24**, 1–31.

Poll, M. (1949). Deuxième série de Cichlidae nouveaux receuillis par la Mission hydrobiologique belge au Lac Tanganika. *Bulletin de l'Institut r. des sciences naturelles de Belgique* **25**, 1–55.

Poll, M. (1952). Révision de la faune ichthyologique du Lac Tanganika. *Annales du Musée r. du Congo belge, Tervuren, Sciences zoologiques* **4**, 141–364.

Poll, M. (1956a). Ecologie des poissons du Lac Tanganika. *Proc. 14th Int. Congr. Zool.*, 465–468.

Poll, M. (1956b). Poissons Cichlidae. *Explor. Hydrobiol. L. Tanganika,* **3**, Fasc. 5B, 619 pp.

Poll, M. (1986). Classification des Cichlidae du Lac Tanganika: Tribus, genres et espèces. *Académie r. de Belgique. Mémoires de la classe des sciences, 8°.* **45**, 1–163.

Reinthal, P. (1989). Morphological analyses of the neurocranium of a group of rock-dwelling cichlid fishes (Cichlidae: Perciformes) from Lake Malawi, Africa. *Zool. J. Linn. Soc.* **98**, 123–139.

Reinthal, P. (1990). The feeding habits of a group of herbivorous rock-dwelling cichlid fishes (Cichlidae: Perciformes) from Lake Malawi, Africa. *Env. Biol. Fishes* **27**, 215–233.

Ribbink, A.J., Marsh, A.C., Marsh, B.A. and Sharp, B.J. (1980). Parental behaviour and mixed broods among cichlid fish of Lake Malawi. *S. Afr. J. Zool.* **15**, 1–16.

Ribbink, A.J., Marsh, A.C. and Marsh, B.A. (1981). Nest-building and communal care of young by *Tilapia rendalli* Dumeril in Lake Malawi. *Env. Biol. Fishes* **6**, 219–222.

Ribbink, A.J., Marsh, B.A., Marsh, A.C., Ribbink, A.C. and Sharp, B.J. (1983). A preliminary survey of the cichlid fishes of rocky habitats in Lake Malawi. *S. Afr. J. Zool.* **18**, 147–310.

Rossiter, A. (1991). Lunar spawning synchroneity in a freshwater fish. *Naturwiss.* **78**, 182–185.

Rossiter, A. (1993). Fishes, fishes, everywhere: Lake Tanganyika and its life. *Env. Biol. Fishes* **37**, 97–101.

Rossiter, A. (1994). Territory, mating success and the individual male in a lekking cichlid fish. In: *Animal Societies: Individuals, Interactions and Organisation* (Ed. by P.J. Jarman and A. Rossiter), pp. 43–55. Kyoto University Press, Kyoto, Japan.

Rossiter, A. and Yamagishi, S. (in press). Intraspecific plasticity in the social system of the lek-breeding cichlid fish, *Cyathopharynx furcifer*. In: *The Ecology of the Fishes of Lake Tanganyika* (Ed. by Kawanabe, H., Yanagisawa, Y. and Hori, M.) Kyoto University Press, Kyoto, Japan.

Roughgarden, J. (1974). Species packing and the competition function with illustrations from coral reef fish. *Theor. Pop. Biol.* **5**, 163–186.

Roughgarden, J. and Feldman. (1975). Species packing and predation pressure. *Ecology* **56**, 489–492.

Sale, P.F. (1977). Maintenance of high diversity in coral reef fish communities. *Am. Nat.* **111**, 337–359.

Sale, P.F. (1978). Coexistence of coral reef fishes—a lottery for living space. *Env. Biol. Fishes* **3**, 85–102.

Sale, P.F. (1980). The ecology of fishes on coral reefs. *Oceanogr. Mar. Biol. Ann. Rev.* **18**, 367–421.

Sands, D. (1983). *Catfishes of the World. Volume II, Mochokidae.* pp. 112. Dunure Enterprises. Ayr, Scotland.

Sato, T. (1986). A brood parasitic catfish of mouthbrooding cichlid fishes in Lake Tanganyika. *Nature* **323**, 58–59.

Sato, T. (1994). Active accumulation of spawning substrate: a determinant of extreme polygyny in a shell-brooding cichlid fish. *Anim. Behav.* **48**, 669–678.

Schoener, T.W. (1974). Resource partitioning in ecological communities. *Science* **185**, 27–39.

Scholtz, C.A. and Rosendahl, B.R. (1978). Low lake stands in Lake Malawi and Tanganyika, East Africa, delineated with multifold seismic data. *Science* **240**, 1645–1648.

Slatkin, M. (1974). Competition and regional coexistence. *Ecology* **55**, 128–134.

Snoeks, J., Rüber, L. and Verheyen, E. (1994). Studies on the taxonomy and the distribution of Lake Tanganyika fishes. *Arch. Hydrobiol.* **44**, 355–372.

Sousa, W.P. (1984). The role of disturbance in natural communities. *Ann. Rev. Ecol. Syst.* **15**, 353–391.

Stiassny, M.L.J. (1981). Phylogenetic versus convergent relationships between piscivorous cichlid fishes from Lake Malawi and Tanganyika. *Bull. Brit. Mus. (Nat. Hist.) Zool.* **40**, 67–101.

Sturmbauer, C. and Meyer, A. (1992). Genetic divergence, speciation and morphological stasis in a lineage of African cichlid fishes. *Nature* **358**, 578–581.

Sturmbauer, C. and Meyer, A. (1993). Mitochondrial phylogeny of the endemic mouthbrooding lineages of cichlid fishes from Lake Tanganyika in eastern Africa. *Mol. Biol. Evol.* **10**, 751–768.

Sugihara, G. (1980). Minimal community structure: an explanation of species abundance patterns. *Am. Nat.* **116**, 770–787.

Taborsky, M. (1984). Broodcare helpers in the cichlid fish *Lamprologus brichardi*: their costs and benefits. *Anim. Behav.* **32**, 1236–1252.

Taborsky, M. (1985). Breeder-helper conflict in a cichlid fish with broodcare helpers: an experimental analysis. *Behaviour* **95**, 45–75.

Taborsky, M. (1987). Cooperative behaviour in fish: Coalitions, kin groups and

recripocity. In: *Animal Societies: Theories and facts* (Ed. by Y. Ito, J.L. Brown and J. Kikkawa), pp. 229–237. Japan Scientific Societies Press, Tokyo.
Taborsky, M. and Limberger, D. (1981). Helpers in fish. *Behav. Ecol. Sociobiol.* **8**, 143–145.
Taborsky, M. (1994). Sneakers, satellites, and helpers: parasitic and cooperative behavior in fish reproduction. In: *Advances in the Study of Behavior*, Volume 23. (Ed. by P.J.B. Slater, J.S. Rosenblatt, C.T. Snowdon and M. Milinsky), pp. 1–100. Academic Press, London.
Takahashi, S. and Hori, M. (1994). Unstable evolutionary stable strategy and oscillations: A model of lateral asymmetry in scale-eating cichlids. *Am. Nat.* **144**, 1001–1020.
Takamura, K. (1983). Interspecific relationships between two aufwuchs eaters *Petrochromis polyodon* and *Tropheus moorei* (Pisces: Cichlidae) of Lake Tanganyika, with a discussion on the evolution and functions of a symbiotic relationship. *Physiol. Ecol. Japan* **20**, 59–69.
Takamura, K. (1984). Interspecific relationships of Auchwuchs-eating fishes in Lake Tanganyika. *Env. Biol. Fishes* **10**, 225–241.
Tiercelin, J.J. and Mondeguer, A. (1991). The geology of the Tanganyika trough. In: *Lake Tanganyika and its Life* (Ed. by G.W. Coulter), pp. 7–48. Oxford University Press, Oxford.
Tiercelin, J.J., Chorowicz, J., Bellon, H., Richert, J.P., Mwanbene, J.T. and Walgenwitz, F. (1988). East African rift system: offset, age and tectonic significance of the Tanganyika-Rulwa-Malawi intercontinental transcurrent fault zone. *Tectonophyiscs* **148**, 241–252.
Trendall, K. (1988a). Recruitment of juvenile mbuna (Pisces: Cichlidae) to experimental rock shelters in Lake Malawi, Africa. *Env. Biol. Fishes* **22**, 117–131.
Trendall, K. (1988b). The distribution and dispersal of introduced fish at Thumbi West Island in Lake Malawi, Africa. *J. Fish Biol.* **33**, 357–369.
Walter, B. (1991). Conflict of interest in a harem: behavioural observations on the snail cichlid *Lamprologus ocellatus* (Steindachner 1909). *Verh. Dtsch. Zool. Ges.* **84**, 333–334.
Walter, B. and Trillmich, F. (1994). Female aggression and male peace-keeping in a cichlid fish harem: conflict between and within the sexes in *Lamprologus ocellatus*. *Behav. Ecol. Sociobiol.* **34**, 105–112.
West-Eberhard, M.J. (1983). Sexual selection, social competition and speciation. *Quart. Rev. Biol.* **58**, 155–183.
Wheeler, A. (1978). *Key to the Fishes of Northern Europe*. Frederick Warne, London.
Witte, F., Goldschmidt, T., Wanink, J.H., van Oijen, M., Goudswaard, K., Witte-Maas, E. and Bouton, N. (1992). The destruction of an endemic species flock: quantitative data on the decline of the haplochromine cichlids of Lake Victoria. *Env. Biol. Fishes* **34**, 1–28.
Worthington, E.B. (1937). On the evolution of fish in the great lakes of Africa. *Int. Rev. Hydrobiol.*, **35**, 304–317.
Worthington, E.B. (1940). Geographical differentiation in fresh waters with especial reference to fish. In: *The New Systematics* (Ed. by J. Huxley), pp. 287–303. Clarendon Press, Oxford.
Worthington, E.B. and Lowe-McConnell, R.H. (1994). African lakes reviewed: creation and destruction of biodiversity. *Environmental Conservation* **21**, 199–213.

Yamaoka, K. (1982). Morphology and feeding behaviour of five species of *Petrochromis* (Cichlidae). *Physiol. Ecol. Japan* **19**, 57–75.

Yamaoka, K. (1983). Feeding behaviour and dental morphology of algae-scraping cichlids (Pisces: Cichlidae) in Lake Tanganyika. *African Study Monogr.* **4**, 77–89.

Yamaoka, K. (1985). Feeding behaviour and dental morphology of Aufwuchs-eating cichlid fishes in Lake Tanganyika. *Verh. Internat. Verein. Limnol.* **22**, 2260.

Yamaoka, K. (1987) Comparative osteology of the jaw of algal-feeding cichlids (Pisces, Teleostei) from Lake Tanganyika. *Rep. Usa Mar. Biol. Inst. Kochi Univ.* **9**, 87–137.

Yamaoka, K. (1991). Feeding relationships. In: *Cichlid Fishes: Behaviour, Ecology and Evolution* (Ed. by M.H.A. Keenleyside), pp. 152–172. Chapman and Hall, London.

Yamaoka, K., Hori, M. and Kurutani, S. (1986). Ecomorphology of feeding in 'goby-like' cichlid fishes in Lake Tanganyika. *Physiol. Ecol. Japan* **23**, 17–29.

Yanagisawa, Y. (1985). Parental strategy of the cichlid fish *Perissodus microlepis*, with particular reference to intraspecific brood "farming out". *Env. Biol. Fishes* **12**, 241–249.

Yanagisawa, Y. (1986). Parental care in a monogamous mouthbrooding cichlid *Xenotilapia flavipinnis* in Lake Tanganyika. *Jpn. J. Ichthyol.* **33**. 249–261.

Yanagisawa, Y. (1987). Social organization of a polygynous cichlid *Lamprologus furcifer* in Lake Tanganyika. *Jpn. J. Ichthyol.* **34**, 82–90.

Yanagisawa, Y. (1991). The social and mating system of the maternal mouthbrooder *Tropheus moorii* (Cichlidae) in Lake Tanganyika. *Jpn. J. Ichthyol.* **38**, 271–282.

Yanagisawa, Y. (1993). Long-term territory maintenance by female *Tropheus duboisi* (Cichlidae) involving foraging during the mouth brooding period. *Ecol. Freshwater Fish* **2**, 1–7.

Yanagisawa, Y. and Nishida, M. (1991). The social and mating system of the maternal mouthbrooder *Tropheus moorii* (Cichlidae) in Lake Tanganyika. *Jpn. J. Ichthyol.* **38**, 271–282.

Yanagisawa, Y. and Nshombo, M. (1983). Reproduction and parental care of the scale-eating fish *Perissodus microlepis* in Lake Tanganyika. *Physiol. Ecol. Japan* **20**, 23–31.

Yanagisawa, Y. and Ochi, H. (1991). Food intake by mouthbrooding females of *Cyphotilapia frontosa* (Cichlidae) to feed both themselves and their young. *Env. Biol. Fishes* **30**, 353–358.

Yanagisawa, Y., Nishida, M. and Niimura, Y. (1990). Sexual dimorphism and feeding habits of the scale-eater *Plecodus straeleni* (Cichlidae, Teleostei) in Lake Tanganyika, *J. Ethol.* **8**, 25–28.

Yanagisawa, Y. and Sato, T. (1990). Active browsing by mouthbrooding females of *Tropheus duboisi* and *Tropheus moorii* (Cichlidae) to feed the young and/or themselves. *Env. Biol. Fishes* **27**, 43–50.

Yuma, M. (1993). Competitive and co-operative interactions in Tanganyikan fish communities. In: *Mutualism and Community Organization: Behavioural, Theoretical, and Food-Web Approaches* (Ed. by H. Kawanabe, J.E. Cohen and K. Iwasaki), pp. 213–227. Oxford University Press, Oxford.

Yuma, M. (1994). Food habits and foraging behaviour of benthivorous cichlid fishes in Lake Tanganyika. *Env. Biol. Fishes* **39**, 173–182.

Building on the Ideal Free Distribution

I.	Summary	253
II.	Introduction	254
III.	The Ideal Free Distribution	255
IV.	The Simplest Ideal Free Model	256
V.	Incorporating Interference	257
VI.	Incorporating Resource Wastage into Continuous Input Models	260
VII.	Incorporating Despotism	260
VIII.	Incorporating Unequal Competitors	262
	A. Continuous Phenotype Unequal Competitor Models	262
	B. The Isoleg Theory	265
	C. Unequal Competitor Distribution Determined by the Form of Competition	268
IX.	Incorporating Kleptoparasitism	269
	A. Modifying the Simple IFD Model to include Kleptoparasitism	269
	B. A Kleptoparasitic IFD Model based on a Functional Response	270
X.	Incorporating Resource Dynamics	273
	A. Continuous Input Models and Standing Crops	273
	B. Decision-making Prey	274
XI.	Incorporating Survival as a Measure of Fitness	275
XII.	Incorporating Perceptual Constraints	277
XIII.	Incorporating Patch Assessment	278
	A. Patch Assessment Models	278
	B. Experimental Investigations of Patch Assessment	282
XIV.	A Summary of IFD Models	284
XV.	Assessing the Success of Ideal Free Distribution Models	289
	A. Testing the Ideal Free Distribution	289
	B. Does the Theory pass the Test?	289
XVI.	Conclusions	300
	Acknowledgements	301
	References	302

I. SUMMARY

The distribution of animals around their environment is one of the corner-stones of ecology. This review aims to show how the question of animal

ADVANCES IN ECOLOGICAL RESEARCH VOL. 26
ISBN 0–12–013926–X

distributions has been approached from the point of view of individual behaviour. The basis for much of the work on this subject has been the ideal free distribution theory, which considers that the suitability of any area of the environment will be a function of the density of competitors occurring there. I review how ideal free theory has been developed and tested empirically since its inception.

The original concept of an ideal free distribution was stated in very general terms and based on a number of assumptions, several of which have been removed by subsequent advances in the theory. The mechanisms of competition have been considered, producing two major classes of models, those dealing with exploitation competition and those which consider competition resulting from interference, including kleptoparasitism and resource monopolization. Empirical work has revealed that foragers frequently differ in their competitive abilities, inspiring a number of theoretical advances, and a corresponding range of predictions. Consideration has also been given to the effects of optimally distributed foragers on resource dynamics and the influence of an individual's energetic state on its optimal strategy.

Numerous empirical studies have been conducted into animal distributions, many of which have been cited as supporting the predictions of the IFD theory. However, in many cases, the evidence supporting a model is weak, indicating the need for experiments specifically designed to test the assumptions and predictions of a particular model. Experimental and field studies are reviewed and their support or otherwise for the various models is assessed. An important contribution of empirical work has been in identifying constraints such as knowledge of resource distribution and ability to discern small differences in patch profitabilities which alter the optimal distribution. Models incorporating patch assessment and perceptual constraints, designed to increase the applicability of the theory are discussed. Finally, consideration is given to the success of ideal free distribution models in predicting animal distributions.

II. INTRODUCTION

'We can now define a habitat distribution which will provide a reference for the study of dispersive populations. This is the ideal free distribution. It rests on assumptions of habitat suitability and the adaptive state of organisms.' Fretwell, 1972.

Why animals are found where they are is one of the major questions facing ecologists. In this review, I concentrate on attempts to address the reasons for the distribution of populations through consideration of decisions made by individuals. This "bottom up" approach is particularly important because the micro-distribution and in some cases the macro-distribution of populations is determined by individual decisions. Therefore, if we are to understand population distributions, we must ultimately consider individuals.

In modelling the distribution of individuals, an assumption is made that natural selection will have given rise to genes equipping individuals with behavioural rules which enable them to optimize their choice of where and when to forage. Since the success of genes is often difficult to ascertain we also assume that an individual's success is an accurate measure of the success of the genes which program it.

Frequently, the main variable affecting which part of the environment provides the best conditions for fitness maximization will be the presence of other individuals. Animals may influence each other's success through various density dependent processes. These include exploitation competition, in which the quantity of available resources is reduced; interference, in which fitness is reduced by factors such as wasting time in interactions with other foragers or disturbing prey; and the influence of density on predation risk. Therefore, animals will be distributed not only in relation to the resources they require, but also in relation to their competitors.

A number of models have been devised to predict the optimal distribution of competitors having some density dependent relationship between their numbers and their fitness in a particular part of the environment. The most successful class of such models can be collectively described as "ideal free distribution" (IFD) models since they are all based on a few key assumptions which are described in the original IFD paper published in 1970. I will describe how the original theory has been developed to include more realistic assumptions, and examine whether the predictions of IFD models are supported by experimental and field studies.

III. THE IDEAL FREE DISTRIBUTION

The term "ideal free distribution" was coined by Fretwell and Lucas (1970). It describes the distribution of animals which are "ideal", meaning that they will go to the patch where their intake rate is highest, and "free" in that they are able to enter any patch on an equal basis with residents, i.e. without restriction or costs in terms of time or energy. Fretwell and Lucas's essential insight was that the "suitability" of a particular part of the environment will decrease with an increase in the density of individuals occurring there. Therefore, if the environment contains several patches of different suitabilities, newly arriving competitors will initially choose to occupy the best patch. However, as their numbers increase, density dependent effects on better patches will lead to previously poorer patches becoming equally suitable. This process occurs across the whole environment and leads to an equilibrium at which all patches are equally suitable.

Conclusions similar to those drawn by Fretwell and Lucas were also reached independently by other biologists investigating resource competition. Orians (1969) predicted that the number of females on a male's territory would be related to the level of resources occurring there, clearly pre-empting the thinking which went to make up the IFD theory. He recognized that as the best patches become crowded the average reproductive success will be expected to drop, and hence that animals in lower quality patches do not necessarily have lower fitness. In the same year, Brown (1969), studying great tits *Parus major* came to similar conclusions. Despite couching his argument in terms of group selection theory, he recognized that "the optimal mix for maximum production requires that some individuals breed in relatively poor habitats", and that the poor habitat and the good habitat would both be exploited in relation to their value. Parker (1970, 1974) used identical logic and assumptions to that in the IFD when he described an "equilibrium position" of the distribution of male dungflies *Scathophaga stercoraria* around cowpats where they search for females. In field observations he found a distribution which conforms well to that predicted by both theories, with male dungflies occurring in proportion to the number of females arriving in patches around a cowpat.

The IFD theory has been used as the basis for numerous models each addressing one or two assumptions of the original theory. There has not been a clear progression, with one assumption being removed after another to produce an increasingly accurate model of the real world. Rather, different authors have tended to use the original IFD as a basis, and have changed that aspect which they consider most wanting.

IV. THE SIMPLEST IDEAL FREE MODEL

In order to produce a testable model, the very general terms of Fretwell and Lucas's theory need to be replaced by specific quantities. The simplest way to do this is to consider a single aspect of habitat suitability, and some linear density dependent effect on it. An example of such a situation can be described if we consider suitability in terms of the intake rate in a patch and produce density dependence by having animals compete for resource items (Parker and Stuart, 1976; Parker, 1978). Increasing competitor density will have a proportional effect on intake rate if resources are simply "shared out" amongst all individuals in a patch. This can occur if patches consist of areas where resource items arrive continuously, and are "utilized" at the same rate as they arrive. This might apply to fish feeding on food items dropping onto the surface of a pool, or males competing for females arriving at a particular patch (e.g. Parker, 1970). Individual intake rate depends only on the input rate into the patch and on how many competitors share the patch. As the number of competitors in the patch increases, each individual gains a smaller and smaller proportion of the resources until the average gain rate becomes so low that the animal would do better to move to a lower quality patch that is subject to less competition. This process occurs over the entire environment until an evolutionarily stable state (ESS) (Maynard Smith, 1982) is reached at which point no individual can do any better by moving from its patch to another patch. At this point all individuals have equal gain rates regardless of which patch they occupy. This situation, known as "continuous input" can be expressed algebraically as:

$$W_i = R_i/n_i \qquad (1)$$

where: W_i = the average gain rate in patch i
R_i = the input rate into patch i
n_i = the number of competitors in patch i

Since gain rates in all patches must be equal, $W_i = R_i/n_i = W_j = R_j/n_j =$ constant for all patches i, j, k, etc. This gives the "input matching rule" (Parker, 1978): $n_i = R_i/c$ (constant), which predicts that the number of competitors in a patch should be proportional to the total input received by that patch. This rule has been supported by numerous studies of continuous input situations. In the field, male dungflies have been shown to match the input of females to a cowpat (Parker, 1978) and grazing catfish (Loricariidae) have been found to distribute themselves such that the standing crop of algae in pools receiving different levels of light is the same (Power, 1983). Numerous experimental studies of distribution between two feeding stations (e.g. Milinski, 1979; Harper, 1982; Godin and Keenleyside, 1984) have also supported input matching. It has also been shown that input matching can occur over time as well as space. Parker and Courtney (1983) and Iwasa et al. (1983) found that male emergence in two species of butterflies matched the mate input rate owing to female emergence as predicted.

V. INCORPORATING INTERFERENCE

In the simple continuous input model, density dependence results solely from exploitation competition. However, although a simple system such as this is convenient for testing the basic prediction of the IFD, it is only one type of foraging situation, and one that is probably relatively uncommon in the wild. Another form of competition is likely to occur if animals search patches for dispersed resources, and are hence limited by search time as

well as resource abundance. This form of competition is called "interference" and can be defined as "a short term, reversible decline in intake
rate as a result of the presence of others" (Goss-Custard, 1980). Interference reduces intake through mechanisms such as time wasted in interactions with competitors or disturbing prey. For instance, oystercatchers
Haematopus ostralegus interfere with each other by stealing prey items,
overt aggression and avoidance of conspecifics (Ens and Goss-Custard,
1984, Goss-Custard *et al.*, 1984). In this type of situation there is now no
certainty that any competitors should be found in lower quality patches.
The level of interference will only dictate the equilibrium distribution when
time wasted in interacting with competitors reduces intake rate, lowering
patch profitabilities and causing dispersal of competitors onto lower quality
sites. Allowing for the effect of interference requires a term which modifies
the influence of competitor density on intake rate. Sutherland (1983) proposed that the "interference constant" m (Hassell and Varley, 1969) be
used such that:

$$W_i = Q_i/n_i^m \quad \text{Where } 0 \leq m \leq \infty \tag{2}$$

In comparison to the continuous input model, intake is now not simply
limited by input rate and competitor density but also by search time.
Therefore, R_i is replaced by Q_i which is a combined measure of search
efficiency and resource density in patch i. Q is equal to the intercept of a
graph of log (intake rate) versus log (competitor density) (Fig. 1) and
hence is related to the maximum intake an individual would achieve in the
absence of competitors (Fig. 1). m is defined as the slope of this graph, so
that the magnitude of the effect of density on intake rate dictates the level
of m.

The strength of interference (the level of m) has important consequences
for competitor distribution. In contrast to continuous input situations, a
patch with half the density of resources relative to the best patch is not
automatically predicted to receive half the number of foragers. If m is zero,
all competitors will go to the good patch. However this is rarely the case.
For instance, Goss-Custard (1980) found that oystercatchers interfered
with each other by fighting and disrupting prey, and it is probably safe to
assume that this is the normal situation for most foraging vertebrates. If
there is some interference ($m > 0$) it will pay some individuals to move to
lower quality patches where interference is lower, and an equilibrium
distribution in which competitors in different quality patches achieve equal
gain rates will again result. The proportion of competitors in patches of
different intrinsic qualities will depend on the level of m. The higher the
level of m, the smaller the proportion of individuals in the better patch.

If we consider competitors to be predators and their resources as prey, it
is apparent that the level of m will also have important effects on prey

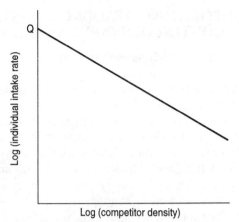

Fig. 1. The definition of Q and m. Q is the intercept of the log graph and m is the slope. This hypothetical relationship between density and intake rate is the basis for the IFD interference models discussed below.

distribution. When m is low, predators are aggregated, which tends to lead to density dependent prey mortality. When $m = 1$ there is a constant predator:prey ratio so mortality is density independent, and predator distribution reflects prey distribution. When $m > 1$ then low density patches will be disproportionately predated, hence predation is inversely density dependent.

There are two predictions implicit in the interference model which are not stated in previous reviews, and which have not always been recognized in studies attempting to apply the theory. Firstly, the equation predicts that if there is any interference at all ($m > 0$), all patches should be used, regardless of competitor density. Secondly, the proportion of competitors in each patch should be constant, and similarly independent of the overall competitor density. The basis for these predictions can be seen if we rearrange the equation $Q_i n_i^{-m} = Q_j n_j^{-m}$:

$$Q_i/Q_j = n_j^{-m}/n_i^{-m},$$

$$\log (Q_i/Q_j)/m = \log(n_i/n_j) \quad (3)$$

Since Q and m are positive constants, the ratio n_i/n_j must also be a constant, and can never equal zero. The model does not consider animals to be indivisible units, whereas, in reality a situation could exist in which the poorest patch is not used because it does not provide an equal intake rate to an entire competitor. However, in large populations these integer effects should not be important. This prediction is the result of the implicit assumption that the relationship between interference and density is linear.

VI. INCORPORATING RESOURCE WASTAGE INTO CONTINUOUS INPUT MODELS

Sutherland and Parker (1992) suggest that simple continuous input situations can be modelled using the same equation as that described for interference competition and $m = 1$. Incorporating m allows the equation to be applied to more complex situations in which some resource items go unexploited or are destroyed owing to increasing numbers of competitors. For instance, competing male tiger blue butterflies *Tarucus theophrastus* sometimes fail to mate with some females, and do so to a disproportionately greater extent at higher male densities (Courtney and Parker, 1985). These situations can be modelled by using $m > 1$, so that intake is depressed by wastage as well as sharing, with greater wastage at higher density. In modelling continuous input using the same equation as used for interference, it is important to recognize that continuous input situations are not simply the point on a continuum of increasing interference, at which $m = 1$. Continuous input is exploitation competition without depletion, and happens to have the same dynamics as an interference situation in which $m = 1$. Although Milinksi and Parker (1991) describe continuous input as being high interference, many ecologists would not use the word interference in this context. Confusion over the difference between continuous input and interference models has led to some authors comparing the results of non-continuous input studies situations with the erroneous prediction that the number of competitors in a patch will always match the density of resources (see Tregenza, 1994). In recognition of the independence of the two constants, I will avoid using m to define the extent that increasing competitor density affects resource wastage and will instead call the constant f:

$$W_i = R_i/n_i^f \qquad \text{Where } 1 \leq f \leq \infty \qquad (4)$$

If $f = 1$ all resources are utilized. Higher values of f represent a greater influence of competitor density on resource wastage.

VII. INCORPORATING DESPOTISM

The ideal despotic distribution (IDD) was proposed by Fretwell (1972), to describe distributions in which animals guard resources, violating the "free" assumption of the IFD. If the residents of a habitat make it dangerous for unsettled individuals to enter, then the average success of non-residents will be lower than the habitat average, meaning they are not free. This assumption seems well justified by studies which have found that incumbent individuals have an advantage in fights. Speckled wood butterflies *Pararge aegaria*, for instance, engage in battles for sunlit patches which are nearly always won by the resident individual (Davies, 1978). Similarly, shrews *Sorex araneus* holding territories have an advantage in competitive interactions over non-residents (Barnard and Brown, 1982). Fretwell based his model on the density limiting hypothesis of Huxley (1934), who envisaged a territory as a rubber disc which can be compressed, but requires increasing force to do so as it gets smaller. The IDD assumes that as the density of territory holders in a patch increases, so does the advantage of holding a territory, making it progressively more difficult for newcomers to settle. This assumption is also supported by observation. For instance, Harris (1964) showed that the centre of a lizard's territory is more vigorously defended than the edge.

In the model, the profitability of a patch to a newcomer is a function of the profitability of the patch to incumbent individuals multiplied by a density dependent factor which reflects the advantage of holding a territory. Competitors entering a habitat choose the patch where their gains will be highest, leading to an equilibrium where the profitability of all patches to newcomers is equal.

Several authors investigating animal distributions have found evidence of despotic behaviour, and higher gain rates for competitors in better patches. For example, Andren (1990) found that in his territorial population of Jays *Garrulus glandarius*, birds holding territories in areas dominated by Norway spruce *Picea abies* had higher breeding success. Similarly, Patterson (1985) found that there was a higher density of convict cichlids *Cichlasoma nigrofasciatum* in better quality habitats, but that some individuals were excluded from these areas. In both these studies (see also "Testing the IFD" Section XV.A. for further examples), observation of territoriality and the lack of alternative models has led to the IDD being cited as a valuable predictor of despotic distributions. However, this is because it comes closer to predicting the distribution than the IFD, which is insufficient grounds for support, since support requires that some unique prediction is fulfilled. In fact, although the IDD is an elegant theory, it is scarcely a predictive model at all. Its only unique prediction is that otherwise equal competitors will differ in success if some hold territories and some do not. Otherwise, it merely states that if animals are despotic, newly arriving individuals will be forced into lower quality patches, and that the higher the density of incumbent individuals the greater will be the depression of patch quality available to new arrivals. The IDD is also limited in its applicability to real situations for the same reasons as the original ideal free distribution, something Fretwell noted: "Like the ideal free distribution the ideal despotic distribution is a useful basis for discussion but because of underlying assumptions can only approximate any real situation."

It is obvious that resource defence will have a profound effect on predator distributions, and the question of optimal territory size is in itself a major topic of research (see Davies and Houston, 1984 for review). However, the effects of territoriality on animal distributions are still poorly understood, and lack the framework for investigation which is provided for non-despotic situations by the various IFD type models. More work in this field would be very valuable, although the combination of intrinsically unequal competitors and the competitive advantage of holding a territory provide a formidable challenge both to modellers and to quantitative experimental work.

VIII. INCORPORATING UNEQUAL COMPETITORS

Individuals within a population are rarely of equal competitive ability. For instance, the largest sexually mature male salmon *Salmo salar* may be as much as 250 times the weight of the smallest (M. Gage, pers. comm.). Such differences will have important implications for distribution and can be divided into two distinct types. They may be short term only, and owing to factors such as age, or animals using different behavioural strategies, that lead to short-, but not long-term differences in competitive ability. Alternatively, differences may be long term, with some phenotypes inherently superior to others.

A. Continuous Phenotype Unequal Competitor Models

Parker (1982) presents a model in which there are two patches and two phenotypes. The phenotypes are assigned "competitive weights" which are measures of their relative abilities. A phenotype with twice the competitive weight of another will gain twice the payoffs in the same patch. This concept is used by Sutherland and Parker (1985, 1992) and Parker and Sutherland (1986), to model predator distributions with unequal competitors of several phenotypes. The payoffs to an individual are dependent on its competitive weight relative to the mean or total competitive weight in the patch. Again, two different classes of model have been developed to deal with situations in which competition is solely through exploitation of continuous input resources and situations in which competition is solely through interference.

1. Unequal Competitor Continuous Input Model

If resources are simply divided between competitors in a patch (as in continuous input situations without interference), then the intake rate of a competitor of phenotype A can be calculated by:

$$W_{Ai} = ((K_{Ai}/K_i)R_i)/n_i \qquad (5a)$$

Where W_{Ai} = the intake rate of an individual of phenotype A in patch i
K_{Ai} = the competitive weight of A in patch i
K_i = the total competitive weight in patch i
R_i = the input rate into patch i
n_i = the number of competitors in patch i

In this model, instead of individual intakes being equal, intake per competitive weight unit is equal across individuals and patches. Therefore, each individual simply receives the proportion of resources equal to its proportion of the total competitive weight in the patch. Since we are dealing only with exploitation competition, animals are considered to differ only in how good they are at gaining a share of incoming resources (their exploitation efficiency). Unlike interference, exploitation efficiency is considered to be independent of competitors. Hence, proportional differences in intake rate between individuals will remain constant across patches with different input rates and competitor densities. This prediction is discussed below and can be described graphically as in Figure 2(a).

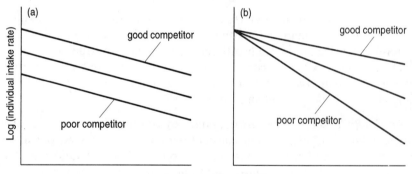

Log (competitor density)

Fig. 2. (a) Phenotype scales intercept, as suggested for continuous input situations. (b) Phenotype scales slope as suggested for situations where different phenotypes suffer different levels of interference. (After Sutherland and Parker, 1985.)

2. Unequal Competitor Interference Model

The same system of competitive weights can be used to predict individual intake rate in situations where resources are assumed to remain at constant density, and competition is solely the result of interference. As in the equal competitor interference model, the effect of competitor density is scaled by the interference constant m. However, because the relative competitive ability of an individual will affect how increased density affects intake, m is scaled by (K_i/K_{Ai}):

$$W_{Ai} = Q_i n_i^{-m}(K_i/K_{Ai}) \qquad (5b)$$

This equation predicts that an individual with a low competitive weight will have its intake reduced more by increased density on a patch than one with a high competitive weight. Therefore lower ranked competitors will suffer more from increased density, and so will do best in low density patches. This means that the relative payoffs of individuals of different phenotype will vary between patches of different quality, and hence different competitor density (Fig. 2(b)).

3. Distributions Resulting from the Models

If logs are taken of equations (5a) and (5b) they can be expressed in terms of the graph equation $y = c + mx$.

$$\log W_{Ai} = \log (K_{Ai}/K_i) + \log Q_i - m \log n_i \qquad (6a)$$

$$\log W_{Ai} = \log Q_i - m(K_i/K_{Ai}) \log n_i \qquad (6b)$$

It can be seen that in equation (6a) phenotype differences affect the intercept whereas in equation (6b) they affect the slope (see Fig. 2). This difference can lead to radically different predictions of distribution.

Depending on whether competitive ability has the same effect across patches, two main types of distribution were found by Parker and Sutherland (1986) and Sutherland and Parker (1992).

(a) Multiple equilibrium distributions. In continuous input situations where relative payoffs remain constant across patch types no one ESS prevails. A range of distributions can occur provided that the ratio between the sum of competitive weights and the input rate in each patch remains constant across all patches. The equilibrium attained depends on the starting conditions.

(b) ESS truncated phenotype distribution. In interference situations, patches with high resource density have more competitors and hence higher interference. Phenotype scales the slope of a log(intake) against log(density) graph (as in Fig. 3), so the higher the density of competitors, the greater the difference in the relative intakes of better and poorer competitors. Better competitors will hence achieve their highest intake in the high resource density, high interference patches. Poor competitors will do best in the low interference patches. This produces a situation in which there is an absolute correlation between competitive ability and patch quality, which can be described as "truncated distribution". There may be phenotypes which have a competitive ability such that at the ESS they do equally well on either of two patches adjacent in the rank order of patch qualities. However, there can only be one "boundary phenotype" for each adjacent pair of patches since any increase or decrease in the competitive weight of a boundary individual will lead to one of the two patches becoming optimal.

If a mutant individual arises (which is not a boundary phenotype), which does not choose the patch correlated with its competitive weight, it will experience reduced gains. If it moves to a higher input patch, it will suffer because its high (m × relative competitive weight) value means that it will be adversely affected by interference and consequently its intake will be reduced. If it moves to a lower input patch it will be unable to benefit sufficiently from the lack of interference to make up for the reduced input and so again its intake will fail.

To summarize, when phenotype has the effect of scaling the *slope* of the equation as in the interference model, then more than one phenotype cannot mix across two patch types. This is a prerequisite for a truncated phenotype distribution, and excludes the possibility of multiple stable equilibria. When phenotype has the effect of scaling the *intercept* of the equation as in the model based on a continuous input situation then several stable equilibria are possible, with the same phenotypes occurring on different patches. A truncated phenotype distribution is ruled out unless relative payoffs change across patches. This could occur if there is some extraneous reason for differential advantage across patches. For example, predators which had learnt to identify a particular type of prey would only find this advantageous in patches in which that prey type occurred. The distribution would then depend on the nature of the variation in relative payoffs.

Sutherland and Parker (1992) examined these conclusions by computer simulation. Using the intercept scaling models, distributions reached an equilibrium dependent on the starting conditions with the number of individuals occurring at each site dependent on the distribution of phenotypes. Using the slope scaling model, the final distribution was independent of the starting conditions. In all cases a truncated phenotype distribution occurred with no more than one phenotype mixing across the same pair of patch types.

B. The Isoleg Theory

A similar conclusion to that drawn from Parker's (1982) two phenotype, differing relative payoffs model was independently reached by Rosenzweig (1974, 1981, 1986). However, his model differs in a number of important ways. Competitors are considered to be members of two different species, each of which has a specific competitive ability which is different from the other species. The environment consists of two habitats of different quality. Both species prefer the richer habitat and both are affected by intraspecific and interspecific competition. The subordinate species is strongly affected by interspecific competition, whereas the only interspecific effect suffered by the dominant species is a slight depression in the value of the poorer patch. Although the theory is expressed in terms of two different species, these are functionally identical to different phenotypes in a single species model.

The theory can be represented as a graph of species habitat selection. As the density of each species changes it will have implications for its own distribution and for that of the other species. On this graph, lines can be drawn to represent the transition points at which members of the species switch from exploiting only the good patch to exploiting both patches, and from exploiting both patches to exploiting just the poorer patch. Rosenzweig (1981) terms these lines 'isolegs' after the Greek *iso*-same and *lego*-choice. An example of an isoleg graph is shown in Fig. 3.

The isoleg model differs from the unequal competitor IFD models discussed so far in that it does not attempt to make quantitative predictions. Also, it considers the absolute size of populations, and limits them with an environmental carrying capacity. Therefore, unlike Sutherland and Parker's unequal competitor IFD models, it predicts that at low overall competitor density, both species will be found on the better patch. As density of either species increases some subordinates will eventually be forced onto the poorer patch, since it is assumed that (as in the unequal competitor IFD with interference), better competitors have a greater advantage on the better patch. If dominant density increases further, all subordinates will be forced onto the poorer patch. Dominants only use the poorer patch when their own density rises sufficiently; subordinates cannot force them onto the poorer patch as it is assumed that their density cannot increase this far.

Pimm *et al.* (1985) tested the predictions of the isoleg theory using 3 species of hummingbirds. This system is remarkably well suited to tests of habitat selection with interspecific competition since:

(i)   The species can be separated into two groups, which do not overlap in competitive ability. Blue-throated hummingbirds *Lampornis clemenciae* are always dominant over Rivoli's *Euglenes fulgens* and Black-chinned *Archilochus alexandri*. There is apparently no dominance relationship between the two subordinate species.

(ii) Their food supply can be manipulated very finely since at the study site they are reliant on sucrose solution provided at feeders, which can be controlled in value (strength of solution) and location.

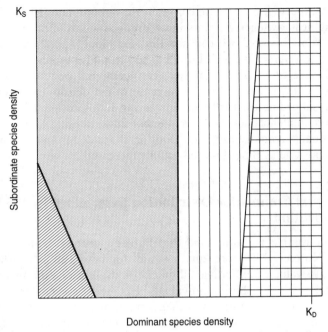

Fig. 3. Isoleg diagram of habitat selection of a dominant and a subordinate species. The subordinate has two isolegs (the thicker lines). In the diagonally shaded region all subordinates select the better habitat. In the stippled region intraspecific and/or interspecific competition leads to both patches being exploited and in the vertically shaded region all subordinates are found in the poorer patch. The dominant either uses only the good habitat to the left of the thin isoleg, or uses both patches to the right (horizontal shading). The K values are the environment's carrying capacity for each species. (After Pimm *et al.*, 1985.)

(iii) The population can be manipulated, both by trapping birds to remove them from the population and by taking advantage of seasonal variations as they migrate into the area.

(iv) The relationship between forager numbers and food resources can be estimated by using a surrogate measure, the time spent at feeders. Since during the summer, feeding is overwhelmingly dominant in hummingbird time budgets, observations of birds feeding was used as an alternative to measuring local densities.

(v) They are easy to identify and observe.

Observations and manipulations of the humming birds using two different concentrations of sucrose in feeders showed habitat selection in accordance with the predictions of the model. Dominant (Blue-throated) hummingbirds switched from feeding at only the good feeders to using both with increased intraspecific density but never used only the poor patch, as expected. Even more convincingly, subordinate (Black-chinned) hummingbirds switched habitats in accordance with the predicted isoleg graph. From Figure 3 we can see that the slope of the first isoleg is predicted to be negative whilst that of the second is predicted to be vertical. In Pimm *et al.*'s study, at low density, increased use of the poorer patch by subordinates resulted from an increase in the density of either dominants or subordinates. However, increasing levels of use of the poorer patch were influenced less and less by subordinate density, so the slope of a line of equal percentage use of the poor patch rotated from being negative to being positive as Blue-throated activity increased.

C. Unequal Competitor Distribution Determined by the Form of Competition

As an alternative to Sutherland and Parker's unequal competitor IFD models, Korona (1989) presents a model for competition in which the competitive value of an individual manifests itself in one on one contests with other randomly chosen individuals. His model assumes that animals are aware of patch qualities and that competitive weights of phenotypes remain constant across patches. In each encounter, the chance of success of an individual A is given by:

Chance of success of A =

$$\frac{competitive\ weight\ of\ A}{comp.\ weight\ of\ A\ +\ comp.\ weight\ of\ adversary} \quad (7)$$

Intake rate is determined as a function of the resources per animal and the success of each individual in its conflicts with other competitors. The model predicts a unique equilibrium distribution, which has three features.

(i) Animals distribute themselves numerically in proportion to patch qualities.
(ii) Each phenotype occurs in the same proportion in each habitat.
(iii) The average gain for all individuals is equal in all patches.

Korona suggests that one on one competition is more likely than each individual simultaneously affecting all others, an implicit assumption in Sutherland and Parker's models. A disadvantage of Korona's model is that it can only be applied to a situation in which resources are simply divided between all the competitors (such as continuous input). However, if density dependence is the result of interference, rather than exploitation, this is unlikely to be the case. This assumption of Korona's model is unfortunate since his prediction that competitive ability will manifest itself through one to one contests would appear to be least likely in continuous input situations, where scrambles for resource items are likely to occur. The biological situations in which one on one competition is most obvious, such as oystercatchers occasionally attempting to steal mussels from each other, all tend to be interference situations, which clearly violate the other assumptions of the model. Nevertheless, Korona's model serves to illustrate the importance of checking assumptions, and indicates an alternative form of competition which may occur when competitors are unequal.

IX. INCORPORATING KLEPTOPARASITISM

A. Modifying the Simple IFD Model to Include Kleptoparasitism

In many animals competition occurs through loss of resource items to other individuals. For instance, Fulmars *Fulmarus glacialis* holding a piece of fish are likely to be subject to repeated attacks by conspecifics and will frequently be forced to relinquish the item (Enquist *et al.*, 1985). This type of interference, termed kleptoparasitism, is distinct from other forms since density on a patch need not affect mean gain rate if competition is only the result of kleptoparasitism. If animals steal resources from each other, and time wasted in fights is negligible compared with search time, it will lead to differences in individual gain rate, but not average gain rate. Parker and Sutherland (1986) investigated this situation using a model in which there is a dominance hierarchy of phenotypes and competitors steal food from their subordinates. In the simplest scenario, kleptoparasitism simply leads to reallocation of resources between individuals which differ in stealing ability. Gains and losses to a competitor increase linearly with the number of individuals respectively above and below it in the hierarchy. This can be expressed algebraically as:

$$W_i = S_i(c + G(n_s) - L(n_d)), \qquad (8a)$$

in which S_i = Capture rate in patch i

c = each capture by the individual

$G(n_s)$ = probable gains from subordinates where there are n_s subordinates

$L(n_d)$ = probable losses to dominants where there are n_d dominants

This simplest relationship between subordinate and dominant density is one in which gains and losses increase linearly with the number of subordinates above and below an individual in the hierarchy, i.e. $G(n_s) = an_s$ and $L(n_d) = an_d$ in which a is a positive constant which is small enough to ensure that payoffs are never negative.

This model can be expanded to include some density dependence, by imagining that a fixed input of resource items R_i is shared between the competitors in a patch as in a continuous input situation. The model assumes that intake is reduced by exploitation competition, in which all competitors are equal, and that subsequently competitive inequalities lead to reallocation of resources:

$$W_i(D) = R_i(c + G(n_s) - L(n_d))/n_i \qquad (8b)$$

Parker and Sutherland ran simulations using both the simple and the density dependent versions of their kleptoparasitism model, and 100 competitors with 10 different ranks. They found that unless patch qualities differed considerably, both models produced unstable distributions, since it pays dominants to move to patches containing subordinates but pays subordinates to move away from patches containing dominants. Payoffs did not remain constant across patches and there was no clear truncation of phenotypes. However, there was a tendency for better phenotypes to be found in better patches more of the time and for low ranking phenotypes to be found in all patch types. As in other models, due to density dependent gains the continuous input model was least sensitive to differences in patch qualities. In the simple model, great differences in patch qualities lead to the poorer patch being avoided by all phenotypes.

Pulliam and Caraco (1984) produced a similar result in a verbal model of the distribution of a subordinate and a dominant individual between two patches. They assumed that individuals had a true estimate of their feeding rate in each of the four possible situations (in either patch with or without the other individual) and that an individual would only change patch if it could expect a higher feeding rate. Travel costs were taken to be negligible. Like Parker and Sutherland (1986) they found that when there is no intrinsic advantage to being together the dominant individual chases the subordinate from patch to patch.

B. A Kleptoparasitic IFD Model Based on a Functional Response

A different approach to modelling competitive distributions has been suggested by Holmgren (1995). His numerical model makes predictions about distribution between two patches using a series of transition equations which dictate the rate at which foragers switch between 3 activities; searching, handling and fighting. The basis of the model can be described graphically as in Figure 4.

The rate of transition between activities is a function of the density of predators engaged in each activity, the time taken to handle each item of prey or to engage in a fight and the rate of encounters with prey. In fights between individuals the higher ranked competitor always wins. Holmgren assumes that there is no patch depletion, and that competition is due only to interference in the form of time lost owing to fighting and kleptoparasitism.

The optimal distribution of competitors was found by starting with an arbitrary distribution between patches and using an iterative procedure which always produced a stable equilibrium, at which no individual can improve its intake by moving patches.

The transition rate structure of the model has two interesting advantages over other types of model. Firstly, interference is generated by interactions between predators, rather than being dictated by the size of a constant which must be imposed from outside. Secondly, the model is very flexible in that it can be used to investigate the effect on the distribution of competitors of differences in various predator attributes such as handling efficiency, searching efficiency and dominance.

When dominance varies between competitors, dominants benefit from the presence of subordinates, which do better in their absence. At low density all animals occupy the best patch, but as density increases the subordinates are the first to begin using the poorer patch. As density increases further some dominants begin using the poorer patch which is made more attractive by the presence of subordinates. This leads to an equilibrium at which both types of predator use both patch types in roughly equal proportion. With more than two predator phenotypes, a group of the most dominant individuals occurs in the best patch, with the remaining predators distributed without any relationship between relative dominance and patch quality.

When predators differ in search efficiency, but suffer equally from interference, inferior searchers are less likely to leave the high density patch since they suffer more from the increased search time associated with the poorer patch. As predator density increases the equilibrium distribution shifts from all competitors in the better patch to all the inefficient searchers in the better patch and all or a proportion of the efficient searchers in the poorer patch, presumably depending on the relative patch qualities. If a range of search efficiency phenotypes is considered there is a truncated distribution with all the best searchers occurring in the worst patch.

Variation in handling efficiency leads to less efficient predators suffering more from an increase in density since it is during prey handling that they are susceptible to kleptoparasitism. Again, at low density all predators are

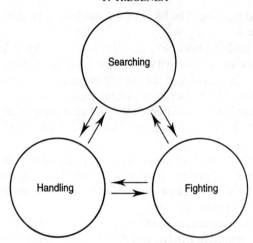

Fig. 4. Diagrammatic representation of Holmgren's model. Predators switch between 3 activities: Searching predators may find prey and become handlers, or
encounter a handling predator and begin fighting. Handling predators may be
attacked, becoming fighters, or complete prey handling and resume searching.
Fights take a set period of time after which the higher ranked individual begins
handling and the lower ranked individual begins searching.

found in the good patch; as density increases the inefficient handlers begin
utilizing the poor patch. There is a density range at which the two phenotypes do not occur in the same patch, followed by some of the "efficients"
moving onto the poorer patch as density increases still further. When this
situation was solved for a range of phenotypes, the better handlers tended
to occur in the better patch, with a second group of intermediate ability
handlers occurring in the poorer patch and the remainder distributed between both.

A disadvantage with numerical simulations is that results are dependent
on the parameter values used, and general predictions which would allow
testing are difficult to come by. An example of how this can reduce confi-
dence in the robustness of a model's predictions comes in Holmgren's
choice of patch qualities. He shows that equal non-depleting patch qualities (standing crops) may require very different input rates in order to be
maintained. However, the choice of qualities for the two patches is one in
which the poorer patch has 95% of the resources of the good patch. A
larger difference in patch qualities leads to all predators aggregating in the
better patch. Numerous field studies have found exploited prey patches
varying in quality by an order of magnitude more than this (e.g. Goss-
Custard, 1970; Fortier and Harris, 1989; Sutherland and Allport, 1994).
Therefore, it seems likely that this model will be most useful as an indicator
of the range of potential distributions that are produced when using differ-

ent, but equally plausible assumptions to those used in other models, rather than a basis for empirical work.

X. INCORPORATING RESOURCE DYNAMICS

A. Continuous Input Models and Standing Crops

The models described so far have only considered one source of density dependence at a time. Resources have been considered either to remain at constant density, with competition resulting from interference, or to be replaced immediately following utilization, with competition resulting from exploitation. Generally, these models have been considered alternatives. However, Lessells (1995) points out that interference models can be regarded as a "snapshot" of depletion or continuous input models. She also shows that if resources are continuously input into patches, the assumption that they are immediately consumed can be relaxed, allowing the standing crop of resources in each patch to be considered. The standing crop consists of those resource items which have arrived in a patch but which have yet to be consumed. It does not need to be more than a single resource item. There can be variation in standing crops even if each item is consumed before arrival of the next, since the time each item is on the patch before consumption is the standing crop. If competitors are equal, all consumers must have equal consumption rates at the stable distribution.

If there is only exploitation competition and no interference, the standing crop must be identical in each patch and hence resource (prey) mortality rate will be density independent. If interference occurs as well as exploitation, it will decrease the intake rate more in better patches, so, since consumption rates must still be equal, there must be a greater standing crop in those patches. Prey mortality rate becomes density dependent since it will increase with resource input rate, but at a decreasing rate, since the standing crop will be higher in patches with higher input rates.

The situation will be complicated further if there is some alternative source of prey mortality, aside from consumption by predators. If there is such alternative mortality, and no interference between competitors, losses due to this will be constant across patches since all patches will have the same standing crop. Alternative mortality can lead to some patches remaining unexploited because their input rates are not sufficiently high to overcome the losses to their standing crop resulting from the alternative mortality. Those patches with intake rates sufficiently high to be exploited will suffer density independent mortality, due to their identical standing crops.

In some situations, there may be interference as well as some source of alternative mortality, for example if competition leads to some resource items being wasted (Sutherland and Parker, 1992). In such situations the equilibrium distribution will depend on the relative strengths of the inter-relationships between consumption rate, standing crop, alternative mortality rate and the number of competitors. The only general predictions which can be made are the same as those for a situation where there is interference but no alternative mortality.

B. Decision-making Prey

The patch selection models discussed so far have taken the predator's point of view and have assumed prey to be immobile. Clearly there are many examples of systems in which this is not the case, such as mosquito larvae *Culex quinquefasiatus* moving to avoid backswimmers *Notonecta hoffmanni* (Sih, 1982) and sunfish *Lepomis macrochirus* changing their distribution in the presence of predatory bass, *Micropterus salmoides* (Werner et al., 1983). The influences of such prey responses are discussed by Sih (1984), who suggests that if a refuge from predation exists and prey are mobile, their response will determine the distribution of both species. More recently, Schwinning and Rosenzweig (1990) have modelled the distribution of populations between an open habitat and one containing refuges. They use a predator–prey system with a top predator, an intermediate predator, which is eaten by the top predator, and a prey item, eaten by both. All individuals of all groups are assumed to be equally mobile.

In their model, there is no competition between predators, and predation rates are determined assuming a type II functional response (Holling, 1966). This leads to unstable distributions, since if either habitat has a higher prey density, all the top predators will do best there. Therefore, they continually migrate onto the better patch, increasing the predation rate there, and causing the prey to leave. This leads to a situation similar to that in Parker and Sutherland's (1986) kleptoparasitism model in which predators pursue prey between patches and the distribution oscillates continually. Overall prey density is assumed to stay constant, which is justified if habitat redistribution occurs on a much faster time scale than changes in population size. Because of the lack of intraspecific competition amongst predators, the system is stable only when there is some mechanism which leads to both habitats providing equal predator intake rates and prey mortality rates. Schwinning and Rosenzweig show that there are three ways in which this can be achieved.

(i) If there is intraspecific competition between prey, it will increase their presence in the open habitat. Without competition, prey only choose the open habitat to escape predation in the refuge and predators only choose the refuge for higher prey density, so the two populations have antagonistic interests.

(ii) The top predator does not respond to small fitness differences, either because it does not perceive them, or because it ignores them. Although this doesn't lead to equal intake rates between patches, the predators perceive that their intakes are the same, which has a similar effect. If the threshold level of the top predator is increased above the level necessary to stabilize its own distribution this subsequently stabilizes the distribution of the intermediate predator and the prey.

(iii) The refuge is so strong that the predators become extinct.

XI. INCORPORATING SURVIVAL AS A MEASURE OF FITNESS

McNamara and Houston (1990) address one of the less explicit assumptions of other IFD models: that predators seek only to maximize their short-term fitness, which is normally measured as the net gain rate of some resource. They point out that variability of food supply and the predation risk incurred during foraging are important determinants of fitness and should be considered when predicting animal distributions. This subject, known as "risk sensitive foraging" is in itself a large field and has been comprehensively reviewed elsewhere (McNamara and Houston, 1992). Consideration is here restricted to McNamara and Houston's use of risk sensitive models to predict optimal distributions.

In their model, fitness is defined in terms of long-term survival during a non-reproductive period rather than in terms of rate of energetic gain. Animals are assigned a level of reserves, above a threshold level L, the animal rests, and below 0 the animal starves to death. It is assumed that animals forage continuously when their reserves are below L (interruptions such as sleep are ignored). Animals may choose between two patches of food with no cost to switching patch. The behaviour of each animal at each point in time is found using a dynamic programming technique (see Ross, 1983). Dynamic programming allows an animal's optimal policy at any point to be found by following it backwards through time. At each step, the animal's "state" (in this case its energy reserves) can be calculated, and from this the optimal patch choice is determined. McNamara and Houston applied their model to several different combinations of patch types and both to a single individual and to a population.

Single animal, no predation, equal mean rates in both patches. Using different patch qualities, when mean net gain is positive in both patches, the low variability patch was chosen. As mean net gain decreases and becomes negative, the state region for remaining food reserves over which it is optimal to choose the high variance patch becomes larger. In effect, the animal reaches a point at which its only hope of survival is to go to the high variance patch and be lucky. Also, as mean net gains decrease, reserves decline, making the high variance patch even more likely to be chosen.

Single animal, one patch is richer and contains predators. In this case, the optimal policy not surprisingly turned out to be to forage in the non-predated patch until reserves dropped below a critical level at which point predation risk is tolerated for increased gain. When food supply is poor, it determines distribution; when it is good, predation takes over.

Population of consumers, no predation, variability difference between patches. To consider the state-dependent decisions of a population of consumers, McNamara and Houston incorporate the effects of exploitation competition into their model. They assume that food items continually appear in a patch and are found at a rate related to their density by the Poisson distribution. Competition manifests itself only through lower prey availability, with no interference effects.

At low prey density, all animals choose the patch with low variation. If density increases but both patches still provide positive gains, then animals with low reserves prefer the more variable patch. As density is increased still further more and more animals choose the more variable patch but not to as great an extent as with one animal, since more animals in the more variable patch means that the less variable patch is a better choice. Density dependent effects are amplified as handling times decrease, hence the tendency to choose the more variable patch is negatively correlated with handling time.

Population of consumers, predation risk in one patch. In this situation it was found that predation was very important in determining distribution at low densities and almost ignored at high densities.

The distributions produced by a state-dependent IFD are a "dynamic equilibrium". This requires a large population since the stable distribution is a result of a continuous flux of animals changing state and behaving accordingly. McNamara and Houston point out that a state-dependent analysis of this type is appropriate for an animal that is simply trying to survive a given period of time, such as an overwintering bird. However, for some animals, feeding more in the present increases future reproduction, which will change the optimal policy.

In contrast to the simple IFD, animals making state dependent decisions do not distribute themselves so as to equalize rates of energy gain. The fundamental assumption of the state dependent IFD, that animals will be influenced by their physiological condition as well as their long-term competitive differences is highly plausible. Therefore the possibility that animals are making state dependent decisions must be considered when investigating distributions, since any mixed ESS is liable to be invaded by state dependent decision makers. This has wide ranging implications for the study of competitive distributions since it is very difficult to separate state dependent and long-term competitive differences in experiments and field studies.

XII. INCORPORATING PERCEPTUAL CONSTRAINTS

All the previous models assume that foragers are omniscient, and that their distribution can be explained through resource distribution and competition. This assumption is clearly unjustified in the majority of situations where animals will have an imperfect knowledge of patch profitabilities. Awareness of patch profitabilities may be limited to two factors. Firstly, being unable to distinguish between the profitability of similar patches, and secondly, being unaware of the profitability of patches which have not been visited. The first problem, described as "perceptual constraint" has been suggested as a cause for departures from an ideal free distribution. Abrahams (1986) presents a simple model in which patch suitability is defined as the total resources in the patch divided by the number of competitors present. This is used in an iterative simulation in which a newly arriving individual chooses the patch providing the highest intake rate after the addition of a further competitor (itself). The new arrival has a perceptual limit which means that it cannot perceive differences between patch profitabilities of less than a certain magnitude. If there is more than one best patch which cannot be distinguished, the individual chooses one of them at random. After all the competitors have entered the environment, they are randomly removed from their patch and have to re-choose a patch under the same constraints. Abrahams considers this to be analogous to the patch "switching" behaviour seen in many studies of competitive distributions. The results of the simulation confirm what one might intuitively suspect.

(i) Perceptual limits can only lead to underuse of better patches and overuse of poorer patches.
(ii) Greater perception limits lead to greater deviations from an IFD.
(iii) Extreme resource distributions lead to better distributions if the perception limit is low, since large differences in patch qualities mean that more animals are able to determine which is better so there are fewer "guesses". However, if perception limits are great then extreme resource distributions lead to a poorer fit to an IFD because guesses are "more wrong" when there is a greater difference between patches.

Abrahams suggests two ways in which distributions resulting from perceptual constraints can be distinguished from those resulting from interference. The first of these is that only the perception limit model predicts higher intake in the better patch. This contrasts with equal competitor IFD models which predict equal intake in all patches. However, if competitors vary, which is probably the rule rather than the exception, there may be higher mean intake in better patches as a result of unequal competitor distribution. The other utilizes the fact that the perceptual constraint simulation is based on absolute numbers whereas the interference IFD uses only relative numbers. If perceptual limits are responsible for a lack of fit to the simple IFD model, then decreased overall resource density or increased competitor numbers will lead to greater deviations since patch differences will be reduced. The interference model does not predict such an effect.

Abrahams points out that it is possible for perceptual limits and interference to operate simultaneously, and putting equations aside, this would certainly appear to be the most likely case. A range of animals have been found to be capable of complex optimal foraging behaviours, including determination of factors such as intake rate (see Stephens and Krebs (1986) for a review). It seems likely that perceptual limits will also have been reduced to a minimum, but it is inevitable that there will be a limit of some sort which will have an effect when patch profitabilities are sufficiently similar.

XIII. INCORPORATING PATCH ASSESSMENT

A. Patch Assessment Models

A second possible reason for violation of the assumption of omniscience is a consequence of the fact that animals cannot know the profitability of patches which have not been visited. They must either visit all the patches in their environment or must estimate patch qualities using as many sources of information as possible. There have been several attempts to model this aspect of predator behaviour, and to test whether the assumption of omniscience is necessary.

Harley (1981) proposed that one way for animals to decide on which patch is most profitable would be to use a simple learning rule called the relative payoff sum (RPS) rule. The basis of the RPS rule is that an animal chooses a particular behaviour (which could be a choice of patch) according to how successful that behaviour has been in the past. It is assumed that decisions are made following a sufficiently large time period for the animal to have experienced payoffs from all possible behaviours. Verbally, the rule states that the probability of displaying a particular behaviour is equal to the sum of the previous payoffs from that behaviour divided by the sum of previous payoffs from all behaviours. When applied to patch choice, the animal chooses to forage on each patch with a probability equal to the gains achieved on that patch in the past divided by the gains achieved on all patches.

Harley adapted this basic theory to postulate a cellular mechanism by which an animal could use the RPS rule. He proposed that each behaviour has a corresponding substance which elicits it, which is synthesized in response to the fitness increase brought by the performance of the behaviour. The substances are continuously degraded so that recent payoff information is most significant in determining behaviour properties. Substances are also continuously replaced to retain a residual concentration, ensuring that no behaviour can be deleted, in the same way that continual degradation ensures that no behaviour can become fixed. This predicts that the probability of choosing a given behaviour is equal to the relative frequency with which it has been rewarded, i.e. it is characterized by the matching law (Herrnstein, 1970).

Harley's RPS rule has been criticized by several people, most notably Houston and Sumida (1987). They present evidence that the RPS rule matches only for certain parameters, and in some situations can be bettered by a non-matching strategy. They also contend that the empirical evidence supporting the RPS rule is weak. This point is also made by Kacelnik and Krebs (1985), who cast doubt on Milinski's (1984) and Regelmann's (1984) applications of the RPS rule in their analysis of stickleback behaviour. Specifically, Kacelnik and Krebs show that the only piece of evidence which allowed testing of whether the fish were using the RPS rule rather than another rule is the predicted probability of switching between patches. In the experiments the observed probability of switching was half as great as the prediction from the RPS rule. Owing to these problems, it must be concluded that although animals must assess their environment and base decisions on that assessment, the exact nature of the rule they use may not be identical to that proposed by Harley. Nevertheless, Harley's model has served as a pioneering heuristic device in the study of patch assessment.

Bernstein, Kacelnik and Krebs (1988) investigated the distribution of predators in a patchy environment using a decision rule similar to Harley's, in that it is also based on a linear function expressing choice of behaviour as a weighted average of past and present experience. They point out that the IFD emphasizes an equilibrium state, rather than a process; individuals move between patches until all achieve the same maximal rate of gain. Interference is modelled in a similar fashion to that proposed by Sutherland (1983). However, in their model, there is a handling time for each prey item so that maximum intake is also limited by the total time available divided by the handling time. Bernstein *et al.* use a learning rule that assumes that predators will abandon their current patch and move to a random alternative patch when local capture rate drops below the estimated capture rate in the environment as a whole. Predators are assumed to be able to detect local capture rate precisely, but to estimate the environmental average on the basis of past experience. The relative weight of past and present capture rate on the animal's behaviour are assumed to be dependent on a "memory factor", such that past experience is slowly devalued relative to present experience. This allows an animal to "update" its estimate of average environmental profitability.

At the beginning of the simulation, each predator is given a personal intake rate threshold value γ. If the predator's intake rate drops below this, it leaves the patch. γ changes at each time step according to an algorithm which gives a weighted mean of immediate past and present intake rate:

$$\gamma_t + 1 = q\alpha + (1 - \alpha)\gamma_t \qquad 0 \leqslant \alpha \leqslant 1 \qquad (9)$$

Where γ_t is the value of γ at time t, q is a measure of the instantaneous profitability of the patch and α is a memory factor.

The migrating predators are randomly distributed between patches at the beginning of the simulation. Prey are immobile. Bernstein *et al.* (1988) found that in the absence of depletion, and assuming that predators are equal, their model produced a distribution very close to the IFD. In fact, predators using a learning rule and migrating blindly approach an IFD more rapidly than omniscient predators which migrate only to a patch with a higher capture rate. This is likely to be an artefact of the mechanism of the simulation, in which all predators migrate synchronously, all crowding into the better patches and reducing their profitability through interference. Nevertheless, this simulation shows that even if realistic assumptions are made about predators' knowledge of their environment, intake maximizing distributions can still be achieved.

Further manipulations to the model using depleting prey showed that prey mortality due to consumption can be density dependent or independent according to the interaction between predator efficiency and speed of learning. Rapid learning leads to density independence of prey mortality because predators track prey distribution. Slow learning leads to the predators' response lagging behind the changes in the environment such that density dependence is produced.

Slow depletion leads to an IFD being established, which in turn causes density dependent prey loss and hence the environment becomes more and more uniform. Fast depletion produces erratic effects since changes in prey distribution occur faster than learning predators can track them. Bernstein *et al.* (1988) suggest that this effect might lead to a systematic difference in temporal and spatial distribution between vertebrate and invertebrate predators if their learning abilities were found to be different.

Bernstein *et al.* (1991) developed their model by examining the influence of travel costs and the structure of the environment, using a simple adjustment to their learning algorithm. Migration is allocated a certain number of time steps, τ during which the predator does not feed. As a consequence it updates its value of γ according to the equation:

$$\gamma_t + 1 = (1 - \alpha)^{\tau}\gamma_t \qquad (10)$$

The simulation was run with varying levels of travel cost. It was found that as travel costs increase, predators become reluctant to leave their current patch even if intake rate is well below what they could achieve elsewhere. This leads to predator distribution departing progressively from the IFD. With depleting patches, prey mortality moves from density dependence, to random, and finally to inverse density dependence. The random stage occurs because at this level of travel costs only predators in very poor habitats move patches. Therefore, at intermediate travel costs, higher prey mortality rates than would be expected without travel costs occur to a greater extent on the medium and higher prey density patches.

The pattern of prey mortality was dependent on the scale over which changes in the environment occurred. In an environment which varied progressively from good to poor or vice versa, predators gained a false impression of the environment as a whole and consequently the pattern of prey mortality was affected. Allowing random dispersal of predators, rather than movement only to neighbouring patches, increased the fit to the IFD as did distributing prey at random. This model predicts that predators in a habitat which may have large systematic variations should move widely, sampling their environment in order to gain a true impression of prey distribution and hence maximize their intake.

Other recent theoretical treatments of patch estimation have focused on Bayesian updating (Oaten, 1977; Green, 1980; Iwasa *et al.*, 1981; McNamara, 1982). The Bayesian process also involves combining past experience with current experience to estimate patch quality. These two sources of information have been called "patch sample" and "pre-harvest" (Valone, 1991). Patch sample information is accumulated during patch use and may include the time spent in the patch, the number of resource items obtained in the patch and the time since last resource capture. Pre-harvest information can include prior knowledge about the distribution of resources in the environment, sensory information and environmental cues that indicate patch quality. Group foragers also have access to a third source of information which has been called "public" information (Valone, 1989), since it is acquired by observation of other group members.

In Bayesian models, foragers go to the patch which their estimations lead them to believe is the most profitable one. This approach has been combined with the IFD by Cézilly and Boy (1991), who simulated a competitor distribution based on the following assumptions.

(i) Individuals are free to enter either of two patches at any time.
(ii) Initially, individuals have no knowledge of resource distribution.
(iii) Switching between patches incurs a cost in terms of time.
(iv) Individuals are able to measure their own success and are aware of the time penalty involved in switching.
(v) An individual's success in a patch relative to other competitors is proportional to its competitive weight.
(vi) Competitors use a Bayesian rule to assess patch qualities and choose the patch with the highest estimated profitability.
(vii) Relative competitive ability is the same regardless of patch.

Patches have a continuous input of resource items arriving with a probability based on the Poisson distribution and with an input ratio of 2:1 between the good and poor patches. Three good and three poor competitors were used. Good competitors had twice the competitive ability of poor ones. Because the foragers initially have no information about patch profitabilities they have an equal probability of going to either patch. It was found that under these conditions the distribution corresponded very closely to the predictions of the IFD. It was also found that better competitors tended to spend more time in the better patch and switched less frequently than the poorer competitors. However, although not mentioned by the authors, the limited number of each type of competitor (3), in itself makes it inevitable that better competitors will spend more time in the good patch. This is because there are only two possible distributions in which the sum of competitive weights on the patches matches their input rates. These are that the good patch has either all the good competitors or two good competitors and two poorer ones. This gives a mean distribution in which there are 2·5 good competitors in the good patch and 0·5 in the poor patch. Therefore the finding that better competitors gain more items in the best patch may be an artefact of the simulation's limited competitor set. Nevertheless, the general conclusion provides more evidence that adaptive complex distributions of competitors can result from simple individual behaviours.

B. Experimental Investigations of Patch Assessment

The practicalities of how foragers estimate patch quality is the subject of a growing body of research. A great deal of work has been done on solitary foragers (e.g. Hunte *et al.*, 1985; De Vries *et al.*, 1989; Marschall *et al.*, 1989; Valone and Brown, 1989; Cuthill *et al.*, 1990; Valone, 1991, 1992), much of which has implications for competitive systems. However, this review is restricted to studies in which animals have been observed in competing situations.

Milinski (1984) concluded that patch choice in his three-spined stickle-backs *Gasterosteus aculeatus* was based on individuals measuring their own foraging rate within a patch. Although this is the most simple method available, it is possible that animals could use "short-cut" methods in order rapidly to maximize their intake whilst minimizing patch sampling. Harper (1982) found that when he fed Mallards *Anas platyrhynchos* at either end of a pond at different rates an IFD was rapidly achieved. However, if he fed at equal rates, but at one end of the pond used pieces of bread which were twice as heavy, the ducks initially overused the poorer patch and took significantly longer to conform to the IFD. This suggests that initially the ducks assessed patch quality by watching the rate at which pieces of bread were thrown. The fact that an IFD was still eventually achieved suggests that visual assessment was followed by personal intake rate assessment, allowing the ducks to correct for the limitations of the visual assessment method.

Croy and Hughes (1991) examined the distribution of 6 fifteen-spined sticklebacks *Spinachia spinachia* between two continuous input feeding stations. They found that like Harper's ducks, the fish were considerably more sensitive to delivery rate than they were to changes in the size of prey items. This suggests that there are certain indicators of patch quality which foragers are able to recognize more readily than others. This is to be expected, since only certain foraging clues are likely to have been present in the animal's evolutionary environment. For instance, it seems more likely that foraging fish would have the choice of two areas of a stream where food flowed past at different rates than two areas where food particles differed in size. Croy and Hughes also found that more switching occurred in the first half of each trial and that better competitors switched more frequently, enabling them to track short-term changes in food profitability. Poorer competitors tended to remain at one patch and concentrate on prey missed by better individuals. This suggests that foragers differing in competitive ability may employ different patch assessment strategies.

Animals foraging in groups have access to a source of patch information denied to dispersed foragers, that is they can use the distribution of conspecifics as an indicator of patch profitabilities. Gotceitas and Colgan (1991) investigated whether three-spined sticklebacks used this shortcut. They found that when individual fish were prevented from physically sampling patches of different food availability, patch choice did not conform to the IFD. Instead, individuals preferred the patch with the greatest number of conspecifics present, but only when the conspecifics were feeding. This suggests that sticklebacks initially use the presence of conspecifics to choose which patch to go to, but require physical sampling in order to choose correctly in the longer term. Pitcher and House (1987) investigated a similar question using goldfish *Carasius auratus*. They found that when food was plentiful, the fish divided between two equal resource patches as expected. However, when food was low the fish preferred to feed near other fish. Anti-predator advantages of group foraging were ruled out since the fish only joined groups which they could see were feeding. The authors suggest that the fish use personal intake to decide whether to remain in a patch, but if intake is low they may join a group in the expectation of future increases in intake rates resulting from group foraging.

It is worth noting that rules of thumb are only valuable when foraging in certain situations. Individuals competing for resources that are patchily distributed in time as well as space will find rules of thumb useful to maximize intake rapidly whilst the distribution is being established. If the number of individuals in each patch is small, this can independently lead to selection for rules of thumb. A new arrival will do best by going to the patch with the highest number of competitors since one additional competitor will have less impact on the overall amount of resource available. It is possible that this is the reason that three-spined sticklebacks choose larger groups of conspecifics in Gotceitas and Colgan's study. However, providing the number of individuals in each patch is large, once an IFD has been achieved any newly arriving individual will do equally well by going to any patch. Therefore, the use of rules of thumb may be rare in animals foraging for resources which do not vary rapidly in time.

XIV. A SUMMARY OF IFD MODELS

As we have seen, to make predictions about distributions it is necessary to ignore many factors which may frequently have important effects. Numerous models have been written in attempts to reduce the number of unjustified assumptions and come closer to reality. However, in order to keep the models simple enough to be mathematically tractable, the removal of one assumption usually has to be bought at the cost of the inclusion of another. Hence, the development of ideal free theory has not led to an increasingly complex formula which makes fewer and fewer assumptions and comes ever closer to mimicking reality. Rather, a large body of work has been produced with different models tackling the removal of only one or two of the major assumptions underlying the theory. In Tables 1 and 2, I firstly summarize these assumptions and then consider which model deals with the removal of each.

It is evident from Table 1 that for all the key assumptions of the original IFD there are foreseeable situations in which these assumptions will be violated. Theoreticians have attempted to address this problem through the development of a range of models that are summarized in Table 2.

Table 1

Assumptions made in ideal free distribution models and likely violations of these

Assumption	Potential 'real' situation
Equal competitive abilities	Variation between individuals in: Susceptibility to interference Search efficiency Handling time "State" (for instance food reserves) Knowledge of resource distribution Cost of travel Resource holding ability Susceptibility to other environmental factors (e.g. predation, temperature) Other priorities (e.g. mate searching, kin selection)
Omniscience	Animals may be constrained by: Having to visit patches to assess their profitability Inability to detect the difference between patches Inability to remember patch profitabilities
No costs to travel between patches	Costs to travel between patches in terms of: Time Energy Risks from predation, physical factors
No interference between competitors	Various reversible factors leading to reduced intake rate with increased competitor density including: Mutual interference—wasting time or resources Kleptoparasitism Despotism
Resources are fixed in space and time	Exploitation leads to patches depleting Prey may move in response to predation New patches may arise
Only rate of resource acquisition affects patch choice	Patches may have associated costs or benefits, such as: Predation risk Physical properties of the environment (e.g. temperature, rate of flow of moving water) Refuges, nest sites
Distribution is dictated entirely by competitors maximizing short-term fitness	Animals may have other priorities, examples being: Mate searching Migrations

Table 2

Summary of ideal free distribution models, including which of the assumptions of
the original theory are removed and resulting predictions

More realistic assumption(s)	Model	Predictions
Exploitation competition of continuous input resources leads to gains decreasing in proportion to competitor density.	Continuous input equal competitor IFD (Parker, 1978).	Gain rate of all individuals is equal in all patches. Stable equilibrium distribution of competitors with competitor density in direct proportion to input rate in each patch.
As above, but resources are not consumed immediately on arrival in patch. Competition is due to exploitation and interference.	Putting resource dynamics into IFD models (Lessells, 1995).	As above, but if there is interference, standing crops of resources are greater in patches with higher input rates.
As above, but including some alternative source of prey mortality other than consumption by predators.	Putting resource dynamics into IFD models (Lessells, 1995).	Poorer patches can remain unexploited, standing crops of resources are greater in patches with higher input rates.
Increased density leads to decreased intake rate through mutual interference.	Interference IFD (Sutherland, 1983).	Fitness of all individuals is equal in all patches. The ratio of number of animals in patches is constant, irrespective of total density. If intake rates are known at different densities, the theory can be used to predict the optimal distribution.
Increased density leads to decreased intake rate through mutual interference. Individuals differ in competitive ability.	Phenotype alters slope of log (intake) against log (competitor density) graph, (Sutherland and Parker, 1985).	No phenotype is found in more than two patch types. Higher quality patches contain most competitive phenotypes. Fitness correlated with both phenotype and patch quality.
Increased density leads to decreased intake rate through mutual interference. Individuals differ in competitive ability.	Phenotype alters intercept of log (intake) against log (competitor density) graphs (Sutherland and Parker, 1985).	A number of stable equilibria are possible dependent on the number of animals present in their relative competitive weights. The statistically most likely equilibrium is one in which number of animals in patch is proportional to input rate (Houston and McNamara, 1988)

Table 2—*cont.*

More realistic assumption(s)	Model	Predictions
Competitors steal resources from one another. Individuals vary in their ability to steal and hold resources.	Kleptoparasitism model (Parker and Sutherland, 1986)	No stable distribution, animals switch patches continuously. Payoffs to each phenotype vary across patches. There is a weak correlation between dominance and patch quality.
Competitors steal resources from one another. Competitors vary in their dominance.	IFD determined by the form of competition (Korona, 1989).	Each phenotype distributes itself independent of other phenotypes and in proportion to patch qualities.
Gains in a patch decrease in proportion to competitor density. Competitors are unable to discriminate between resource patches differing in profitability less than a threshold amount.	Perceptual constraints model (Abrahams, 1986).	Relative to the predictions of the continuous input IFD, model predicts under-use of the better patch. This is increased when perceptual limits are great. Large differences between patches may decrease under-use of better patch, but will increase if perceptual limits are also large.
Predator gains increase with prey density according to a type II functional response and decrease. Prey move to minimize predation risk.	3 population predator–prey system (Schwinning and Rosenzweig, 1990).	No stable distribution unless one habitat contained a strong refuge, competition between prey was strong or the predators were unable to perceive the difference between habitats of similar profitability (perceptual constraints).
Increased predator density leads to decreased gains in patch due to depletion. Predators have variable energy reserves and are able to choose between patches of different variability of food supply on the basis of their risk of starvation.	State dependent IFD (McNamara and Houston, 1990).	At low density, all animals choose the less variable patch. As density increases more and more animals choose the more variable patch.
As above but with predation in one patch.	State dependent IFD.	Animals avoid the depredated patch at low density but use it more and more as density increases.

Table 2—*cont.*

More realistic assumption(s)	Model	Predictions
Competitors steal resources from one another. Competitors vary in their dominance.	A kleptoparasitic IFD based on a functional response Holmgren (1995)	Distributions change with population density. Generally, both predators use both patch types in roughly equal proportion.
Competitors vary in their searching efficiency.	A kleptoparasitic IFD based on a functional response Holmgren (1995)	Distributions change with population density. At intermediate density, inefficient searchers occupy the better patch, while efficient searchers use the poorer patch.
Competitors vary in their handling efficiency.	A kleptoparasitic IFD based on a functional response Holmgren (1995)	Distributions change with population density. At intermediate density, better handlers tend to occur in the better patch, and poorer handlers in the poorer patch.
Animals must learn the profitability of different areas of the environment. Mutual interference.	Bernstein, Kacelnick and Krebs (1988)	Animals distributed as predicted by Sutherland's interference IFD despite using a simple learning rule.
Animals must learn patch profitability and there is a time cost to travel. Mutual interference.	Bernstein, Kacelnik and Krebs (1991).	There is an approximate fit to the interference IFD, but poorer patches are overused more if patch quality varies progressively across the environment.
Animals switch between patches.	Houston *et al.* (1995)	There is a compromise between equal intake rates in patches and equal competitor numbers. This leads to slight overuse of the poorer patch.

It is evident from the breadth of theoretical work into competitive distributions that great improvements have been made in the biological reality of IFD models. Nevertheless, the fact that each model tends to consider only a limited reduction in assumptions shows that there is still the potential for further advances.

XV. ASSESSING THE SUCCESS OF IDEAL FREE DISTRIBUTION MODELS

A. Testing the Ideal Free Distribution

The predictions and assumptions of ideal free distribution models have been tested in many studies. Those prior to 1985 are reviewed by Parker and Sutherland (1986), whose table I have modified by including comment on which model explains the results of experiments and observations (if one does) (Table 3). I have also added the results of work published after 1985. In the table, "IFD" is used to mean the simplest model, i.e. the continuous input model or Sutherland's interference equation, depending on which form of competition is occurring. I have noted only the most simple model which explains the results of the experiment. If a study is deemed to provide support or to be consistent with a model's predictions and assumptions then the results can be assumed to be inconsistent with other theories. The difference between providing "support" and being "consistent" is not clear-cut. I have considered that an experiment or study cannot be deemed to support a model unless it fulfils some quantitative predictions which is distinct from those of other models.

B. Does the Theory Pass the Test?

In theory, it should be possible to apply ideal free type theories to the distribution of any animal. However, it is apparent from Table 3 that even situations chosen specifically for the approximation to the assumptions of the theory do not produce an overwhelming level of support for the predictions of the models they test. There are two possible explanations for this: either current theory is inappropriate for the situations examined, or researchers have failed accurately to characterize the distributions they are studying. Although there have been a number of published errors in applying ideal free theory (Tregenza, 1994), we can largely rule out this second explanation. However, there are profound difficulties in studying animal distributions. These are worth considering, since they are likely to be responsible for the shortage of empirical work specifically designed to test the predictions of particular models.

1. Difficulties in Studying Competitive Distributions

(i) The value of resources is often difficult to assess. Worse still, there may be several different resources being exploited, so that the one being studied may not be the only relevant factor. This means that the

Table 3
A summary of empirical investigations into competitive distributions including
consideration of whether the results support any particular model

Reference	Species	Result	Support for a model of predator distributions?
Non-continuous input situations			
Andren (1990)	Jay *Garrulus glandarius*	Highest reproductive success found in dense forest, territoriality results in unequal resource partitioning	Consistent with ideal despotic distribution. Evidence is non-quantitative
Buxton (1981)	Shelduck *Tadorna tadorna*	Spend less time feeding in areas of dense prey	No convincing fit to any model
Fortier and Harris (1989)	Marine fish larvae	Post-larval stages of copepod feeders distributed in proportion to resource density (not input rates)	May support IFD but intake rates not measured
Goss-Custard *et al.* (1982)	Oystercatcher *Haematopus ostralegus*	Greater spreading out across patches occurred as density increased. Also, immatures moved off better patches. As birds matured they used better patches	Conflicts with IFD prediction that all patches should be used regardless of density. Competitors unequal
Goss-Custard *et al.* (1984)	Oystercatcher	Average food intake differs consistently between mussel beds	Consistent with unequal competitor models
Goss-Custard *et al.* (1992)	Oystercatcher	Greater spreading out across patches occurred as density increased	Conflicts with IFD prediction that all patches should be used regardless of density
Jakobsen and Johnsen (1987)	Waterflea *Daphnia pulex*	Both food patches depleted to same level. Proportionally more use of better patch at lower food density	Conflicts with IFD predictions due to change in proportional patch use with forager density

Table 3—*cont.*

Reference	Species	Result	Support for a model of predator distributions?
Korona (1990)	Flour beetle *Tribolium confusum*	With low travel costs the number of ovipositing beetles corresponded to the amount of flour in each patch	May support IFD but no measure of fitness of "suitability" is measured
Messier *et al.* (1990)	Muskrat *Ondatra zibethicus*	Different survival rates between patches. Some individuals gain higher fitness via resource monopolisation	Consistent with ideal despotic distribution
Møller (1991)	Passerine birds	In species where patch size correlates with reproductive success, younger individuals found in smaller patches	Supports unequal competitors (truncated distribution) models
Monaghan (1980)	Herring gull *Larus argentatus*	Average food intake about five times higher in better areas of rubbish tip	Consistent with unequal competitor models
Morris (1989)	White-footed mouse *Peromyscus leucopus*	Fitness of mice (by various measures) was not significantly different between three habitats	Supports IFD
Nishida (1993)	Coreid bug *Colpula lativentris*	Males in bigger aggregations had higher mate acquisition probability despite identical competitive ability	No support for any model, may be due to perceptual constraints
Perusse and Lefebvre (1985)	Feral pigeon *Columba livia*	Smaller food patches lead to greater dispersal across patches. Large patches lead to grouped sequential exploitation	Conflicts with IFD prediction that all patches should be used regardless of density
Pusenius and Viitala (1993)	Field vole *Microtus agrestis*	Inferior habitat used when increased density lowered reproductive success in preferred habitat	Conflicts with IFD prediction that all patches should be used regardless of density

Table 3—*cont.*

Reference	Species	Result	Support for a model of predator distributions?
Sibly and McCleery (1983)	Herring gull	Average food intake consistently higher on open tip than elsewhere	Consistent with unequal competitor models
Sutherland (1982)	Oystercatcher	Average food intake differs between parts of cockle bed	Consistent with unequal competitor models
Talbot and Kramer (1986)	Guppy *Poecilia reticulata*	Proportion of fish in a habitat correlated with food supply, but no fit to IFD	No convincing fit to any model, may be due to poor patch assessment
Thompson (1981)	Shelduck	Time spent feeding similar between areas with different prey densities	No convincing fit to any model, may be due to poor patch assessment
Thompson (1984)	Lapwing and gold plover *Vanellus vanellus* and *Pluvialis apricaria*	Rate of food intake greater in fields where prey is most abundant. Also, increase in density led to a change in proportional use of patches	No convincing fit to any model, may be due to poor patch assessment or to unequal competitors
Zwarts and Drent (1981)	Oystercatcher	Food intake in one patch constant in years of different mussel availability	Weak support for IFD, since forager density is related to resource density
Continuous input situations			
Abrahams (1989)	Guppy	As proportion of food input into a patch was increased proportion of fish using patch increased	Supports IFD
Alatalo *et al.* (1992)	Black grouse *Tetrao tetrix*	Males in larger leks gain more mates. Poorer males go to smaller leks	Supports unequal competitor models

Table 3—*cont.*

Reference	Species	Result	Support for a model of predator distributions?
Courtney and Parker (1985)	Tiger blue butterfly *Tarucus theophrastus*	Density of mate searching males on different sized bushes fits ideal free better than random distribution	Supports IFD
Croy and Hughes (1991)	Fifteen-spined stickleback *Spinachia spinachia*	Fish distributed in proportion to food when rate of input was source of patch variation	Consistent with unequal competitor models—size differences in success
Davies and Halliday (1979)	Toad *Bufo bufo*	Density of mate searching males in pond and edge consistent with obtaining equal numbers of females	Consistent with unequal competitor models—size differences in success
van Duren and Glass (1992)	Cod *Gadus morhua*	Fish spent more time in better food patch than IFD prediction, but only better competitors did better there. Poor competitors gained very little food	Significant deviation from unequal competitor IFD may be due to small group of fish and very low success of poorer competitors
Gillis and Kramer (1987)	Zebra fish *brachydanio rerio*	Deviations from IFD seen only at high densities— therefore not due to despotism, but to both interference and exploitation competition occurring at higher densities	"Ideal interference distribution".[1] Conflicts with IFD models due to change in proportional patch use with density
Gillis *et al.* (1993)	Canadian trawler	Distribution of vessels led to equalization of catch rates between areas	Consistent with unequal competitor models— differences in success[2]

Table 3—*cont.*

Reference	Species	Result	Support for a model of predator distributions?
Godin and Keenleyside (1984)	Cichlid fish *Aequidens curviceps*	Average food intake same in two sites, despite individual differences	Consistent with unequal competitor models
Harper (1982)	Mallard *Anas platyrhynchos*	Average food intake same in two sites, despite individual differences	Consistent with unequal competitor models
Inman (1990)	Starling *Sturnus vulgaris*	Over-use of better food patch. Unequal relative payoffs, birds more likely to be found in patch where their success is highest	No support for any model, may be due to despotism
Järvi and Pettersen (1991)	Salmon parr *Salmo salar*	Mature parr are superior competitors. Under-use of better food patch, even allowing for unequal competitive abilities	No support for any model. May be due to perceptual constraints, incorrect assessment of competitive abilities or despotism
Lamb and Ollason (1993)	Wood-ant *Formica aquilonia*	The number of ants using continuous input sucrose patches reflected the relative input rates to two sites	Supports IFD
Milinski (1979, 1984, 1986)	Three-spined stickleback *Gasterosteus aculeatus*	Average food intake same in two sites, despite individual differences. Ratio of phenotypes in each patch is the same as in the environment as a whole	Supports unequal competitor models
Otronen, M. (1993)	*Dryomyza anilis*	Large males are more successful. No correlation between male size and location, unequal gains across patches.	No support for any model, optimal distribution is spatial and temporal

Table 3—*cont.*

Reference	Species	Result	Support for a model of predator distributions?
Parker (1970, 1974)	Dungfly *Scathophaga stercoraria*	Average mating success with newly arriving females same in different sites, but large males more successful at take-overs, which occur mainly in one site	Supports unequal competitor models, bigger males more successful and tend to occur on dropping (Borgia, 1980)
Patterson (1985)	Convict cichlid *Cichlasoma nigrofasciatum*	Territorial. Higher density in preferred habitat. During settlement, poorer habitat used when nest sites still remained in better habitat	Consistent with ideal despotic distribution
Power (1983, 1984)	Loricariid catfish	Standing crop of algae in sunny and shaded pools is equal as a result of foraging by catfish	Consistent with IFD (see Section X, Lessells, 1995)
Recer *et al.* (1987)	Mallard	Forager number correlated with input rate on two patches	Supports IFD
Sutherland *et al.* (1988)	Goldfish *Carasius auratus*	Mean rank of fish inversely proportional to number of fish in site but fish gained more food in better site. No relationship between rank and time in better site found	Supports unequal competitor model, with some deviation possibly due to sampling or perceptual constraints
Thornhill (1980)	Lovebug *Plecia nearetica*	Larger males at bottom of swarm, smaller males at top. Mating success increases with size and position in swarm	Supports unequal competitor models
Utne *et al.* (1993)	Two-spotted goby *Gobiusculus flavescens*	Input matching only occurred when two patches equal. Greater ratios of input between sites led to increased deviations.	Conflicts with IFD, consistent with perceptual constraint model

Table 3—cont.

Reference	Species	Result	Support for a model of predator distributions?
Whitham (1980)	Aphid *Pemphigus betae*	Average number of offspring per individual equal on different sized leaves despite individual differences	Supports unequal competitor models

Notes:
[1] Gillis and Kramer suggest that the term "ideal interference distribution" should be used to describe deviations from the IFD due to various factors associated with high predator densities or despotism. They do not suggest any specific model associated with this name.
[2] It is not obvious whether this is a continuous input situation or not, since fish clearly are replaced by reproduction, although this may be seasonal, with depletion occurring over much of the year.

number of individuals expected in a particular patch cannot be estimated and the distribution cannot be defined.

(ii) In the field it is often hard to identify the location of all competitors. Frequently, competitors are not obvious, since they employ many strategies to avoid predation and physical stresses which makes them difficult to locate. Ideal free distributions occurring under stones and in clumps of long grass are difficult to observe.

(iii) Since most situations involve unequal competitors, characterizing the distribution requires individual recognition, which is often difficult. If individuals cannot be recognized, then the observer cannot say which animals are where, and hence cannot describe the distribution except in the simplest terms.

(iv) Apart from individual recognition, it is necessary to assess the *relative* abilities of unequal competitors in order to make quantitative predictions. This is often difficult, since it requires that the same individuals are present in the same location for a long period.

(v) In the field it is hard to assess which animals are being successful. This is often partly a consequence of difficulties in recognition of individuals but may also be because in many situations it is difficult to identify when an animal has succeeded in securing a resource.

(vi) When using the models of Sutherland and Parker in non-continuous input situations it is impossible to determine the level of the interference constant, m, without measuring individual intake rates at different densities. This makes quantitative predictions, from even an elegant theory such as this, difficult except in the most accommodating circumstances.

(vii) It is difficult to monitor a situation 24 hours a day. It is likely that optimal distributions are achieved over time as well as space. Temporal and spatial IFDs have been described, but little theoretical work has been done on distributions in four dimensions. Animals may be able to choose patches based on their future suitability as well as their present suitability. This could be a stable strategy if resources tended to appear unpredictably at different times in different places, and if animals were able to exploit them more efficiently by being present when they arrived.

2. Factors Affecting the Optimal Distribution not Considered by Current Theory

The systems studied in past investigations into the predictions of IFD models will have been chosen to avoid the problems discussed above. Therefore the reasons why only 12 out of 48 studies reviewed here are designated as providing support for a particular model are presumably to do with the theory, rather than difficulty in testing it. There are a number of likely reasons for the limited success of contemporary IFD models.

(i) The assumption that animals are omniscient is clearly unjustified. In order to determine patch profitabilities, animals have to assess their environment. This activity has various associated costs and benefits. Costs include time and energy spent moving between patches, and during assessment, predation risk, physical dangers (e.g. desiccation) and the possibility that it will not be possible to return to a patch. Possible benefits are searching for mates at the same time as searching for food (Abrahams, 1989) and inbreeding avoidance through movement. This is discussed in greater detail in sections VII and VIII.

(ii) Competition may occur in several ways, each of which requires a different theoretical approach. Currently, only the models for continuous input and pure interference are sufficiently quantitative to allow their predictions to be tested in a rigorous fashion. Despotism is suggested in five studies, but since the ideal despotic model's only unique prediction is that holding territories in itself will lead to greater success, none of the studies can be designated as supporting the model, since none of them investigate this prediction. Furthermore, although different types of competition are not mutually exclusive, current models tend to only consider one form of density dependence.

(iii) Although differences in competitor phenotype are considered in recent models, quantifying competitive ability in a single constant may frequently be too great a simplification. Animals may have one

set of relative abilities in one foraging situation and a different ranking in a situation which suits different individuals better. Similarly,
individuals may vary in their competitive abilities over time.

(iv) Factors other than competition for a single resource may affect habitat choice. The optimal distribution will be altered by factors such as
the presence of other resources, predation, physical constraints, and
relatedness between competitors.

(v) It is likely that the complex learning rules necessary for perfect patch
choice will not have evolved in all species. This may be the result of
costs in terms of nerve tissue or simply because evolution does not
lead to perfection.

(vi) In situations where patches are small, the fact that animals are discrete units may limit the distribution. For instance, a "good" patch
might have enough resources to support three top competitors, but
not enough for four. The surplus resource might be exploited by an
inferior competitor which might otherwise be predicted to occur on a
poorer patch.

(vii) Any particular model is only one potential relationship between the
terms it comprises. This relationship may not be applicable to all, or
indeed any, real situations. It is worth noting that of those 21 studies
in which interference was considered to be the source of density
dependence, five are inconsistent with the Sutherland and Parker
interference IFD theory because proportional patch use changed with
competitor density. This suggests that in many situations the relationship between density and the influence of interference, upon which
these models are based may not be applicable.

An interesting feature of continuous input IFD experiments is that many
have produced distributions which conform to the equal competitor
model's numerical predictions, despite the fact that the participants differed in their competitive ability and intake rates. A simple competitive
weights model (one where ratios of phenotypic payoffs stay constant across
patches) would predict a range of potential equilibria, comprising all distributions in which the ratio of competitive weight to resource input in each
patch was constant. Only in one of these equilibria does the average payoff
in each patch remain constant: that in which the ratio of different competitor phenotypes is the same in each patch and the habitat as a whole. The
fact that this distribution was apparently over-represented suggests that
there is some reason why distributions of this type are more likely to occur
than any of the other equilibria. A number of suggestions have been made
to explain this situation.

(i) Sutherland and Parker (1985) point out that the distribution most
frequently seen in experiments is the most likely outcome if the equi-

librium is selected by chance. They also postulate (Parker and Sutherland, 1986) that if competitors distribute themselves randomly with respect to phenotype, and match numbers with input rate this will lead to the observed distribution. This could occur if individuals were quick at assessing the input rate of a patch and the competitor density but much poorer at assessing their relative competitive weight.

Houston and McNamara (1988) analysed this theory in more detail using an approach based on statistical mechanics and found that the average of all potential distributions gave a weighted mean payoff to better and poor patches of 1·54:1·45. This means that for large populations there will be a slightly higher proportion of good competitors in the better location than in the habitat as a whole, rejecting Parker and Sutherland's idea except for small groups. However, Milinski and Parker (1991) point out that in practice, experiments will probably not be able to distinguish between 1·54:1·45 and 1·5:1·5, so the possibility that the observed distributions are a result of statistical probabilities remains.

(ii) Competitors could distribute themselves with respect to their own competitive rank only, so that each rank is distributed in an ideal free fashion ignoring all other ranks (N.B. Davies quoted in Milinski and Parker, 1991).

(iii) Korona (1989), reviewed earlier, provides a third possibility. His model produces a unique distribution in which each phenotype occurs in the same proportion in each habitat, as is frequently observed. Also, his model does not require that animals use a "short-cut" patch assessment method to explain their distribution. The simplest way in which an optimal distribution could be reached is via competitors assessing their personal intake rate and moving to the patch where it is highest. This method might be superseded by a direct patch or competitor assessment method such as is suggested in the previous two explanations. However, although it is likely to allow more rapid intake rate maximization, there is no guarantee that a short-cut assessment method will have evolved. Also, there are some situations in which patch assessment without sampling will be difficult.

(iv) Holmgren (1995) predicts a number of alternative distributions, depending on how competitors vary in their competitive ability. His simulations predict that if they differ in dominance then an equilibrium similar to that most frequently observed will be produced. Like Korona's model, Holmgren does not invoke any more sophisticated environmental assessment than personal intake rate.

In a recent paper, Kennedy and Gray (1993) state that, "in its current form, the IFD does not accurately predict the distribution of foraging

300   T. TREGENZA

animals". Their review criticizes much past work on competitive distributions and claims to provide a new perspective on deviations from the IFD theory's predictions. However, as Milinski (1994), points out, their paper contains a series of misconceptions and most of what they consider to be new insight has been in the literature for many years.

Kennedy and Gray claim that the IFD predicts that the proportion of competitors in a patch will match the proportion of resources occurring there. As we have seen, this is only the case for an interference situation in which the value of the interference constant m happens to be exactly 1, and does not apply to the majority of natural foraging scenarios. They also claim that their re-analysis of data from past distribution studies is a new finding since it shows that there is consistent overuse of poorer patches relative to the predictions of simple IFD models ("undermatching"). However, as discussed in Sections VII and VIII there has already been a considerable amount of work on this subject. Further, contrary to their claims, Kennedy and Gray's method of detecting overuse of the poorer patch is not novel, having been used by Fagen (1987).

Kennedy and Gray discuss possible reasons for undermatching. Because they expect that competitors will automatically match resources even without interference, they incorrectly predict that as interference increases there will be more overuse of the better patch (overmatching). This is contrary to the predictions of the ideal free distribution in an interference situation, where increasing interference leads to the distribution shifting from all competitors being found in the best patch with zero interference to undermatching with very high ($m > 1$) interference. I will not review all the errors in Kennedy and Gray's brief review, but would refer the reader to Milinski (1994).

XVI. CONCLUSIONS

For decades, ecologists have described animal distributions without a comprehensive understanding of how they occur at an individual level. In the last twenty-five years attempts have been made to remedy this situation through investigation of the behavioural mechanisms that ultimately lead to population distributions. The basis for much of this work has been the concept that the most important factor affecting the suitability of any particular part of the environment is the presence of other competitors. The original ideal free distribution theory has been refined and expanded in many directions through a large body of theoretical and experimental work. This has revealed both the complexity inherent in competitive distributions and the potential for individual behaviours to have significant effects at a population level.

The assumptions of equal competitive ability, omniscience, movement without costs and a single type of competition have all be shown to be inappropriate in many situations. When unequal competitors are considered, a range of new distributions are predicted including phenotypes being truncated across patches of different quality. Lack of omniscience and costs to movement also affect predicted distributions, tending to lead to greater use of poorer patches, with the magnitude of this effect dependent on the composition of the environment.

Several different forms of competition have been identified, and a corresponding range of models and predictions formulated. Exploitation competition has mainly been considered in a situation in which resource items arrive continually on a patch, although iterative models have also thrown some light on how depletion of fixed resources may affect forager distribution. A range of models based on interference competition have been proposed. These differ in the mechanism of interference considered, such as time wasted in interactions with other foragers or kleptoparasitism. Also, since there is no pre-defined relationship between competitor density and the strength of interference, different models assume different relationships between these factors and produce correspondingly different results. Further empirical work is badly needed to investigate the relationships between factors such as competitor density, resource distribution and intake rates, since current models are based largely on theory rather than observation. Also, although different forms of competition frequently occur together in nature, current theory tends to consider one or the other in isolation, so further theoretical and practical work would be valuable here too.

It is unlikely that a "grand unified theory" of animal distribution will ever be possible since it is difficult to assess the relative importance of the numerous factors involved. Nevertheless, the development of individual level models has provided numerous insights into short term (i.e. less than a single generation) population dynamics. It now remains for ecologists to combine classical population dynamics with these behavioural models.

ACKNOWLEDGEMENTS

I would like to thank Geoff Parker, Dave Thompson and Nina Wedell for numerous valuable comments on earlier versions of the manuscript and Bill Sutherland for suggestions on content. My warmest thanks are also due to Kate Lessells, without whom this review would have been greatly inferior. Funding was provided by NERC studentship GT4/91/TLS/30.

REFERENCES

Abrahams, M.V. (1986). Patch choice under perceptual constraints: a cause for departures from the IFD. *Behav. Ecol. Sociobiol.* **10**, 409–415.

Abrahams, M.V. (1989). Foraging guppies and the IFD: The influence of information on patch choice. *Ethology* **82**, 116–126.

Alatalo, R.V., Höglund, J., Lundberg, A. and Sutherland, W.J. (1992). Evolution of black grouse leks: Female preferences benefit males in larger leks. *Behav. Ecol.* **3**, 53–59.

Andren, H. (1990). Despotic distribution, unequal reproductive success, and population regulation in the Jay *Garrulus glandarius* L. *Ecology* **71**, 1706–1803.

Barnard, C.J. and Brown, C.A.J. (1982). The effects of prior residence, competitive ability and food availability on the outcome of interactions between shrews (*Sorex araneus* L.) *Behav. Ecol. Sociobiol.* **10**, 307–312.

Bernstein, C., Kacelnik, A. and Krebs, J.R. (1988). Individual decisions and the distribution of predators in a patchy environment. *J. Anim. Ecol.* **57**, 1007–1026.

Bernstein, C., Kacelnik, A. and Krebs, J.R. (1991). Individual decisions and the distribution of predators in a patchy environment II. The influence of travel costs and structure of the environment. *J. Anim. Ecol.* **60**, 205–225.

Borgia, G. (1980). Sexual competition in *Scatophaga stercoraria*: size- and density-related changes in male ability to capture females. *Behaviour* **75**, 185–206.

Brown, J.L. (1969). The buffer effect and productivity in tit populations. *Am. Nat.* **103**, 347–354.

Buxton, N.E. (1981). The importance of food in the determination of the winter flock sites of the shelduck. *Wildfowl* **32**, 79–87.

Cézilly, F. and Boy, V. (1991). Ideal free distribution and individual decision rules: a Bayesian approach *Acta. Œcol.* **12**, 403–410.

Courtney, S.P. and Parker, G.A. (1985). Mating behaviour of the tiger blue butterfly (*Tarucus theophrastus*); competitive mate-searching when not all females are captured. *Behav. Ecol. Sociobiol.* **17**, 213–221.

Croy, M.I. and Hughes, R.N. (1991). Effects of food supply, hunger, danger and competition on choice of foraging location by the fifteen-spined stickleback, *Spinachia spinachia* L. *Anim. Behav.* **42**, 131–139.

Cuthill, I.C., Kacelnik, A., Krebs, J.R., Haccou, P. and Iwasa, Y. (1990). Starlings exploiting patches: the effect of recent experience on foraging decisions. *Anim. Behav* **40**, 625–640.

Davies, N.B. (1978). Territorial defence in the speckled wood butterfly (*Pararge aegaria*), the resident always wins. *Anim. Behav.* **26**, 138–147.

Davies, N.B. and Halliday, T.R. (1979) Competitive mate searching in common toads, *Bufo bufo. Anim. Behav.* **27**, 1253–1267.

Davies, N.B. and Houston, A.I. (1984). Territory economics. In: *Behavioural Ecology: An Evolutionary Approach*, 2nd edition (Ed. by J.R. Krebs and N.B. Davies), pp. 148–169. Blackwell, Oxford.

DeVries, D.R., Stein, R.A. and Chesson, P.L. (1989). Sunfish foraging among patches: the patch-departure decision. *Anim. Behav.* **37**, 455–464.

van Duren, L.A. and Glass, C.W. (1992). Choosing where to feed: the influence of competition on feeding behaviour of cod, *Gadus morhua* L. *Fish. Biol.* **41**, 463–471.

Enquist, M., Plane, E. and Roëd, J. (1985) Aggressive communication in fulmars (*Fulmarus glacialis*) competing for food. *Anim. Behav.* **33**, 1007–1020.

Fagen, R. (1987). A generalised habitat matching rule. *Evol. Ecol.* **1**, 5–10.

Fortier, K. and Harris, R.P. (1989). Optimal foraging and density dependent competition in marine fish larvae. *Marine Ecology* **51**, 19–33.

Fretwell, S.D. (1972). *Populations in a seasonal environment.* Princeton University Press, Princeton, New Jersey.

Fretwell, S.D. and Lucas, H.J. Jr. (1970). On territorial behaviour and other factors influencing habitat distribution in birds. *Acta Biotheor.* **19**, 16–36.

Gillis, D.M. and Kramer, D.L. (1987). Ideal interference distributions: population density and patch use by zebrafish. *Anim. Behav.* **35**, 1875–1882.

Gillis, D.M., Peterman, R.M. and Tyler, A.V. (1993). Movement dynamics in a fishery: Application of the ideal free distribution to spatial allocation of effort. *Can. J. Fish. Aquat. Sci.* **50**, 323–333.

Godin, J.G.J. and Keenleyside, M.H.A. (1984). Foraging on a patchily distributed prey by a cichlid fish (Teleosti Cichlidae): A test of the IFD theory. *Anim. Behav.* **32**, 120–131.

Goss-Custard, J.D. (1970). The responses of Redshank (*Tringa totanus* (L.)) to spatial variations in the density of their prey. *J. Anim. Ecol,* **39**, 91–113.

Goss-Custard, J.D. (1980). Competition for food and interference amongst waders. *Ardea* **68**, 31–52.

Goss-Custard, J.D. and Durell, S.E.A.Le V. Dit. (1982). Use of mussel *Mytilus edulis* beds by Oystercatchers *Haematopus ostralegus* according to age and population size. *J. Anim. Ecol.* **51**, 543–554.

Goss-Custard, J.D., Clarke, R.T. and Durell, S.E.A.Le V. Dit. (1984). Rate of food intake and aggression of oystercatchers *Haematopus ostralegus* on the most and least preferred mussel *Mytilus edulis* beds of the Exe estuary. *J. Anim. Ecol.* **53**, 233–245.

Goss-Custard, J.D., Caldow, R.W.G. and Clarke, R.T. (1992). Correlates of the density of foraging oystercatchers *Haematopus ostralegus* at different population sizes. *J. Anim. Ecol.* **61**, 159–173.

Gotceitas, V. and Colgan, P. (1991). Assessment of patch suitability and ideal free distribution: the significance of sampling. *Behaviour* **119**, 65–76.

Green, R.F. (1980). Bayesian birds: a simple example of Oaten's stochastic model of optimal foraging. *Theor. Pop. Biol.* **18**, 244–256.

Harley, C.B. (1981). Learning the evolutionarily stable strategy. *J. Theor. Biol.* **89**, 611–633.

Harper, D.G.C. (1982). Competitive foraging in mallards "ideal free" ducks. *Anim. Behav.* **30**, 575–584.

Harris, V.A. (1964). *The Life of the Rainbow Lizard.* Hutchinson Tropical Monographs. Hutchinson and Co., London. 174 pp.

Hassell, M.P. and Varley, G.C. (1969). New inductive model for insect parasites and its bearing on biological control. *Nature* **223**, 1133–1136.

Herrnstein. R.J. (1970). On the law of Effect. *J. Exp. Anim. Behav.* **21**, 159–164.

Holling, C.S. (1966). The functional response of predators to prey density and its role in mimicry and population regulation. *Mem. Entomol. Soc. Can.* **45**, 5–60.

Holmgren, N. (1995). The ideal free distribution of unequal competitors—predictions from a behaviour based functional response. *J. Anim. Ecol.* **64**, 81–102.

Houston, A.I. and McNamara, J.M. (1988). The IFD when competitive abilities differ: approach based on statistical mechanics. *Anim. Behav.* **36**, 166–174.

Houston, A.I. and Sumida, B.H. (1987). Learning rules, matching and frequency dependence. *J. Theor. Biol.* **126**, 289–308.

Houston, A.I., McNamara, J.M. and Milinski, K. (1995). The distribution of animals between resources: a compromise between equal numbers and equal intake rates. *Anim. Behav.* **49**, 248-251.

Hunte, W., Myers, R.A. and Doyle, R.W. (1985). Bayesian mating decisions in an amphipod *Gammarus lawrencianus* Bousfield. *Anim. Behav.* **33**, 366-372.

Huxley, J.S. (1934). A natural experiment on the territorial instinct. *Brit. Birds.* **27**, 270-277.

Inman, A.J. (1990). Group foraging in starlings: Distributions of unequal competitors. *Anim. Behav.* **40**, 801-810.

Iwasa, Y., Higashi, M. and Yamamura, N. (1981). Prey distribution as a factor determining the choice of optimal foraging strategy. *Am. Nat.* **117**, 710-723.

Iwasa, Y., Odendaal, F.J., Murphy, D.D., Ehrlich, P.R. and Launer, A.E. (1983). Emergence patterns in male butterflies: a hypothesis and a test. *Theor. Pop. Biol.* **23**, 363-379.

Jakobsen, P.H. and Johnsen, G.H. (1987). Behavioural response of the water flea *Daphnia pulex* to a gradient of food concentration. *Anim. Behav.* **35**, 1891-1895.

Järvi, T. and Pettersen, J.H. (1991). Resource sharing in Atlantic salmon: A test of distribution models on sexually mature and immature parr. *Nord. J. Fresh. Res.* **66**, 89-97.

Kacelnik. A. and Krebs, J.R. (1985). Learning to exploit patchily distributed food. In: *Behavioural Ecology—ecological consequences of adaptive Behaviour* (Ed. by R.M. Sibly and R.H. Smith), pp. 189-205. Blackwell, Oxford.

Kennedy, M. and Gray, R.D. (1993). Can ecological theory predict the distribution of foraging animals? A critical analysis of experiments on the ideal free distribution. *Oikos* **68**, 158-166.

Korona, R. (1989). Ideal free distribution of unequal competitors can be determined by the form of competition. *J. Theor. Biol.* **138**, 347-352.

Korona, R. (1990). Travel costs and the IFD of ovipositing female flour beetles, *Tribolium confusum. Anim Behav.* **40**, 186-187.

Lamb, A.E. and Ollason, J.G. (1993). Foraging wood-ants *Formica aquilonia* Yarrow (Hymenoptera, Formicidae) tend to adopt the ideal free distribution. *Behav. Proc.* **28**, 189-198.

Lessells, C.K. (1995). Putting resource dynamics into continuous input ideal free distribution models. *Anim. Behav.* **49**, 487-494.

Marschall, E.A., Chesson, P.L. and Stein, R.A. (1989). Foraging in a patchy environment: prey-encounter rate and residence time distributions. *Anim. Behav.* **37**, 444-454.

Maynard Smith, J. (1982). *Evolution and the theory of Games.* Cambridge University Press, Cambridge.

McNamara, J.M. (1982). Optimal patch use in a stochastic environment. *Theor. Pop. Biol.* **21**, 269-288.

McNamara, J.M. and Houston, A.J. (1990). State dependent ideal free distributions. *Evol. Ecol.* **4**, 298-311.

McNamara, J.M. and Houston, A.I. (1992). Risk sensitive foraging: A review of the theory. *Bull. Math. Biol.* **54**, 355-378.

Messier, F., Virgl, J.A. and Marinelli, L. (1990). Density-dependent habitat selection in muskrats: a test of the ideal free distribution model. *Oecologia* **84**, 380-385.

Milinski, M. (1979). An evolutionary stable feeding strategy in sticklebacks. *Z. Tierpyschol.* **51**, 36-40.

Milinski, M. (1984). Competitive resource sharing: an experimental test of a learning rule for evolutionarily stable strategies. *Anim. Behav.* **32**, 233–242.

Milinksi, M. (1986). A review of competitive resource sharing under constraint in Sticklebacks. *J. Fish Biol.* **29**, 1–14.

Milinski, M. (1994). Ideal free theory predicts more than only input matching—a critique of Kennedy and Gray's review. *Oikos.* **71**, 163–166.

Milinski, M. and Parker, G.A. (1991). Competition for resources. In: *Behavioural ecology an evolutionary approach, vol. 3.* (Ed. by J.R. Krebs and N.B. Davies), pp. 137–168. Blackwell, Oxford.

Møller, A.P. (1991). Clutch size, nest predation, and distribution of avian unequal competitors in a patchy environment. *Ecology* **72**, 1336–1349.

Monaghan, P. (1980). Dominance and dispersal between feeding sites in the Herring gull (*Larus argentatus*). *Anim. Behav.* **28**, 521–527.

Morris, D.W. (1989). Density dependent habitat selection: Testing the theory with fitness data. *Evol. Ecol.* **3**, 80–94.

Nishida, T. (1993). Spatial relationships between mate acquisition probability and aggregation size in a gregarious coreid bug, (*Colpula lativentris*): A case of the ideal free distribution under perceptual constraints. *Res. Pop. Ecol.* **35**, 45–56.

Oaten, A. (1977). Optimal foraging in patches: A case for stochasticity. *Theor. Pop. Biol.* **18**, 244–256.

Orians, G.H. (1969). On the evolution of mating systems in birds and mammals. *Am. Nat.* **103**, 589–603.

Otronen, M. (1993). Male distribution and interactions at female oviposition sites as factors affecting mating success in the fly *Dryomyza anilis* (Dryomyzidae). *Evol. Ecol.* **7**, 127–141.

Parker, G.A. (1970). The reproductive behaviour and the nature of sexual selection in *Scatophaga stercoraria* L. II. The fertizilation rate and the spatial and temporal relationships of each sex around the site of mating and oviposition. *J. Anim. Ecol.* **38**, 205–228.

Parker, G.A. (1974). The reproductive behaviour and the nature of sexual selection in *Scatophaga stercoraria* L. IX. Spatial distribution of fertilization rates and evolution of mate search strategy within the reproductive area. *Evolution.* **28**, 93–108.

Parker, G.A. (1978). Searching for mates. In: *Behavioural Ecology: An Evolutionary Approach, 1st edition.* (Ed. by J.R. Krebs and N.B. Davies), pp. 214–244. Blackwell Scientific Publications, Oxford.

Parker, G.A. (1982). Phenotype-limited evolutionary stable strategies. In: *Current problems in sociobiology.* (Ed. by Kings College sociobiology group). Cambridge University Press, Cambridge.

Parker, G.A. and Courtney, S.P. (1983). Seasonal incidence: adaptive variation in the timing of life history stages. *J. Theor. Biol.* **105**, 147–155.

Parker, G.A. and Stuart, R.A. (1976). Animal behaviour as a strategy optimizer: evolution of resource assessment strategies and optimal emigration thresholds. *Am. Nat.* **110**, 1055–1076.

Parker, G.A. and Sutherland, W.H. (1986). Ideal free distributions when individuals differ in competitive ability: Phenotype limited ideal free models. *Anim. Behav.* **34**, 1222–1242.

Patterson, I.J. (1985). Limitation of breeding density through territorial behaviour: experiments with convict cichlids, *Cichlasoma nigrofasciatum.* In: *Behavioural Ecology—ecological consequences of adaptive Behaviour* (Ed. by R.M. Sibly and R.H. Smith), pp. 393–405. Blackwell, Oxford.

Perusse, D. and Lefebvre, L. (1985). Grouped sequential exploitation of food patches in a flock feeder, the feral pigeon. *Behav. Proc.* **11**, 39–52.

Pimm, S.L., Rosenzweig, M.L. and Mitchell, W. (1985). Competition and food selection: field tests of a theory. *Ecology* **66**, 798–807.

Pitcher, T.J. and House, A.C. (1987). Foraging rules for group feeders: Area copying depends upon food density in shoaling goldfish. *Ethology* **76**, 161–167.

Pusenius, J. and Viitala, J. (1993). Demography and regulation of breeding density in the field vole, *Microtus agrestis*. *Ann. Zool. Fenn.* **30**, 133–142.

Power, M.E. (1983). Grazing response of tropical freshwater fishes to different scales of variation in their food. *Env. Biol. Fish.* **9**, 103–115.

Power, M.E. (1984). Habitat Quality and the distribution of algae-grazing catfish in a Panamanian stream. *J. Anim. Ecol.* **53**, 357–374.

Pulliam, H.R. and Caraco, T. (1984). Living in groups: is there an optimal group size? In: *Behavioural Ecology an Evolutionary approach*, 2nd edition (Ed. by J.R. Krebs and N.B. Davies), pp. 122–147. Blackwell Scientific Publications, Oxford.

Recer, G.M., Blanckenhorn, W.U., Newman, J.A., Tuttle, E.M., Withiam, M.L. and Caraco, T. (1987). Temporal resource variability and the habitat matching rule. *Evol. Ecol.* **1**, 363–378.

Regelmann, K. (1984). Competitive resource sharing—a simulation model. *Anim. Behav.* **32**, 227–232.

Rosenzweig, K.L. (1974). On the evolution of habitat selection. In: *Proceedings of the First International Congress of Ecology* pp. 401–404. Centre for Agricultural Publishing and Documentation, Wageningen, Netherlands.

Rosenzweig, M.L. (1981). A theory of habitat selection. *Ecology* **62**, 327–335.

Rosenzweig, M.L. (1986) Hummingbird isolegs in an experimental system. *Behav. Ecol. Sociobiol.* **19**, 313–322.

Ross, S.M. (1983). *Introduction to Stochastic Dynamic Programming*. Academic Press, New York.

Schwinning, S. and Rosenzweig, M.L. (1990). Periodic oscillations in an ideal-free predator-prey distribution. *Oikos* **59**, 85–91.

Sibly, R.M. and McCleery, R.H. (1983). The distribution between feeding sites of herring gulls breeding at Walney island. *U.K. J. Anim. Ecol.* **52**, 51–68.

Sih, A. (1982). Foraging strategies and the avoidance of predation by an aquatic insect, *Notonecta hoffmanni*. *Ecology* **63**, 786–796.

Sih, A. (1984). The behavioral response race between predator and prey. *Am. Nat.* **123**, 143–150.

Stephens, D.W. and Krebs, J.R. (1986). *Foraging theory*. Princeton University Press, Princeton, New Jersey.

Sutherland, W.J. (1982). Spatial variation in the predation of cockles by oyster-catchers at Traeth Melynog, Anglesey. II. The pattern of mortality. *J. Anim. Ecol.* **51**, 491–500.

Sutherland, W.J. (1983). Aggregation and the "ideal free" distribution. *J. Anim. Ecol.* **52**, 821–828.

Sutherland, W.J. and Allport, G.A. (1994). A spatial depletion model of the interaction between bean geese and wigeon with the consequences for habitat management. *J. Anim. Ecol.* **63**, 51–59.

Sutherland, W.J. and Parker, G.A. (1985). Distribution of unequal competitors. In: *Behavioural Ecology—ecological consequences of adaptive Behaviour* (Ed. by R.M. Sibly and R.H. Smith), pp. 255–274. Blackwell, Oxford.

Sutherland, W.J. and Parker, G.A. (1992). The relationship between continuous input and interference models of ideal free distributions with unequal competitors. *Anim. Behav.* **44**, 345–355.

Sutherland, W.J., Townsend, C.R. and Patmore, J.M. (1988). A test of the ideal free distribution with unequal competitors. *Behav. Ecol. Sociobiol.* **23**, 51–53.

Talbot, A.J. and Kramer, D.L. (1986). Effects of food and oxygen availability on habitat selection by guppies in a laboratory environment. *Can. J. Zool.* **64**, 88–93.

Thompson, D.B.A. (1981). Feeding behaviour of wintering shelduck in the Clyde Estuary. *Wildfowl* **32**, 88–98.

Thompson, D.B.A. (1984). Foraging economics in flocks of plovers and gulls. Unpublished PhD thesis, University of Nottingham.

Thornhill, R. (1980). Sexual selection within mating swarms of the lovebug, *Plecia nearetica* (Diptera: Bibionidae). *Anim. Behav.* **28**, 405–412.

Tregenza, T. (1994). Common misconceptions in applying the ideal free distribution. *Anim. Behav.* **47**, 485–487.

Utne, A.C.W., Assnes, D.L. and Giske, J. (1993). Food, predation risk and shelter: An experimental study on the distribution of adult two-spotted goby *Gobiusculus flavescens* (Fabricius). *J. Exp. Mar. Biol. Ecol.* **166**. 203–216.

Valone, T.J. (1989). Group foraging, public information, and patch estimation. *Oikos* **56**, 357–363.

Valone, T.J. (1991). Bayesian and prescient assessment: foraging with pre-harvest information. *Anim. Behav.* **41**, 569–577.

Valone, T.J. (1992). Information for patch assessment: a field investigation with black-chinned hummingbirds. *Behav. Ecol.* **3**, 211–222.

Valone, T.J. and Brown, J.S. (1989). Measuring patch assessment abilities of desert granivores. *Ecology* **70**, 1800–1810.

Werner, P.A., Gilliam, J.F., Hall, D.J. and Mittelbach, G.G. (1983). An experimental test of the effects of predation risk on habitat use in fish. *Ecology* **64**, 1540–1550.

Whitham, T.B. (1980). The theory of habitat selection: examined and extended using *Pemphigus* aphids. *Am. Nat.* **115**, 449–466.

Zwarts, L. and Drent, R.H. (1981). Prey depletion and the regulation of predator density: oystercatchers (*Haematopus ostralegus*) feeding on mussels (*Mytilus edulis*). *Feeding and Survival Strategies of Estuarine Organisms* (Ed. by N.V. Jones and W.J. Wolff), pp. 193–216. Plenum Publishing Corporation, London.

The Role of Recruitment Dynamics in Rocky Shore and Coral Reef Fish Communities

DAVID J. BOOTH and DEBORAH M. BROSNAN

I.	Summary	309
II.	Introduction and Historical Perspective	310
III.	Settlement and Recruitment of Marine Organisms	314
IV.	Are Marine Systems Open?	316
V.	Recruitment: Patterns and Processes	329
	A. The Extent of Recruitment Variability in Marine Systems	329
	B. Factors Affecting Recruitment Strength	332
VI.	The Influences of Recruitment on Patterns of Population and Community Structure	336
	A. Recruitment and Population Structure	337
	B. Recruitment Effects on Community Structure and Dynamics	342
	C. The Role of Recruitment Variability	351
	D. Summary of Models, Predictions and Tests	358
VII.	Management Implications, Conservation Issues and Future Directions	360
	A. Management and Conservation	360
	B. Future Directions	363
VIII.	Concluding Remarks	364
	Acknowledgements	365
	References	365

I. SUMMARY

The paradigm that communities are composed of populations of species at equilibrium, first developed for terrestrial birds and mammals, is inadequate for open populations (larvae dispersing to different habitats) such as marine organisms. Recent conceptual and empirical advances in the study of recruitment of young into populations of such organisms have demonstrated the critical role of limited recruitment in determining population structure. However, the effects of recruitment at the community level are less clear, since community-level interactions may obscure signals

ADVANCES IN ECOLOGICAL RESEARCH VOL. 26
ISBN 0–12–013926–X

from recruitment. In addition, recruitment effects have only been established for a restricted range of common and conspicuous elements of the community. We evaluate evidence that recruitment dynamics of component species can exert a strong effect on processes at a community level. We use rocky shores and coral reef fish communities as exemplar systems since considerable work has been done in these areas. Both systems share some important ecological characteristics, and we wished to bring together two areas that have progressed in parallel with each other for the past 15–20 years. We conclude that models already developed for recruitment at a population level, coupled with models of processes at a community level, may be used to build a predictive base for the study of recruitment and community structure. More rigorous field testing of hypotheses is needed, which considers both recruitment in isolation and its interaction with other forces affecting community structure. Given recent and increasing threats to marine communities by the advancing tide of human population, an understanding of recruitment processes and effects will have implications beyond academic application.

II. INTRODUCTION AND HISTORICAL PERSPECTIVE

Throughout the 1950s and 1960s community ecologists were strongly influenced by theories of density dependent population regulation. This led to an explanation of population and community structure based on competition and predation. Up until the early 1980s, most studies of community structure of organisms were based upon assumptions of equilibrium conditions, with density dependent predation or competition for space or food considered to be major structuring forces (e.g. Cody, 1968; Diamond, 1978). Early community studies were conducted on "closed" communities (e.g. studies by Elton, 1958; MacArthur and MacArthur, 1961; MacArthur et al., 1966), where offspring remain close to their natal site. In such systems, populations and communities would be expected to be strongly influenced by local fecundity patterns. However, beginning with studies of insects in the 1950s (see Andrewartha and Birch, 1954), density independent explanations for variations in community structure which highlight the role of stochastic environmental perturbation were developed. Recently, non-equilibrial models emphasizing such density independent forces have become more popular (e.g. Connell, 1978; Buss and Jackson, 1979; Talbot et al., 1978). Driving this change has been the recognition that communities containing organisms such as insects, marine invertebrates and coral reef fishes differ in biologically important ways from terrestrial birds and mammals. The majority of animal (and probably plant) species, including most marine organisms, have dispersive offspring (propagules such as larvae or spores) or disperse as adults. They are therefore considered open, with replenishment of local populations from elsewhere. This means that local fecundity may have no effect on appearance of new individuals into the local population. Recognition of the potential effects of recruitment variability on populations and communities has led to a flurry of studies, so that a new paradigm has developed (see Sale, 1991).

Much research on recruitment in marine organisms has centred on two systems: fishes on coral reefs and invertebrates and algae on temperate rocky intertidal shores. Work has often progressed in parallel or on different tangents in both systems, with little attempt at cross-referencing. Our purpose here is to synthesize some of the relevant information from both systems so that it may be more accessible. In addition, while the role of recruitment in populations has been well established, (see below) its role in community dynamics has been relatively unstudied (but see Hawkins and Hartnoll, 1983; Hartnoll and Hawkins, 1985; Underwood and Denley, 1984; Sutherland, 1987; Menge and Farrell, 1989; Menge, 1991). An additional purpose of this paper is to explore the role of recruitment variability in communities. Rocky intertidal and coral reef fish communities have much in common ecologically. For example, both have traditionally been viewed as open, space-limited systems (e.g. Roughgarden, 1986) and subject to disturbance (e.g. Dayton, 1971, 1975; Connell, 1979; Paine and Levin, 1981; Sale, 1978).

It is important to note that there are certain constraints inherent in our comparison. For instance, we compare a tropical and highly speciose (coral reef fish) with a less geographically restricted system (rocky shores), which is subject to more seasonal fluctuations. (Note also that most work on rocky shores has been carried out in temperate latitudes.)

The potential importance of recruitment to adult populations of marine organisms has been recognized for at least 75 years. Hjort (1914 and discussion in Richards and Lindeman, 1987; Cushing, 1975), and Hatton (1938) proposed that larval supply may affect community structure of marine organisms. Recruitment effects have been incorporated into fisheries models, and stock-recruitment models have been used to regulate fisheries for many years (e.g. Beverton and Holt, 1957; Cushing, 1975). Here, recruitment has a very specific meaning: entry of juveniles into the adult (fishable) population although some studies (e.g. Lasker, 1985, for various clupeoid fishes, Cohen et al., 1988, for cod and haddock) specifically address the causes of variable larval recruitment and their consequences to the fishery. Much of the focus of fisheries management has been on the relationship between adult stock size and recruitment. Exploited adult stocks may be depleted to such a level that recruitment overfishing may result (e.g. Cushing, 1975). Partly due to the stochastic nature of recruitment, the consequences of overfishing have proved very difficult to

312   D.J. BOOTH AND D.M. BROSNAN

predict or prevent in temperate fisheries. Models predicting recruitment from adult stock size (e.g. Ricker, 1954, Beverton and Holt, 1957) even when "tweaked' to incorporate random variation owing to environment (see Cushing, 1975) have proved of limited use. These problems are even greater in tropical, multi-species fisheries, where larval durations are less well known, and long-term data sets are few, (e.g. see Russ, 1991).

In intertidal communities and non-resource coral reef fish communities, however, recruitment generally refers to number of new juveniles entering a site or community (see below), and the ramification of recruitment variation have been discussed and debated for many years. Thorson (1950, 1966) noted that many marine species have a planktonic phase and a "settled" juvenile or adult phase. He also observed that such species are subject to large-scale fluctuations in abundance. He further realized that recruitment was dependent on pre-settlement events, settlement and post-settlement factors (which he referred to as the first to third levels of selection). However, these early insights were not incorporated into the ecological theory of the day.

Ecologists studying coral reef fish up until the 1970s generally focused on space limitation and consequent competition as major determinants of community structure (e.g. Smith and Tyler, 1975). Reviews of coral reef fish ecology in the 1970s (e.g. Ehrlich, 1975) gave precedence to studies of competition consequences such as habitat and food partitioning, with little mention made of the ecology of larval or juvenile states of the organisms, despite their well established importance in temperate fisheries and inter-tidal systems. Studies of recolonization of reef sites and artificial reefs by fishes (e.g. Talbot et al., 1978; Bohnsack and Talbot, 1980; Lassig, 1982) revealed the potential importance of recruitment to the development of coral reef fish communities. "Rediscovery" of larval stages by reef fish ecologists in the early 1970s was followed by important advances in methods used to sample ichthyoplankton (Doherty, 1987a; Leis, 1991; Dufour, pers. comm.).The swiftness of the paradigm shift is seen in the contrast between comments of Anderson et al. (1981), who asserted that recruitment does not affect community structure, and two years later by Munro (1983), who stated dogmatically that as a general rule, populations of reef fishes are recruitment-limited. More recently, however, Sale (1990) called recruitment study in coral reef fishes the "supply-side bandwagon", and there has been a trend for more pluralistic approaches, which simultaneously consider the effects of both recruitment and post-recruitment processes (see Hixon, 1991; Jones, 1991). The focus in studies of coral reef fish recruitment has usually been at the level of the population or guild (sensu Root, 1967). Indeed, most studies have been biased toward several taxonomic groups (especially pomancentrids, labids and chaetodontids). For example, a literature search that we conducted identified 55 papers dealing with the effects of recruitment on population structure in fishes, with 20% concerning territorial pomacentrid damselfish. Compared with many intertidal studies, coral reef studies tend to be restricted to few taxa and trophic levels, perhaps owing to sampling difficulties and greater species diversity (and more rare species) than in temperate intertidal systems.

The importance of larval stages and settlement in marine intertidal species was recognized in the wealth of new, and largely European, literature published mainly in the 1950s and 1960s. This early work focused on factors affecting the settlement process itself, rather than its demographic consequences. Emphasis was on physical and chemical factors (Barnes, 1955; 1956a; Crisp, 1974 (and references therein) 1976a,b; Crisp and Meadows, 1962; Knight-Jones and Stevenson, 1950; Knight-Jones and Crisp, 1953; Knight-Jones, 1951, 1953, 1955,); larval behaviour, including territoriality, gregariousness and spawning behaviour (e.g. Barnes, 1957; Crisp, 1961, 1974 (and references therein); Crisp and Spencer, 1958; Meadows and Campbell, 1972); and factors affecting dispersal, particularly in relation to the role of larval life histories in dispersal patterns, and the colonization of new habitats (e.g. Barnes, 1956b; Barnes and Barnes 1977; Crisp, 1955, 1958, 1974; Southward and Crisp, 1954, 1956). Other studies focused on variations and fluctuations in settlement and density (e.g. Barnes, 1956b; Crisp, 1958, 1976a,b; Southward, 1967). It is not within the scope of this paper fully to explore factors affecting settlement, and the reader is directed to the comprehensive reviews of Crisp (1974), Meadows and Campbell (1972), and more recent papers referenced in Table 1 of this review. Lewis (1964) speculated that fluctuations in population densities and species composition on rocky shores were due to the constant rain of larvae and spores, and the relatively short life of intertidal species (in highly competitive environments). He also noted that this was most evident for "planktotrophic species, on shores where settlement was high, growth rates rapid and competition severe".

Recent emphasis on the role of recruitment in intertidal invertebrates arose from a more general interest in the effects of non-equilibrial processes on populations and communities. In the 1970s, non-equilibrial models were introduced and applied successfully to marine intertidal communities. Studies such as those of Harger (1970), Dayton (1971), Suchanek (1979, 1981), Sousa (1979a,b, 1984a, 1985), Paine and Levin (1981) and theoretical advances such as that of Connell's (1978) intermediate disturbance hypothesis (developed in part from coral reef studies) served to highlight this new approach. Nevertheless, all of the above studies share the common tactic assumption that recruitment is not limiting, and there is an abundant supply of larvae to occupy and recolonize available space. Experimental and theoretical emphasis in the role of larval recruitment in structuring populations and communities began in the 1980s (e.g. Denley and Underwood,

1979; Hawkins and Hartnoll, 1982; Kendall *et al.*, 1982; Underwood and Denley, 1984; Roughgarden *et al.*, 1984, 1985; Wethey, 1984, 1986).

Depending on your viewpoint, rediscovery of larval stages by ecologists in the early 1970s stimulated either a "new field" or simply "old wine in new bottles" (Underwood and Fairweather, 1989). They point out that many recent studies overlook earlier knowledge and references to recruitment limitation: the heavily empirical, natural historical European tradition was always aware of the importance of recruitment variation (see above). However, current research in recruitment has a much stronger theoretical and predictive base than earlier work, despite the many questions that remain unanswered and even unaddressed.

In this paper, we first define settlement and recruitment, and discuss the uses and misuses of these terms in the current literature. Unclear definitions in the ecological literature have lead to confusion in the past (e.g. the various meanings of "competition": Birch, 1957). Next, we evaluate evidence that rocky intertidal and coral reef fish species can indeed be considered open (larvae recruiting from a distant source).

Following this we will review the dynamics of recruitment, highlighting the differences and similarities between the systems. We will next consider the consequences of these recruitment patterns to population and community structure, by drawing empirical examples and summarizing existing conceptual models. Finally, these models and new insights (derived from other studies) are synthesized into composite models, and predictions and tests stemming from them are suggested. The paper concludes by addressing the practical/management implications of recruitment and by recommending future research directions. It is not our purpose to attempt a comprehensive review of recruitment, since a number of thorough studies already exist (e.g. coral reef fishes: Richards and Lindeman, 1987; Doherty and Williams, 1988; Mapstone and Fowler, 1988; Doherty, 1991). There are no equivalent comprehensive reviews for rocky intertidal communities, the reader is directed to Morse, 1992 (review of the effect of algae on invertebrate recruitment) and to references below (e.g. Connell, 1985; Menge and Farrell, 1989), for further insights. Instead, our goal is to draw the attention of workers in other fields to significant findings in these systems, and to promote exchange of ideas between rocky intertidal and coral reef fish researchers.

III. SETTLEMENT AND RECRUITMENT OF MARINE ORGANISMS

Larvae of benthic marine organisms typically disperse from natal sites and later arrive at suitable benthic habitat, after which they rapidly associate with substrate in the process of *settlement*. High mortality occurs in the plankton, either through predation (e.g. Gaines and Roughgarden, 1985), starvation (Hjort, 1914), offshore larval transport (Roughgarden et al., 1988; Farrell et al., 1991) or inability to find suitable habitat before settlement competency is lost (Strathmann and Branscombe, 1979; Victor, 1991). Mortality has been estimated at over 99% in the days before settlement (Fig. 1), although few studies have quantified actual rates (Gaines and Roughgarden, 1985). This is partly due to difficulty in matching the spatial scales of reproduction and settlement. Settlement can involve actually attaching to the substrate (e.g. barnacles, algae) or using the substrate as shelter (e.g. fishes). Usually, settlement is accompanied by some physiological changes such as colour change, sometimes of significant enough nature to be called metamorphosis (e.g. coral reef fish: Victor, 1991; Leis, 1991; invertebrates: Durante, 1991). The newly settled larva may be immediately visible to an observer, or may not appear for several days (e.g. Keough and Downes, 1982; Kingsford, 1988) especially if larvae or spores are microscopic or settle in cryptic habitats such as crevices or algal fronds or holdfasts (e.g. Bowman and Lewis, 1977; Lewis and Bowman, 1975; Connell, 1985), or if they bury in the sand (e.g. some wrasses: Victor, 1991). For these reasons, and because settlement often occurs at night, documentation of the actual settlement event is uncommon (see e.g. Olson, 1985; Sweatman, 1985b; Stoner, 1990; Doherty, pers. comm.; Booth, 1991a).

While settlement is clearly a biological phenomenon, *recruitment* requires more of an operational definition. Almost invariably, the days/weeks that follow settlement are ones of high mortality, with Type-III survivorship (Deevey, 1947) the rule. One definition of recruitment that may be biologically meaningful is the entry into the benthic population of the recruit after this period of high mortality (i.e. following the second kink in the survivorship curve shown in Figure 1). Three main factors account for low survivorship at this phase.

(i) In species that undergo metamorphosis, some simply cannot successfully complete the transformation.
(ii) Abiotic and biotic disturbances. These include environmental stress e.g. dislodgement by water movement, heat or desiccation (Connell, 1961; Strathmann and Branscombe, 1979; Strathmann, 1987; Menge, 1991; Southward, 1991), bulldozing (Dayton, 1971; Hawkins, 1983; Branch, 1985; Menge, 1991) and algal whiplash (Dayton, 1971; Menge, 1978; Hawkins, 1983).
(iii) Biotic interactions. Examples include predation (intended), and accidental ingestion (e.g. Connell, 1961; Keough and Downes, 1982; Doherty and Sale, 1986, planktivorous fish "wall of mouths": Hamner et al., 1988; Hawkins et al., 1989; Menge, 1991; Watanabe and Harrold, 1991; Caley, 1993).

In contrast to settlement, there are almost as many definitions of recruitment as there are studies that measure it, and many of the definitions may be arbitrary with respect to the life cycle of the organism (see Table 1). This may be appropriate, since the definition may depend on the questions being addressed. Logistic constraints on sampling usually mean that actual settlement is rarely monitored (but see Kobayashi, 1989; Breitburg, 1991) and so recruits are usually counted after some period post-settlement. Most intertidal studies define recruitment as "presence after a set interval". However, time intervals are highly variable and range from days to months. In reality, time interval is often determined by sampling constraints. In general, settlement of barnacle larvae can be measured easily, and most of the studies that distinguish between settlement and recruitment processes in marine intertidal invertebrates have been carried out on intertidal barnacles. For example, Connell (1985) defined settlement as the total number of barnacle larvae that attached to a plot daily. Total number surviving at the end of a specified period was taken as recruitment. By contrast, Caffey (1985) defined settlement in barnacles as the number surviving at the end of a one-month period. Note that based on our definitions, Caffey's "settlement" is actually recruitment (definition 6, Table 1). Given the extremely high mortality of new recruits (coral reef fish: up to 10% per day for the first few days, e.g. Aldenhoven, 1986; Doherty and Sale, 1986; Victor, 1986; Eckert, 1987; Booth, 1991a; 13% per day for mussel larvae: Jorgenson, 1981), census interval can greatly affect estimates of settlement rate. The general lack of consistency of census interval among studies, even of similar organisms, thus often precludes meaningful comparison (Booth, 1991a). Methods involving back-calculation of settlement time from older specimens (e.g. using ageing structures such as otoliths; Victor, 1982, 1986; Pitcher, 1988a) offer a possible solution, but have important biases as well (see Holm, 1990; Thorrold and Milicich, 1990; Doherty, 1991; Booth, unpublished).

IV. ARE MARINE SYSTEMS OPEN?

Generally, it is considered that most benthic organisms have open life cycles, that is, the number of new immigrants arriving in any local area (through settlement) is unrelated to (decoupled from) fecundity of the local population (Fig. 2). For coral reef fishes, only one exception is known—the damselfish *Acanthochromis polyacanthus* (Robertson, 1973). Note also that some intertidal species have direct development (e.g. some nudibranchs, sponges, and littorine and thaiid gastropods). Traditionally, larvae are viewed as passive propagules whose dispersal patterns are at the whim of the ocean currents. The discovery that many species have a long planktonic phase (e.g. 3–6 weeks in coral reef fish larvae, up to 6 weeks in the mussel *M. edulis*) bolstered the idea that marine larvae and algal

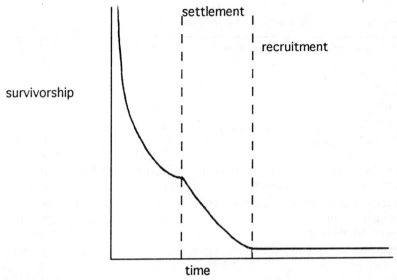

Fig. 1. Hypothetical survivorship curve for intertidal species, and coral reef fish. Early survivorship is low (see text) and survivorship generally follows the classical type III model (Deevey, 1947). Survivorship drops further at settlement owing to many factors, including inability to metamorphose (see text for details). To distinguish between settlement and recruitment, we define recruitment as occurring after the second dip in the mortality curve. Thus recruitment encompasses early planktonic mortality and settlement associated mortality.

propagules can be carried great distances, and recruit in areas far from the parent stock. Under this model, settlement may be density independent (i.e. unaffected by adult interactions). Furthermore, if adult populations are recruitment limited, then a study of recruitment will yield more information on local demography than a study of local fecundity patterns.

Whether or not the members of the community have open life cycles is therefore of critical importance in evaluating potential effects of recruitment. The decoupling of recruitment and adult fecundity is due to the highly variable nature of the pelagic environment and the presumed passivity of larvae and algal propagules. This conclusion evolved as a reaction to an earlier failed notion from the fisheries literature that a meaningful stock-recruitment relationship could be used to predict catch rates of commercial fishes (Cushing, 1975). Recent recruitment models for marine intertidal species are based on the "open system" assumption (e.g. Gaines and Roughgarden, 1985; Possingham and Rougharden, 1990). However, the open system hypothesis is rapidly becoming dogma (Sale, 1990 "bandwagon"), despite some recent challenges (see below). The openness of a population is difficult to demonstrate, and although the open system hypothesis is attractive, supporting evidence is still weak and fragmentary.

Table 1
Some definitions of recruitment used in studies of fishes and marine invertebrates

Context/Example	Definition	References
1. Entry into fishable/ adult population	Fisheries	Beverton and Holt, 1957; Ricker, 1975
2. Appearance on benthic habitat a. direct from plankton (i.e., synonymous with settlement)	Reef fishes	Levin, 1993
b. from adjacent seagrasses	Caribbean reef fishes	Shulman, 1985
3. Presence at end of settlement season (usually late summer)	Coral reef fishes	Williams, 1983; Doherty, 1981, 1983a, b, 1987a, b, 1991; Jones, 1988, and many others
4. Presence on certain day after full moon	Coral reef fishes, invertebrates	Robertson, 1988; Keough and Downes, 1982
5. Presence after passing through an early mortality gauntlet	Barnacles	Connell, 1985; Sewell and Watson, 1993
6. Survival to an age when settlers become prey items	Barnacles	Caffey, 1982
7. Attainment of a certain size	Fisheries Barnacles	Richards and Lindeman, 1987; Southward, 1991
8. Presence after a set interval (biologically arbitrary)	Marine intertidal invertebrates	Keough and Downes, 1982; Keough and Butler, 1983; Caffey, 1985; Connell, 1985; Jernakoff, 1985a, b; Bowman, 1986; Chabot and Bouget, 1988; Roughgarden et al., 1985; Farrell, 1989, 1991; Farrell et al., 1991; Judge et al., 1988; Raimondi, 1988; Bushek, 1988; Menge and Farrell, 1989; Stoner, 1990; Hurlbut, 1991a,b; Menge, 1991; Bertness et al., 1992.
	Coral reef fish	Jones, 1990; Booth, 1992.

How do we detect whether populations are open? Direct detection would come from following organisms from release of gametes or spores to final settlement, perhaps distant in time and space. It has been possible only in special circumstances. Stoner (1990) was able to follow the large larvae from a tropical ascidian from their release to settlement. However, the distances of dispersal were less than 100 m. In most cases, indirect approaches for following larvae offer the only tractable solutions to unbiased studies of population structure. These include: tracking of large water masses suspected to contain larvae (e.g. Gaines and Roughgarden, 1987; Rumrill, 1987, 1988; Black and Gay, 1987; Black et al., 1990; Farrell et al., 1991; Sewell and Watson, 1993); sampling of recruits in the water column and onshore (Sousa, 1984b; Reed et al., 1988; Sammarco and Andrews, 1989; Farrell et al., 1991; Gaines and Bertness, 1992), and studies of local genetic and morphological differentiation (e.g. Koehn et al., 1976; Levinton and Suchanek, 1978; Shaklee, 1984; Avise and Shapiro, 1986; Grosberg and Quinn, 1986; Todd et al., 1988; Behrens-Yamada, 1989; Gaines and Bertness, 1992). All these techniques have advantages and disadvantages. For instance, sampling larvae in the water column does not firmly identify their origin. Similarly, monitoring recruits on the shore can lead to difficulties in interpretation. Arrival of a new species at a particular site implies open recruitment, but for species already present the appearance of a new individual can be explained by either open or local recruitment. Genetic studies generally require polymorphic loci, and even then local differentiation may result from restricted gene flow or strong local selection (e.g. Gosling, 1992 and references therein). Recent advances in microsatellite DNA analysis might offer some new solutions (Queller et al., 1992).

The evidence for population openness from different taxonomic groups is conflicting, and so we will first review the evidence from each community. We will then argue that much of the confusion results from an inappropriate model and confounding effects of scale. We suggest that a modified metapopulation approach may resolve these difficulties and form the basis for future studies of dispersal and population structure.

In coral reef fishes, genetic-based studies assume that, if populations are open, then genetic relatedness between resident adults and incoming recruits will be low. Some studies support this model. Avise and Shapiro (1986) found this to be the case for Anthias squamipinnis in the Red Sea. Soule (in Ehrlich, 1975) found less latitudinal variation in allozymes among Great Barrier Reef damselfish that were assumed to be wide dispersers (pelagic larvae) than for more limited dispersers (young brooded)—most fish examined were apparently dispersing widely, and so conformed with an open population assumption. Shaklee (1984) found negligible genetic differentiation between pomacentrid fishes (Stegastes fasciolatus) over the 2500 km Hawaiian archipelago, with almost complete homogeneity among samples at 8 polymorphic loci, as would be expected for widely dispersing, panmictic open populations. As with recruitment studies, most research into genetics of dispersal in coral reef fishes has concentrated on pomacentrids.

In contrast, some recent studies have demonstrated that larval coral reef fishes may be retained on reefs, and that some sites may even be self-recruiting. Meekan (unpubl. data) has found that a common damselfish at Lizard Island on the northern Great Barrier Reef, in Australia (henceforth abbreviated GBR) shows a local correlation between egg production, near-shore larval abundance and recruitment, and that local egg production was a good predictor of settlement intensity. This suggests the possibility of at least partial closure of that population, although he was not able to verify the actual source of the larvae. Black and Gay (1987) and Black *et al.* (1990) developed field-based oceanographic models which suggest local retention of water masses adjacent to GBR reefs, although they assumed that larvae behaved as passive water molecules within these masses (cf. Leis, 1991)!

Mechanisms for retention of larvae near the point of dispersal include tidal fronts and local current gyres, and enclosed bays (e.g. Lobel, 1988; Gaines and Bertness, 1992; Sewell and Watson, 1993 and Table 2). Lobel (1988) hypothesized that such gyres could return Hawaiian reef fish larvae to or near to natal reefs and so maintain populations on the small and very isolated islands of the archipelago. However, such evidence is generally weak and circumstantial, and needs to be verified by actual sampling of larvae inside such gyres. Kingsford *et al.* (1991) concluded that tidally induced fronts around reefs could increase the effective detection area of the reef to incoming larvae, and may also act to retain larvae produced locally.

On a cautionary note, a recent study by Planes and Doherty (unpubl. data) measured spatial patterns in allele frequencies of some polymorphic loci in the damselfish *Acanthochromis polyacanthus* on the Great Barrier Reef. One locus (AaT-2), which correlated well with colour patterns of the fish, showed similar ratios of its 3 alleles between reefs off Cairns and Townsville, a distance of 300–500 km. This pattern may suggest gene flow between these populations, and perhaps open population status for this species. However, as stated earlier, *A. polyacanthus* is the only known larval brooder among coral reef fishes, and so should be the archetypal candidate for closed population status among reef fishes! This serves to remind us that data on polymorphism should be combined with other information (e.g. direct observations of life history) in evaluating the population biology of marine species.

Dispersal in marine algae has been the focus of much attention (Chapman, 1974, 1986; Sousa, 1984b; Hoffman, 1987; papers in Santelices, 1990; Hoffmann and Santelices, 1991). Many results indicate that open recruitment is not the rule and that recruitment of many intertidal algal species is predominantly local (but see Menge *et al.* (1993) for contrasting data). Factors affecting dispersal of algal spores include size and weight of spores, and phototrophic responses. Consequently some spores tend to occur predominantly in slow-moving water bodies such as the boundary layer, while others are found in faster moving current and surge layers. Experimental

Table 2
Some of the major factors affecting recruitment rates of benthic marine organisms

Factor	Process	Examples
A. Physical Oceanographic	upwelling,	Carr, 1991; Farrell et al., 1991; Pineda, 1991.
	tidal fronts, slicks, gyres and eddies	Kingsford and Suthers, in press. Lobel, 1988; Shanks, 1983, 1986; Cowan, 1985; Shanks and Wright, 1987; Sewell and Watson, 1993.
	large-scale current patterns and ESNO events, wind driven patterns	Barnes and Powell, 1953; Crisp, 1955, 1958; Lewis, 1963; Lewis and Bowman, 1975; Barnes and Barnes, 1977; Bowman and Lewis, 1977, 1986; Wethey, 1984, 1986; Hawkins and Hartnoll, 1982; Connell, 1985; Reed et al., 1988; Sutherland, 1990; Farrell et al., 1991; Gaines and Bertness, 1992
	exposure (higher recruitment on exposed shores)	Sutherland, 1987, 1990; Busek, 1988; Roughgarden et al., 1985; Farrell, 1989; Menge, 1991; Southward, 1991; but see Rumrill, 1987, 1988
Substrate	affects settlement and survival	Ryland, 1959; Crisp and Meadows, 1962; Crisp, 1974; Meadows and Campbell, 1972; Lewis and Bowman, 1975; Chabot and Bouget, 1988; Raimondi, 1988; Stephens and Bertness, 1991; Morse, 1992; Pawlick, 1992; Brosnan upubl. but see Caffey, 1982; Booth, 1992
Disturbance	desiccation, cyclones, storms	Connell, 1961; Dayton, 1971; Hawkins, 1981; Hawkins and Hartnoll, 1983; Keough and Downes, 1982; Hartnoll and Hawkins, 1985; Reed et al., 1988; Menge, 1991
B. Biological Biotic disturbance	whiplash bulldozing	Dayton, 1971; Menge, 1978; Hawkins and Hartnoll, 1982, 1983; Branch, 1981, 1986

Table 2—*cont.*

Factor	Process	Examples
Predation	in the water column, and at settlement	Thorson, 1946; Hancock, 1973; Bayne, 1976; Menge and Lubchenco, 1981; Jernakoff, 1983, 1985a; Gaines and Roughgarden, 1987; Roughgarden *et al.*, 1985; Roughgarden, 1986; Doherty and Sale, 1986; Lively and Raimondi, 1987; Fairweather, 1988; Santelices and Martinez, 1988; Santelices, 1990; Hoffmann and Santelices, 1991; Menge, 1991; Hixon and Beets, 1993
Adult spawning	periodicity	Barnes, 1956a,b, 1957; Barnes and Barnes, 1977; Crisp and Spencer, 1958; Bowman and Lewis, 1977, 1986; Hawkins and Hartnoll, 1982; Doherty, 1983b; McFarland *et al.*, 1985; Victor, 1986; Robertson *et al.*, 1988
	strategy (demersal vs. pelagic modes)	Thresher, 1991
	proximity to spawning adults	Sundene, 1962; Anderson and North, 1966; Dayton, 1973; Paine, 1979; Sousa, 1984b; Reed *et al.*, 1988
Larval behaviour	active swimming	Breitburg, 1991
	substrate preference	Barnes, 1955; Ryland, 1959; Crisp and Meadows, 1962; Bayne, 1965; Dayton, 1971; Meadows and Campbell, 1972; references in Crisp, 1974, 1976a,b; Wethey, 1984, 1986; Southward and Barnes, 1979; Southward, 1991; Nelson, 1981; Cooper, 1982, 1983; Peterson, 1984a,b; Eyster and Pechenik, 1987; Strathman, 1987; Chabot and Bouget, 1988; Le Tourneux and Bouget, 1988; Raimondi, 1990; Paine and Levin, 1981; Keough and Downes, 1982; Morse, 1992; Pawlick, 1992
	predator avoidance	Young and Chia, 1981; Johnson and Strathmann, 1989
	gregarious settlement	Knight-Jones and Stevenson, 1950; Knight-Jones, 1953; Meadows and Campbell, 1972; Crisp, 1974, 1976a,b; Barnett and Crisp, 1979; Wethey, 1984, 1986; Peterson, 1984a,b; Eyster and Pechenik, 1987; Raimondi, 1990; Southward, 1991

Table 2—*cont.*

Factor	Process	Examples
	phototropism vertical migration delayed metamor- phosis	Reed *et al.*, 1988 Davis and Olla, 1987 Bayne, 1965, 1976; Day and McEdward, 1984; Victor, 1991
Larval condition	starvation	Hjort, 1914; Hancock, 1973 Bayne, 1976; Suthers, 1989

studies in algal recruitment patterns have been primarily of two types: the first involves positioning settlement slides to trap algal spores in the water column at various distances from shore; the second involves clearing areas and counting the density of new recruits and their distance to nearest adult plant (these experiments often form part of disturbance studies, e.g. Sousa, 1984a). Drawbacks of these techniques have been discussed above.

Experiments involving clearing space and monitoring recruitment suggest that many algal species recruit locally. For example, kelp species may recruit short distances: 10 m for *Alaria esculanta* (Sundene, 1962); 1·5–7 m for *Postelsia palmaeformis* (Dayton, 1973; Paine, 1979). Adult density affects dispersal distance in the kelp *Macrocystis pyrifera*, distance increasing from 5 m to 40 m as the density of adults rises (Anderson and North, 1966). Sousa (1984a) found that while most species on rocky shores disperse over a few metres, a few opportunistic species, such as *Enteromorpha*, dispersed more widely. *Enteromorpha* spores have been found on slides placed 35 km from the nearest population (Amsler and Searles, 1980). The preponderance of these studies show that most algal dispersal is local. However, complemen- tary studies on recruits in the water column seem to contradict this pattern (Hoffman, 1987; Santelices, 1990; Hoffman and Santelices, 1991). Algal spores are found at great distance (out to sea) from adult populations on and near shore. Similarly, seaweeds including kelps, can be found on remote islands or on new structures such as oil platforms (Chapman, 1986; Druelh, 1981). These results suggest a more open population structure for algae.

There may be both intra- and inter-specific differences in algal recruit- ment. For instance a study by Reed *et al.* (1988) on dispersal and recruitment patterns in kelps and filamentous brown algae suggests that propagule behaviour influences dispersal. Both algal groups produce spores of similar size and shape, but filamentous brown algal spores are positively phototac- tic, and kelp spore are negatively phototactic. Kelp recruitment onto settle- ment slides decreased rapidly with distance from adult plants, and was significantly lower at 3 m. Variation in recruitment was due to variable rates of settlement and early post-settlement mortality. In contrast, filamentous brown algae had much greater dispersal distance and there was no signifi-

D.J. BOOTH AND D.M. BROSNAN

cant decrease in recruitment up to 500 m from adult plants. Episodic disturbances, such as storms, were also found to increase algal dispersal distances.

Invertebrate larvae have long been held to disperse over long distances. Ideas of open population structure were deduced largely from an understanding of reproductive patterns, and from laboratory studies showing that the planktonic phase is long for many species. Studies also showed that many larvae have a competency phase, and cannot settle prior to this stage. For example, the mussel *Mytilus edulis* is unable to metamorphose before a shell length of around 260 μm (McGrath *et al.*, 1988). Some species can prolong their competency phase if a suitable habitat is not immediately available—for example, *M. edulis* can continue to live and grow in the plankton up to a shell length of 350 μm (Bayne, 1965; Sprung, 1984). But ultimately, larvae lose the ability to settle and metamorphose if they do not find a suitable site within a specified time of reaching competency (Bayne, 1965; Strathmann, 1987; Pechenek, 1984). Cooler temperatures may indirectly increase dispersal distances, by retarding larval development rates, and consequently increasing time spent in the plankton. Species-specific differences in dispersal are also known (e.g. Hawkins and Hartnoll, 1983). Dispersal distances have often been calculated using the product of larval development time (from laboratory studies) and projected current speeds (usually assumed to be constant and unidirectional). Not surprisingly, these calculations often suggest that larvae can travel over great distances. These calculations are simplistic and do not account for larval position and behaviour in the water column, changes in current patterns, or the presence of eddies and reduced circulation bays. Consequently, results from calculations and field experiments are often at odds.

Recently more sophisticated techniques have been used to study larval transport. For example, Gaines and Roughgarden (1987) used satellite imagery to track movement and survival of barnacle larvae in a large body of water off California. In the same area (Farrell *et al.*, 1991) used satellite imagery and field studies (plankton tows and settlement plates on shore) to evaluate barnacle recruitment events. Recruitment pulses were strongly associated with periods of onshore water movement, which occurred during cessation of upwelling and relaxation of alongshore winds. One of the species studied, *Chthamalus fissus*, is at its northern geographic range in Monterey Bay. Recruitment of this species was associated with onshore and poleward flowing water. Therefore recruits either came from the local bay or were released from more southerly populations. Appearance of recruits near the entrance of the bay prior to their arrival in the bay interior suggests that recruitment is primarily open and from outside the bay (Farrell *et al.*, 1991). In addition, flushing rates in Monterey Bay average 6 days, which is considerably less than time to competency for barnacle larvae. Other studies have also focused on the importance of the inter-play between wind, oceanographic processes and recruitment (for instance, Bennell, 1981; Hawkins and Hartnoll, 1982; Kendall *et al.*, 1982; Shanks, 1983, 1986; Shanks and Wright, 1987; Gaines and Bertness, 1992; Sewell and Watson, 1993).

Relatively few studies have looked at the morphological and genetic differentiation of marine invertebrate species. Behrens-Yamada (1989) showed that there were greater differences in growth and reproductive timing between populations of directly developing gastropods (with presumably closed populations) than between populations of species with planktonic development. Similar patterns have been found for differences in shell shape (Behrens-Yamada, pers. comm.). In genetic studies, Grosberg and Quinn (1986) found that ascidian larvae settle close to adult kin. Havenhand *et al.* (1986) used four polymorphic loci for the Scottish intertidal mollusc *Adalaria proxima* to determine that actual larval transport was considerably less than that expected on the basis of larval culture alone. Levington and Suchanek (1978) in a study of polymorphic enzymes in the mussel *Mytilus californianus* found that total variation along the West Coast of the USA from Alaska to California was less than intra-shore variation.

Additional information on disperal can be gained from studies of natural selection and gene flow in marine species. Extensive studies of this type have been carried out on the mussel *M. edulis*. The first large scale genetic survey was carried out along the east coast of north America by Koehn *et al.* (1976). In this study, three common alleles for LAP were found. LAP frequency was homogenous in oceanic samples collected over 800 km from Virginia to Cape Cod. Similarly, at another polymorphic locus, AP, allele frequency was homogenous from Nova Scotia to Virginia. These patterns support an open population model. However, the frequency of LAP[94] declined and relative frequency of LAP[96] increased in less saline waters of Long Island Sound (Koehn *et al.*, 1976; Hilbish and Koehn, 1985). A similar pattern of decline in LAP[94] was noted in the Gulf of St. Lawrence and northern Newfoundland. In Long Island Sound, larvae exhibiting oceanic LAP frequencies migrate into the Sound waters every year, but genotype dependent mortality results in high mortality of non LAP[94] larvae. This pattern suggests that selection can swamp gene flow.

At a larger geographic scale, allele frequencies in *M. edulis* are homogenous within Northwest Europe, and within the east coast of North America, but there is noticeable transoceanic differentiation at a number of loci (Gosling, 1992). These results imply either two open populations restricted to either side of the Atlantic ocean, or strong selection across the Atlantic.

As has been shown above for algae, there may be variability in degrees of openness. For instance, Sutherland and Ortega (1986) found that barnacles (*C. fissus*) recruit sporadically onto shores in Costa Rica (high pulses are associated with El Nino years). However, the offspring of these barnacles do not recruit back onto the shore, and may be carried offshore by currents (their subsequent fate is unknown). This population clearly follows an open structure. Sammarco and Andrews (1989) placed coral settlement plates in a regular spiral array outward from Helix reef, central GBR. By simultaneously monitoring coral larval settlement on these plates and surface currents, they concluded that pocilloporid spat produced on the reef were largely self-recruiting, owing to water masses with high retention and low flushing rates. However, they concluded that some spat dispersed widely, and so gene flow among reefs was significant enough to prevent allopatric speciation. This raises the possibility that a population may be mainly closed for the purposes of population demography and ecology. It must be remembered that the evolutionary consequences of even small amounts of gene flow are important.

In summary, some studies suggest open population structure; others, even for the same taxonomic groups, imply local recruitment. How can these conflicting results be reconciled? One possibility is that the dichotomy between open and closed systems is false. It may be more appropriate and useful to ask: under what conditions will most propagules recruit locally, and under what conditions is long distance dispersal likely? A population may show local and open recruitment at different times and in different geographic areas. For example, barnacles whose larvae are trapped in an enclosed bay may have a closed population (see situations studied by Bennell, 1981; Kendall et al., 1982; Gaines and Bertness, 1992). A similar situation may exist for a population of the sea star Pisaster ochraceus in an enclosed inlet of British Columbia Canada (Sewell and Watson, 1993). Adjacent populations of the same species, on a coastline with a predominantly longshore current will be open (Farrell et al., 1991). Temporal variability and episodic events can also change the pattern of recruitment to increase or decease the relative importance of local or distant recruitment (e.g. Hawkins and Hartnoll, 1982, 1983; Reed et al., 1988).

Much of the evidence above suggests that most species have a leptokurtic distribution of dispersal distances (references above). This implies that the majority of propagules disperse short distances but some travel much farther. If this is the case, then the discrepancy between conclusions on algal dispersal, based on shore and water column methodologies, is simply an artefact of the technique and scale. Based on this idea, we would expect to find spores both near and offshore, although the relative abundance of spores from a single population would be higher close to the parent population. Similarly, in some invertebrates and fish, most larvae may recruit locally (i.e. the population appears closed), but some larvae may disperse to, and colonize, distant vacant habitat (i.e. show degrees of openness). Whether or not one characterizes the population as closed or open will therefore depend on where sampling is carried out. Figure 2 illustrates this concept. This model is essentially a metapopulation model (see below).

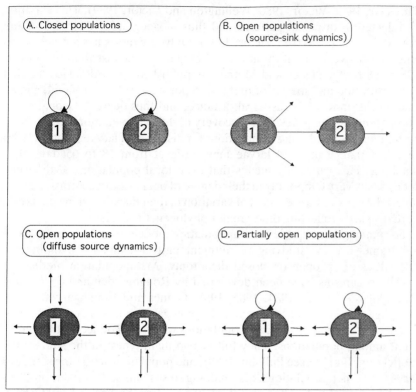

Fig. 2. Open populations, closed populations, and metapopulations: degrees of openness for hypothetical populations of marine organisms. Arrows indicate direction of propagule dispersal. A. Fully closed (self-seeding populations). B. Fully open populations, where all larval influx is from distant sources. In this example, dispersal is directional, some populations (here population 1) act as source populations for others (here population 2, the sink population). This also represents the classic source-sink metapopulation model. C. Fully open populations, where all propagules arriving at a habitat are from many distant source populations. D. Mixed open/closed populations, where populations 1 and 2 are seeded by a combination of their own offspring and from distant sources. The relative importance of distant and local propagule influx may vary seasonally and spatially. Note that B–D can also be considered metapopulations.

In general, Fairweather (1988) predicts that species with shorter planktonic duration (e.g. algae, colonial animals) may disperse less, so have a tighter stock/recruitment relationship than other groups (e.g. barnacles and fish). Within taxa, considerable variation can exist in larval duration. Coral reef fishes have competency periods in the plankton (range of days over which successful settlement occurs) that vary from under 20 to over 100 days

328 D.J. BOOTH AND D.M. BROSNAN

post-hatching, but most are in the range 15 to 30 days (Brothers and
Thresher, 1985; Victor, 1986; Wellington and Victor, 1989). This variation
is reflected in dispersal patterns and thus biogeographic distribution and
population openness (see Victor, 1991). Only in wrasses has a relationship
emerged between larval duration and biogeographic distribution (Victor,
1986). Doherty, Planes and Mather (unpublished) provide evidence that
larval duration in 7 species of reef fish (5 pomacentrids, a caesionid and an
acanthurid) may affect dispersal distance, and thus degree of openness of
populations of reef fishes. The discovery of delayed metamorphosis among
certain reef fishes also has implications for dispersal of larvae. Victor (1986)
reported that acanthurid larvae may settle at from 18 to 65 days after
hatching. This variability means that even local populations of the same
species may vary temporally in their degree of openness. In contrast, benthic
invertebrates show a great deal of variation in their duration in the plankton,
perhaps partly reflecting their greater phylogenetic diversity.

For many marine species, metapopulation models offer a more appropriate framework for studying recruitment patterns, and for avoiding the
difficulties of an open or closed dichotomy. Metapopulation models for
marine organisms have been developed by Roughgarden and co-workers
(Roughgarden et al., 1984, 1985, 1988; Gaines and Roughgarden, 1985;
Roughgarden, 1986; Possingham and Roughgarden, 1990). A metapopulation is a set of spatially distinct populations which are connected by recruitment among populations. They follow two main types; in the source-sink
model (e.g. Fig. 2B, see Pulliam, 1988), one population acts as the source of
larvae for other populations, and in doing so sustains sink populations (this
situation is exemplified by barnacles in Costa Rica (Sutherland and Ortega,
1986)). In the second type of metapopulation, there is no single source
population, rather there is dispersal between populations (e.g. Fig. 2C,D).
Local extinctions and colonization events characterize this model. At any
one time, a population may become extinct and the area subsequently
recolonized by immigration of new recruits of the same species. Metapopulation models are enjoying much attention and success in conservation
biology, where they are used to study dispersal across fragmented habitats.
Rocky shores and reefs may essentially be viewed as a series of spatially
separated fragments of suitable habitat surrounded, literally, by an inhospi-
table sea of unsuitable settlement sites.

In marine intertidal communities, the sea palm Postelsia palmaeformis
may exemplify a metapopulation. Local persistence of this species depends
on a particular disturbance regime (Paine, 1979). Most sporophyte and
gametophyte recruitment is local (Dayton, 1973; Paine, 1979). However
some spores recruit to more distant shores, and new populations arise in
recently disturbed exposed shores (Brosnan pers. obs.). Although populations will establish on exposed intertidal rocks, and will persist for at least a

couple of generations, an inadequate disturbance regime, and distances from 10–500 m from other populations, will ultimately lead to extinction (Paine, 1979). However the key point is that, dispersal and recruitment ensures that the species persists at a large spatial scale, even though there are cycles of colonization and extinction at a local scale. This leads to a spatial and temporal mosaic of interconnected populations, any of which have a set probability of extinction at a particular time, and to a mosaic of available space, which has a probability of colonization. For *P. palmaeformis* the relative importance of open and closed recruitment patterns will be determined by dispersal abilities and disturbance patterns.

In conclusion, since patterns of larval retention may vary taxonomically, spatially and seasonally, we need to know more about connectedness among larval habitats (e.g. Dight *et al.*, 1988). An integrated approach is necessary, combining modelling, oceanography, genetic, ecological and behavioural studies (see Pulliam, 1988; Fairweather, 1991; Gaines and Bertness, 1992). Possibilities for testing the relationship between genetic differentiation and dispersal include a comparison of the population genetics of fish assemblages in atoll lagoons with various degrees of closure. For example, Galzin *et al.* (1994) have described 7 atolls within French Polynesia which range from fully open lagoons to permanently closed lagoons, with minimal historical connection with the surrounding ocean. A simultaneous study of population genetics of several common species and local oceanography at each atoll may allow the relationship between degree of closure of the population and spatial genetic differentiation to be examined.

V. RECRUITMENT: PATTERNS AND PROCESSES

A. The Extent of Recruitment Variability in Marine Systems

Few long-term studies of recruitment have been attempted, but they have found a great deal of variation at all scales. Both tropical reef fishes and temperate intertidal organisms display highly seasonal patterns of recruitment. Comprehensive treatments of recruitment variability in coral reef fishes are given in reviews by Doherty and Williams (1988) and Doherty (1991). Generally, recruitment intensity is highest in the summer months for coral reef fishes (e.g. Walsh, 1987; Doherty and Williams, 1988; Doherty, 1991; Williams, 1991), with a secondary peak sometimes in early spring (Walsh, 1987). Within the period of major annual recruitment, recruitment may oscillate on the basis of lunar (Doherty, 1983a,b; McFarland *et al.*, 1985; Robertson, 1992; Booth and Beretta, 1994), semi-lunar (Ochi, 1986), or episodic patterns (nearly all yearly recruitment in one day: Victor, 1983, 1986; Milicich, 1988). Alternatively, recruitment may occur apparently at

330   D.J. BOOTH AND D.M. BROSNAN

random within the summer period (Ochi, 1986; Booth, 1992). Nearby habitats may or may not receive synchronous recruitment pulses of a single species or a group of related taxa (e.g. Luckhurst and Luckhurst, 1978; Williams, 1983; Doherty, 1983a,b; Booth, unpubl. data). Williams and Sale (1981) found that recruitment synchronicity was less among more distantly related taxa, although Booth and Beretta (1994) found considerable variation in a suite of closely related species in the US Virgin Islands. Recruitment of both species of *Chromis* damselfish monitored was lunar periodic, and a summer phenomenon. Of 5 species of *Stegastes* damselfish, 4 showed predominantly summer recruitment, while one (*S. planifrons*) recruited throughout the year. None of these species exhibited lunar periodicity to their recruitment. Interestingly, recruitment of several of these *Stegastes* species is lunar periodic off the coast of Panama (Robertson, 1992).

Studies on marine intertidal species indicate that recruitment varies by orders of magnitude at many spatial and temporal scales. Many studies find that recruitment to rocky shores is inconsistent and unpredictable on a temporal basis. Shores that rank highest in recruitment one year will not necessarily retain that ranking in subsequent years (e.g. Caffey, 1985; Keough, 1983; Gaines and Roughgarden, 1985; Judge *et al.*, 1988; Farrell *et al.*, 1991; Bertness *et al.*, 1992; Menge *et al.*, 1993, pers. comm.). Others note that some sites have consistently high or low settlement, and maintain their ranking through time (Connell, 1985; Keough and Butler, 1983). Rumrill (1988) found predictable and intense asteroid settlement in Barkely Sound Canada, on shores protected from direct surf. Often, predictability and consistency seem to increase as scale decreases (Caffey, 1985). In contrast, Raimondi (1990) found that variation decreased with increasing spatial scale. Similarly, Doherty (1991) noted for GBR reef fishes, predictability increases at larger geographic scales. Connell (1985) noted that recruitment variability of barnacles was greater where adult barnacles were less abundant.

The degree of seasonality is variable for recruitment of invertebrate larvae. For instance, in the temperate zone, peaks of recruitment often occur in spring and early summer (Gaines and Roughgarden, 1985), but other species recruit mainly in autumn (Southward, 1991). In New England, USA, recruitment is correlated with the onset of warm long days (Menge, 1991). Intense pulses of recruitment, when larvae swamp all available space, seem common in the temperate zone (Menge, 1991). However, smaller peaks of recruitment do occur throughout the year in many species (Peterson, 1984a,b; Farrell, 1989; Brosnan unpubl. data). There are also large differences in reproductive behaviour between boreal and warm temperate marine species in North America (Hawkins *et al.*, 1992). Boreal species usually produce one brood annually and in tight synchrony with spring diatom blooms; by contrast warm temperate species produce multiple broods per year, mostly in late spring (Hawkins *et al.*, 1992). In the tropics, there may be less seasonality (Sutherland and Ortega, 1986; Sutherland, 1987, 1990; Menge, 1991) for marine intertidal organisms. Nonetheless, at least in tropical systems, peaks of recruitment of intertidal species seem less predictable than for coral reef fish.

Is there a temperate/tropical difference in recruitment pattern? Obviously coral reef fish studies are necessarily biased towards the tropics, whereas most intertidal studies have been carried out in temperate latitudes. Could temperate/tropical differences be artefacts of latitudinal biases between coral reef fish and intertidal invertebrate studies. For reef fishes in general, there is little information on the seasonality of recruitment in temperate areas. In some well-studied areas, conspicuous species may be viviparous, and larvae appear to be retained locally (e.g. embiotocids of the west coast USA). Carr (1991) found that recruitment was linked to major upwelling events for 9 species of rockfish (*Sebastes* spp.) off the Californian coasts. Jones (1984) found that recruitment of *Pseudolabrus celidotus* in New Zealand was a summer phenomenon, with recruitment rates comparable to tropical labrids. Thresher *et al.* (1989) found that settlement pulses of small fishes of rocky reefs off Tasmania coincided with phytoplankton blooms. Authors who have worked in both temperate and tropical shore environments (e.g. Menge and Lubchenco, 1981; Menge *et al.*, 1983, 1986a,b; Lubchenco *et al.*, 1984; Sutherland and Ortega, 1986; Sutherland, 1987, 1990; Menge, 1991) note that recruitment of intertidal invertebrates is generally lower in the tropics than in temperate latitudes. Menge (1991) found that recruitment accounted for 39–87% of the variation in sessile invertebrate density in Panama compared with up to 11% in New England. He found that recruitment of one species, the barnacle *C. fissus* (the highest recruiter in the community), was variable at all scales. Apart from these few studies, we have little information on the variability of recruitment at latitudinal scales. It is possible that tropical/temperate recruitment intensity gradients may be reversed for reef fish and rocky intertidal species. However, we must await further information before rigorous latitudinal comparisons can be made.

Despite the huge amount of variability in recruitment (see Table 3), are there any consistent patterns? Unfortunately, few studies measure recruitment in the same way, or at the same temporal or spatial scales, so different measures of variability are used (e.g. numbers of recruits, variance estimates or qualitative measures). This is a serious problem. Minchinton and Scheibling (1993) found that estimates of barnacle recruitment were strongly influenced by sampling frequency. This makes comparison across studies difficult. Nevertheless, one pattern that many authors have documented for intertidal invertebrates is that recruitment is higher on more wave-exposed sites (Caffey, 1985; Connell, 1985; Gaines and Roughgarden,

1985; Raimondi, 1990; Menge, 1991), and that change in large-scale oceano-graphic events can alter recruitment patterns (e.g. Sutherland, 1990; Farrell *et al.*, 1991). This may also be true for coral reef fishes (e.g. Leis, 1986; Leis and Goldman, 1987; Milicich, 1988). Consistent patterns for coral reef fish include: predominantly summer recruitment (see McFarland and Ogden, 1985, and Booth and Beretta, 1994, for exceptions), cross-shelf predictable distributions of "specialized" vs. "unspecialized" larvae across the GBR (see Williams, 1991), and lunar peaks of recruitment (with numerous exceptions, some listed above).

In comparing both systems, there appear to be seasonal pulses of recruitment for at least some species in reef fish and temperate shore communities. Many fish show a semi-lunar pulse not reported for intertidal species (e.g. Walsh, 1987). Both systems are characterized by high spatial and temporal variability in recruitment. Spatial variability seems to occur across all scales from the single shore or reef, to the larger geographic (temperate-tropical) scales.

In summary, there is considerable variation in temporal and spatial patterns of recruitment. Wide-scale variation is found in both temperate and tropical latitudes. Often recruitment is unpredictable and high recruitment at one site does not guarantee high recruitment at adjacent sites. In intertidal communities (unlike reef fishes) recruitment in the tropics may be lower than in temperate shores, but few studies have compared tropical and temperate sites. However, many studies are short-term. Long-term and integrated studies of variation and causes of variation are necessary for better understanding of patterns of recruitment. Methods such as otolith back-calculation reconstruction of recruitment patterns between widely-separated sites (Victor, 1982, 1986; Pitcher 1988a) offer hope of measuring differences in recruitment of coral reef fishes at widely separate sites.

B. Factors Affecting Recruitment Strength

Recruitment patterns are the result of a suite of biological and physical factors that are rarely predictable. They include everything from larval choice at settlement to oceanographic patterns. For example, El Nino years have been shown to significantly alter recruitment patterns of barnacles in California and Costa Rica (Sutherland and Ortega, 1986). In some studies, oceanographic patterns are cited as the dominant factor (e.g. Shanks 1983, 1986; Williams *et al.*, 1984; Connell, 1985; Gaines and Roughgarden 1985; Shanks and Wright, 1987; Kingsford *et al.*, 1991; Pineda, 1991), in others combinations of factors (Kingsford, 1990; Menge, 1991). We will not dwell on all these factors in detail (some are discussed in other sections of this paper). Instead, in Table 2 we present some of the many factors that have been shown to affect recruitment timing and inten-

Table 3

Examples of recruitment variability, at temporal and spatial scales for coral reef fish and intertidal species. (Note that not all studies gave the extent of variability at both scales, and so this table is incomplete.)

Species	Spatial variability/locality	Temporal variability	References
1. Intertidal species			
Balanus glandula	Central California, at Hopkins Marine Station: 2 sites, KLM rocks (nearshore) and Pete's rock (seaward) near Hopkins Marine Station. Recruitment was 20 fold higher at Pete's rock than at KLM.	Recruitment monitored from 1982–1984. Recruitment rates to all sites in 1983 were 2–4 times rates for 1982, and 1984.	Gaines and Rough-garden, 1985
Chthamalus fissus	Pacific coast of Costa Rica, 10 sites, representing 5 different microhabitats. On one shore, densities per site ranged from 23/cm^2 to 339/cm^2.	Censused 1–3 month intervals from 1983–1985. Recruitment rates of 0·1–3 cm^2/day.	Sutherland, 1990
Crassostrea virginica	Galveston Bay, Texas. Study had 4 nearshore sites (<10 m from shore) and 4 offshore (35–55 m from shore). Settlement on hard substrata ranged from 5–62 larvae per plate, nearshore and 0–7 offshore.	Settlement monitored daily Mar–July 1986. All oyster settlement occurred within one week (13–19 May).	Bushek, 1988
Balanus eburneus	Recruitment higher at offshore sites. Between Mar–July 1986. 0·2–190 larvae/plate settled on offshore plates and 0–25·8 settled on nearshore plates.	Recruitment occurred during the entire 5 month sampling period.	Bushek, 1988

Table 3—*cont.*

Species	Spatial variability/locality	Temporal variability	References
Semibalanus glandula	Bodega Bay exposed point and sheltered harbour. Recruitment dissynchronous over scales of several kilometres.	1985–1987. Temporal variability in recruitment in the open coast varied by 5 orders of magnitude (mean rate 3586/m²/mo), and by 4 orders of magnitude in harbour sites (mean rate 3370/m²/mo).	Judge *et al.*, 1988
Kelps, and filamentous brown algae (FBA)	Santa Barbara California, short distance dispersal, two sites 5 km apart. For long distance an additional site 65 km distant from the first site. Recruitment decreases by an order of magnitude from 1–3 metres from adult plant.	1986–1987. Settlement density of FBA averaged 4 sporelings/mm² from 0–10 m from adults. Kelps, *Pterogophora* averaged 8 sporelings/mm² close to adults and <4/mm² 10 m from adult plants. For *Macrocystis*, densities were 4/mm² and 0·5/mm² respectively.	Reed *et al.*, 1988
Chthamalus anisopoma	Northern Gulf of Mexico, exposed reefs near Punta Penasco. Scales ranged from 1 cm–10 km. Differences at largest scale ranged from 24·9–35·5 barnacles/cm². At the smallest scale differences ranged from 5·71–14·79 barnacles/cm².	1984, 1986, study monitored from 12 hours to 1 month. Most settlement occurred during daylight tides. Temporal variability represents 84·5% of the variance in 1984–1985 and 91·2% in 1985–1986.	Raimondi, 1990
Semibalanus balanoides	Rhode Island, two bays 25 km apart. Average settlement at low tidal heights was 390 cyprids/day compared with 34 cyprids/day at high tidal heights.	Ongoing study for 8 years. Settlement occurred earlier at the high recruitment shore, and recruitment rate was four times higher.	Bertness *et al.*, 1992
2. Coral reef fishes	If location unstated, read Great Barrier Reef		

Table 3 *continued*

Species	Spatial variability/locality	Temporal variability	References
Dascyllus albisella	Hawaii, Kaneohe Bay 1989 7·3 (range 1·9–19)	One patch reef in Kaneohe Bay over 3 years: 1987—25 recruits. 1988—152 recruits. 1989—228 recruits.	Booth, 1991b, 1992
Caribbean reef fishes	10 fold variation in recruitment in quadrats <400 m apart.		Luckhurst and Luckhurst, 1978
2 pom-acentrids	10 fold variation among patch reefs 400 m apart.		Williams, 1979 in Doherty and Williams, 1988
1 pom-acentrid	12 fold difference over 25 hrs at 2 sites		Doherty, 1983
Caribbean (various)	highest variation for schooling haemulids.		Shulman, 1984
Thalassoma bifasciatum	Caribbean	3–7 fold variation in recruitment over 4 years	Victor, 1983
Thalassoma bifasciatum	Caribbean	no correlation between 1981/ 1982 magnitude of 10 successive peaks (r = 0·19 n.s. n = 10)	reanalysed from Victor, 1986
pomacentrids	Central GBR	multi-species pulses in same few days over 3 years, but density of recruits varied by up to 10 fold over the 3 yrs.	Williams, 1983
Pomacentrus wardi		Accounted for 20% of early recruitment in one year, but 0% the next.	Williams, 1983
pomacentrids, labroids, apogonids, blenniids, eleotrids		Timing of recruitment highly variable over 4 summer months, during 3 years of study.	Sale, 1985
Harvested siganids	Guam	19 tonnes in 1972. <0·1 tonne in 1973.	Kami and Ikehara, 1976;
P. wardi	Similar timing of pulse at sites 4 km apart.		Pitcher, 1988b

Table 3—cont.

Species	Spatial variability	Temporal variability	References
5 *Stegaster* spp. 2 *Chromo* spp.	St. Thomas, Caribbean 3-week lag between sites	20-fold variations in weekly recruitment; *Chromis* spp. show lunar variation	Booth and Beretta, 1994
T. bifasciatum	Caribbean low coherence within 1 km², but high coherence between sites 2 km apart.		Victor, 1984
Labrids	South GBR variation among 7 reefs (100 km range) greater than variation within lagoons of the 7 reefs (2 km diam.).		Eckert, 1984
P. wardi	25 fold difference among 7 lagoons 100 fold difference among 7 reef slopes.		Sale *et al.*, 1984
Labroides dimidiatus	35 fold difference among 7 lagoons— no significant difference among 7 reef slopes.		Sale *et al.*, 1984
Thalassoma lunare	20 fold difference among 7 lagoons. No significant difference among 7 reef slopes.		Sale *et al.*, 1984

sity in coral reef and marine intertidal communities. Not surprisingly, given the similarities of the two systems, many of the same factors operate in both systems.

VI. THE INFLUENCES OF RECRUITMENT ON PATTERNS OF POPULATION AND COMMUNITY STRUCTURE

Several models have been developed that incorporate effects of recruitment processes on population structure. A *population* is defined as a reproductively isolated unit within a species (e.g. Mayr, 1963), although open populations with source-sink dynamics do not conform closely to this definition.

A *community* is "all species at all trophic levels at a particular location in a given habitat" (e.g Menge and Farrell 1989; Begon *et al.*, 1990). Community structure is then the combination of distribution, abundance, trophic interactions, and diversity of component species of community (e.g. Menge and Sutherland 1987). Sale (1991) defines community structure as "the patterns in species composition and species' abundances present in a defined habitat".

There is a distinct difference in approach to community models and studies between rocky intertidal ecologists and coral reef fish ecologists. For coral reef fish, subsets of a community (often called assemblages or guilds) are usually considered in isolation (e.g. the territorial damselfish guild of Sale, 1975). Of 68 studies considering community structure in coral reef fishes, identified in a bibliographic database (ASFA 1982–1992), 55 considered only fishes in their definition of a community. By contrast, intertidal community-based studies typically focus on many trophic levels (e.g. Jernakoff, 1983; Moran *et al.*, 1984; Mengel and Farrell, 1989) although many ignore mobile consumers like fish (but see Edwards *et al.*, 1982; Menge, 1982; Menge *et al.*, 1983, 1986a,b; Lubchenco *et al.*, 1984; Gaines and Roughgarden, 1987) and shorebirds (but see for example Frank, 1982; Hockey and Branch, 1984; Marsh, 1986a,b; Dumas and Witman, 1993). However, both intertidal and coral reef ecologists tend to focus on conspicuous space-dominants. In general, it seems that models developed for rocky intertidal invertebrates generally consider a wider range of species and trophic levels than reef fish models. This may reflect the history of development of both fields.

In our discussion of recruitment models and studies, we will separate population and community models, although most models will be addressed to varying degrees in both sections.

A. Recruitment and Population Structure

1. Coral Reef Fishes

Despite the much earlier recognition of the role of larval dynamics on adult populations of commercial fishes (see Cushing, 1975), it wasn't until the early 1980s that recruitment was suggested to affect coral reef fish adult populations. Doherty's (1981, 1983a) recruitment limitation model for coral reef fishes hypothesized that despite post-recruitment processes (such as competition, predation), patterns of abundance apparent at settlement will be reflected in adult demography. Evidence against space limitation on coral reefs and absence of adult populations from apparently suitable habitat are consistent with this model. So too are some studies of GBR damselfishes involving direct monitoring of cohorts of recruits through tagging, manipulative experiment and otolith back-calculation (see Doherty, 1991 for references). Problems in testing this hypothesis are the longevity of coral reef fishes (some live more than ten years), and so recruitment fluctuations may not show up directly ("storage effects" see Chesson, 1981; Warner and Chesson, 1985; Warner and Hughes, 1988). Also, processes that occur through juvenile stages can dampen the recruitment/adult demography relationship (e.g. Jones, 1988; Booth, 1991b, 1995). Therefore, the most direct evidence for recruitment limitation may come from populations where recruitment is episodic and where adults are short-lived (e.g. Victor, 1986, but see below).

A widespread case of coral mortality in the Caribbean and Atlantic due to El Nino conditions (Wellington and Victor, 1985), allowed a large-scale natural experiment of the relative importance of space-limitation and other factors (such as recruitment limitation) to be assessed. Soon after the coral death, algal cover increased dramatically, but the densities of herbivorous reef fishes (acanthurids, young scarids, pomacentrids) did not increase. This was indirect evidence that space and food were not limiting for these fishes, and that recruitment probably was.

Few experimental tests of recruitment limitation have been attempted. Jones (1990) studied the relationship between recruit density and adult density in the damselfish *Pomacentrus amboinensis* by simultaneously monitoring them on 16 small patch reefs in One Tree Island lagoon, southern GBR. On reefs where recruitment was artificially set at 0, 0·5, and 1 recruit per metre square per year over the 3-year study, adult densities were positively related to recruitment rate. However, a treatment of higher recruitment rate ($2/m^2/yr$) did not yield higher adult densities. This result was supported by monitoring unmanipulated recruitment rates and adult densities on 16 additional reefs. Again, the relationships levelled out at higher recruitment densities. This study suggests that recruitment limitation occurs below threshold levels of recruitment, and so may vary spatially and yearly, with recruitment rate.

A recent study of the age structure of 10 yearly cohorts of 2 species of damselfish across 7 reefs in the Capricorn-Bunker group of the Great Barrier Reef (Doherty and Fowler, 1994; Doherty and Fowler, unpubl. manu.) provides some of the most compelling evidence for the recruitment limitation hypothesis. They found that differences in inter-annual recruitment among sites could account for differences in age-frequency distributions for the 2 species (the damselfishes *Pomacentrus moluccensis* and *P. nagasakiensis*) among the 7 reefs over 10 years (i.e. recruitment variation (see Table 3) could account for temporal and spatial patterns of adult abundance). This test of recruitment limitation is one of the few long-term studies of coral reef fish population structure (but see below).

For Caribbean reef fishes, it appears that recruitment processes are more complex. Many species settle on the seagrass beds that commonly abut reefs in that region (e.g. Shulman and Ogden, 1987). After a short period, recruits migrate to nearby reef habitat. Since presumably post-settlement mortality due to predation occurs in the seagrass habitat, apparent effects of settlement on population (and community) structure may be dampened (Shulman and Ogden, 1987). However, studies of the wrasse *Thalassoma bifasciatum* (Victor, 1986) and several surgeonfishes (Robertson *et al.*, 1988) have supported the recruitment limitation hypothesis in the Caribbean. In contrast, inconsistent with recruitment limitation, while Hixon and Beets (1993) found uneven distributions of juvenile and adults of several fish species among 5 small artificial reefs, new recruits were evenly distributed among reefs (Hixon and Beets, in prep.).

An important shortcoming for many of the studies of recruitment limitation results from the protocol adopted for measuring recruitment. For instance, both Jones (1990) and Doherty and Fowler (1994) define recruitment as presence of young-of-the-year past the end of the summer recruitment period (definition 3, Table 1). Fishes censused using this protocol may have survived on the reef for several months after settlement, and been subjected to the many sources of mortality and redistribution that operate at this time (see earlier). Therefore, they do not allow resolution of the relative importance of larval processes (primary recruitment limitation (*sensu* Victor, 1986)) and early post-settlement processes on adult population structure.

2. Marine Intertidal Species

Population models for marine intertidal organisms have been pioneered by Roughgarden and co-workers (Roughgarden *et al.*, 1984, 1985, 1988; Gaines and Roughgarden, 1985; Possingham and Roughgarden, 1990). These models focus on the demographic and population dynamic processes of space-limited open populations (the classic view of rocky intertidal communities). Most of these models take a metapopulation approach. The model of Roughgarden *et al.* (1985) directly addresses the question of demographic consequences for populations resulting from differential recruitment. Specific predictions of the model include that where recruitment is limiting, settlement and recruitment densities will be positively correlated with adult densities. Population structure (e.g. size and age class) will be more uniform and there will be fewer oscillations, as the population approaches a steady state. By contrast, when recruitment is high, the model predicts that abundances will oscillate (due to time lags associated with recruitment in a rapidly growing species), cohorts and age-classes will be scattered in a spatial mosaic on the shore, and other factors such as competition and predation will be important. As a result final densities will not reflect settlement or recruitment patterns. A test of the model on rocky shores near Monterey (Gaines and Roughgarden, 1985; Roughgarden,

1986) supported these predictions. At high recruitment densities, interspecific competition resulted in hummocking and subsequent loss of barnacles. In addition, the predatory seastar *Pisaster ochraceus* was attracted to the area and consumed barnacles. This resulted in a periodic cycle of bare space, settlement, predation/competition leading to more bare space. The study also confirmed that settlement onto vacant space is proportional to the fraction of space available. Intermediate recruitment levels led to a switch between low recruitment and high recruitment dynamics. Roughgarden *et al.* (1985) suggest that, based on results from other studies, periodic cycles in barnacle cover may have been ongoing at least since the early 1960s in central California.

Does variation in recruitment affect adult density? Few studies follow settlers through to the adult state. Consequently direct information on the role of recruitment is mostly lacking. Indirect (correlative) studies are more frequent. In Connell's (1985) study of density dependence in barnacles, he found that adult and recruit density were positively correlated when recruitment was light, but adult and recruit density were uncorrelated when recruitment was heavy. Similar results were found by Gaines and Roughgarden (1985) who also suggested that, in general, low recruitment results in low adult densities, while high recruitment produces high adult densities. In Panama, recruitment density was the best predictor of adult density for five of six invertebrate species studied, and recruitment explained 55–97% of the total variance in adult densities of sessile bivalves and barnacles (Menge, 1991). The exception to this was the barnacle *Chthamalus fissus*, which had the highest recruitment of any species in Panama (Menge, 1991). Since then other studies have confirmed the relationship proposed by Gaines and Roughgarden (1985), including Raimondi (1990) for *Chthamalus anisopoma* in the Gulf of California, and Sutherland (1987, 1990) for *Chthamalus fissus* in Costa Rica.

Not all the evidence supports the predicted relationship between recruitment and adult densities. Menge (1991) found that although the relationship between recruitment and adult densities of *Semibalanus balanoides* was significant in New England (where recruitment density is high) the relationship was weak. Further, results for fucoids suggested that recruitment and adult densities were negatively correlated (Menge, 1991). In a review of 39 species (focusing mainly on barnacles), Menge and Farrell (1989) evaluated the relationship between recruitment and adult density. They found significant positive correlations in 19 of these studies. Of these 19, 10 had high recruitment, 5 had intermediate recruitment and 4 had low recruitment. For the remaining 20 studies where recruitment and adult density were uncorrelated, 12 had high recruitment, 6 had intermediate recruitment, and 2 had low recruitment. Of the studies that showed a significant relationship, the causal factor in the relationship was determined as recruitment in 16 cases, predation in 9, intraspecific competition in 3, interspecific competition in 2, and larval behaviour in 2. Recruitment and larval behaviour were determined alone to affect some species. Predation and competition were never cited as sole regulating factors—they acted in concert with recruitment to affect adult density. However, as a note of caution, Menge and Farrell (1989) add that statistical significance does not demonstrate biological significance.

In determining the relationship between settlement, recruitment and adult populations, many factors must be considered. Interactions between settlement density and abiotic factors can affect final densities. These factors may vary spatially and seasonally (e.g. Hartnoll and Hawkins, 1985; Roughgarden et al., 1988). Many coral reef fish and marine organisms reproduce and settle more than once per year. Physical factors, and factors affecting growth, may mean that density dependence impacts on one cohort but not another. For example, in parts of the UK, the limpet Patella vulgata, spawns and settles numerous times between spring and autumn (Bowman, 1986). Larvae settle into crevices and later, when they have attained a certain size, crawl out into tide pools. In an experimental study, Bowman (1986) found no correlation between an August–September settlement density, and number of recruits emigrating to tide pools. However, the number of settlers and emigrants were positively correlated when recruits emigrated to tide pools before the onset of winter. Winter frost kills many limpet recruits (Bowman, 1986). Thus, season and timing of settlement is as important as settlement density.

Larval behaviour can also affect settlement and recruitment patterns of a species. For instance, gregarious settlement can influence distribution patterns, and change the intensity of intra-specific interactions (e.g. Knight-Jones, 1951, 1953, 1955; Barnes, 1957; Meadows and Campbell, 1972; Lewis, 1964; Crisp, 1974, 1976a; Kobayashi, 1989; Breitburg, 1991): Southward (1991) found that in gregarious species of barnacles (Chthamalus spp) (juveniles settle close to adults), there was a significant positive relationship between density of adults from previous cohorts on the shore and recruitment densities. However, settlement still only accounted for less than 20% of total variance in adult densities, indicating that post-recruitment factors may be more important. In contrast, Raimondi (1990) found that at high recruitment levels, the effects of gregarious settlement were obscured. For coral reef fishes, some direct studies of larval behaviour immediately before settlement have suggested that behaviour can obscure the larval supply signal. Breitburg (1989, 1991) observed that larvae of the naked goby Cobisoma bosci gather in nearshore shoals prior to settlement. She concluded that aggregated settlement was an artefact of the clumped larval distribution, and not site preferences of settlers.

In summary, recruitment limitation of adult populations may vary annually and intra-annually, so that in some years density dependent post-settlement processes may predominate (e.g. Jones, 1990). Detection of recruitment limitation can be difficult, especially when the species is long-lived and recruitment is much lower than adult density. More studies need to follow organisms through and past maturity, either by direct observations (e.g. Jones, 1990) or by using simulation models that incorporate field data on settlement and juvenile growth and survival (e.g. Warner and Hughes, 1988; Holm, 1990; Booth, 1991b).

B. Recruitment Effects on Community Structure and Dynamics

Variation in recruitment can significantly affect the importance and frequency of species interactions within communities. For instance, Menge and Sutherland (1987) predict that species interactions (competition and predation) will be more important at high recruitment levels. In the following sections we explore the effects of recruitment on intensity and scale of such interactions.

1. Competition and Recruitment

Competition for space has often been regarded as a prime factor in both rocky intertidal and coral reef fish communities (e.g. Connell, 1961; Ehrlich, 1975; Roughgarden, 1986). Influential models such as the intermediate disturbance hypothesis (Connell, 1978), the predation hypothesis (Paine, 1966), and the lottery hypothesis (Sale, 1974) are based on the premise of competition in some form.

The lottery hypothesis developed by Sale (1975, 1978) for coral reef fishes holds that space is saturated on the reef. It assumes that there is an abundant supply of larvae in the water column, but that larval species composition varies stochastically over space and time. There is no competitive hierarchy; instead whichever larvae settle first pre-empt the space. Therefore, the assumptions of Sale's model are that:

(i) space on reefs is a limiting resource;
(ii) residents can retard settlement or at least establishment of larvae, and
(iii) when habitat becomes vacant, all species of larvae in the vicinity of the new space are equally likely to occupy it in proportion to their relative abundance.

Sale's model was developed to explain the stochastic species composition of a guild of territorial pomacentrid damselfishes among local patch reefs at Heron Island, southern GBR. No rigorous tests of the lottery hypothesis have been made for fishes, although Sale, 1975) found that pomacentrid assemblages on patch reefs were consistent with lottery predictions (but see Robertson and Lassig (1980) for an alternative interpretation). It is also possible that variable and limiting recruitment without space-limitation could equally well explain Sale's results. Despite the shortcomings of the lottery model, it has served to highlight the significance of recruitment and has stimulated more recent studies.

Although we found no studies from marine intertidal communities that directly tested the lottery hypothesis, some studies tended to support it indirectly. Indirect evidence supporting the hypothesis comes from a marine intertidal study of seasonal limpet exclusions in the UK by Hawkins (1981): different algal assemblages developed in exclusion plots, depending on seasonality and available propagules. Other sets of studies that also indirectly support the lottery hypothesis include the work of Sutherland (1974) and Sutherland and Karlson (1977) on a fouling community in North Carolina, studies of Keough and co-workers (Keough and Butler, 1983; Keough, 1984a,b; Connell and Keough, 1985) on the fouling community in *Pinna* shells in Australia, and work on a kelp community in California (Foster, 1975a,b). Sutherland and Karlson's study was on a shallow sub-tidal fouling community which developed on experimental plates in South Carolina, east coast USA. Although a competitive hierarchy existed, with a sessile tunicate *Styela* as the dominant species, there was considerable inter-plate variation in community structure, due to space monopolization of the first colonists. Keough *et al.*'s studies on communities which settled on *Pinna* shells also support the lottery hypothesis. On a large scale, community structure was predictable, but species composition differed between shells, due to recruitment differences. Foster (1975a,b) has argued that recruitment patterns in algal communities are determined more by stochastic and seasonal events.

Despite the limited field evidence for the lottery model, there is other, albeit restricted, evidence that recruitment can interact with competition in reef fishes. Unoccupied reefs generally have higher recruitment rates than occupied reefs (e.g. Molles, 1978; Talbot *et al.*, 1978), consistent with space-limitation hypotheses, and several studies have shown that residents of some species can facilitate or inhibit recruitment. Shulman *et al.* (1983) showed that very small artificial reefs seeded with adult *Stegastes* damsel-fish and snappers had lower recruitment than empty reefs. Shulman (1985) showed that smallest and most recent recruits suffer most from adult aggression on *Stegastes* spp., with resultant higher mortality. Sweatman (1983, 1985a,b), Jones (1987) and Booth (1992) have shown that conspecific and congeneric reef fishes can both positively and negatively affect settlement, growth and survival of new settlers.

A major implicit assumption of the lottery hypothesis was that there is an abundant supply of larvae. Recent studies suggest that this assumption

344   D.J. BOOTH AND D.M. BROSNAN

can be unwarranted (see Mapstone and Fowler, 1988; Doherty, 1991). Doherty's (1981) recruitment limitation model (above) has also been extrapolated to a community level (e.g. Doherty, 1983a; Munro 1983). It suggests that component species are each recruitment-limited, and so post-settlement interactions such as symbioses, predator/prey, and competition have no significant effects on community structure. Therefore, the model predicts that emergent properties of communities (*sensu* Wilson and Botkin, 1990) are modifying, but not limiting. Doherty's model, in its extreme form, may operate only where population densities are extremely low (and may apply during some seasons/years but not others). In support of this model (but equally of the lottery model), Sale and Douglas (1984) found that communities of coral reef fishes on small patch reefs were collections of independent species, since recruitment and mortality are independent of adult density, and so the "communities" may have been predominantly recruitment-limited.

The effect of recruitment variability on competition has been discussed in the model of Menge and Sutherland (1987) (see below), and has been experimentally tested (e.g. Sutherland and Ortega, 1986; Quinn, 1988). A basic prediction is that when recruitment is low, the relative importance of competition is reduced. Evidence in support of this proposition is given by Menge and Farrell (1989). Because few studies contained quantitative recruitment measurements, the relative importance of competition could be compared at only three sites of intermediate wave exposure and high recruitment. At high recruitment sites, competition was significantly higher (32·1%) than at low recruitment sites (9·0%). There are various other studies supporting Menge and Sutherland's view (including Menge and Lubchenco, 1981; Keough 1983, 1984a,b; Menge *et al.*, 1983, 1986a,b; Lubchenco *et al.*, 1984; Connell, 1985; Gaines and Roughgarden, 1985; Sutherland and Ortega, 1986; Sutherland, 1990).

Roughgarden (1986) suggested that although both coral reef fish and marine communities on hard substrates (including intertidal and subtidal habitats) are open and space-limited systems, space can be partitioned in fish communities but not in marine invertebrate communities. Two species of fish can overlap spatially, provided that they use the area of overlap in a sufficiently different way. By contrast, sessile invertebrates are abundant on many marine rocky shores, and because of direct attachment to the substrate, space requirements cannot overlap. However, this overlooks the point that many intertidal species (sessile and mobile) rely on, or use, secondary space as a settlement and recruitment site (Suchanek, 1979, 1981; Witman and Suchanek, 1984; Lee and Ambrose, 1989; Brosnan, 1990, 1992; Dittman and Robles, 1991). For example, Lee and Ambrose (1989) found that a higher proportion of the intertidal barnacle population was on secondary space (mussel shells) than on primary substrate. Thus, although sessile invertebrates are more intimately associated with the substrate, they can partition it on the basis of primary, secondary and even tertiary space.

2. Predation and Recruitment

At the most fundamental level, persistence of a predator population is dependent on recruitment of the prey. Variation in recruitment can affect the nature of predator–prey interactions (Moran et al., 1984; Dethier and Duggins, 1988; Fairweather, 1988). In closed population models, predators have a strong effect on prey densities. However, in open populations, these dynamics will be altered (see below), and predator–prey interactions will not follow the classical predictions. A key parameter in these systems will be the mobility of a predator, and its ability to emigrate from areas of low prey density.

Prey recruitment rates should affect predator reproductive rates, so that areas with high recruitment rates show the highest rates of predation (Menge and Sutherland, 1987). In addition, mobile predators may aggregate to high prey densities, intensifying the effects of predation in these areas. For example, high barnacle recruitment in Costa Rica (Sutherland and Ortega, 1986) and California (Roughgarden et al., 1985) leads to immigration of predators, and high predation rates. However, when predators are mobile and can emigrate from areas of low prey density, there may be a threshold prey density below which predation is absent. For instance, Fairweather (1988) found that when the predatory whelk Morula marginalba was present, barnacle recruits could be driven to local extinction, even though they recruited in high density. In contrast, low recruitment sites, without predatory whelks, ended up with a higher density of adults. Whelks tend to migrate from areas with low barnacle density.

Predation has been shown to interact with recruitment in affecting coral reel fish assemblages. Stimson et al. (1983) monitored the effects of removing muraenid eel predators on prey fishes on small patch reefs in Kaneohe Bay, Hawaii. Densities of several fish species (Dascyllus albisella and Chaetodon miliaris) increased over the year following eel removal, apparently through recruitment, although the authors were not able to demonstrate that eel removals lead to a drop in their abundance. Shulman (1985) found that recruitment and predation on recruits affected fish species composition on reefs in the US Virgin Islands. Hixon and Beets (1993) found a significant negative relationship among reefs between predator abundance and maximum prey-species richness, and concluded that high local species diversity is maintained despite, rather than because of, predation. Caley (1993) removed several species of small piscivores from small (1 metre diameter) patch reefs formed from dead corals, while leaving other reefs as controls.

Total abundance and species richness of non predatory recruits and older resident fishes were higher by the second year of the experiment on predator-removal reefs, suggesting that predators, by removing recruits, can play a role in structuring communities. Note that these studies were conducted at the scale of the patch reef. There is no clear evidence from studies at the whole-reef scale that predator abundance affects density of prey fish species (Russ, 1991; Sean Connell, pers. comm.). Studies in freshwater systems have demonstrated an interaction between predation and recruitment, which affects community structure. Tonn *et al.* (1992) found that predation indirectly affected community structure in a freshwater lake by affecting recruitment of lake perch, a numerically dominant fish. Perhaps such a situation is less likely in diverse coral reef systems.

3. Non trophic Interactions and Recruitment

Non trophic interactions (other than competition for space) have tended to be ignored by ecologists. Nevertheless they are ubiquitous and important. Non trophic interactions can be both facilitatory and inhibitory (Brosnan, 1990, 1992; Stephens and Bertness, 1991; Morse, 1992). For instance over three hundred species use mussel beds as a habitat (Suchanek, 1979, 1981). For these species, abundance of mussels may be the limiting factor affecting their recruitment. Many other intertidal species will not settle directly onto bare space, but require other species as settlement sites (e.g. Paine and Levin, 1981; Peterson, 1984a,b; Brosnan, 1993). Morse (1992) notes the critical and extensive role of algae in the recruitment of larval invertebrates. Similarly fish diversity may be dependent upon reef structural diversity. For instance, the abundance of Caribbean and Red Sea reef fishes is correlated with habitat complexity (Luckhurst and Luckhurst, 1978; Roberts and Ormond, 1987).

Non trophic interactions may be relatively more important under conditions of low recruitment. The limiting factor for recruitment of one species may be the previous recruitment of its settlement-substrate species. The orthodox view that species interactions (such as competition) are relatively unimportant when recruitment is low (Roughgarden *et al.*, 1985; Menge and Sutherland, 1987; Menge and Farrell, 1989; Sutherland, 1990), does not hold for non trophic interactions. For instance, under low recruitment conditions, competition between barnacles is unimportant (Sutherland, 1987, 1990). However, low recruitment of species such as barnacles or mussels, which facilitate establishment of other species, can lead to increased competition between epibionts that settle on barnacles and mussels, and between species that settle within mussel beds. In addition, in the above scenario, it is interesting to note that fluctuations in recruitment of the host species can lead to changes in the importance of competition between "facilitated" species (e.g. epibionts) even though there is no change in recruitment of facilitated species themselves. If recruitment of host species (e.g. mussel) declines, then competition between epibionts can increase. Similarly, if host recruitment increases, then competition between epibiont species can decrease, despite the fact that there has been no change in epibiont recruitment levels. The current framework, used to describe the importance of interactions such as competition and predation, does not easily translate to descriptions and studies of non trophic interactions.

4. Succession and Recruitment

Succession is obviously dependent upon recruitment. The very concept of progressive arrival of species implies recruitment from an outside source. The rate of succession can therefore depend upon the availability of recruits. In closed populations with limited dispersal or in open populations with limited recruitment, succession may be slowed. Potentially, succession may be most rapid in areas with abundant supplies of recruits. For example, succession in a marine intertidal community was faster at a high recruitment site (Farrell, 1989, 1991).

Connell and Slatyer (1977) have identified three models of successional change: facilitation, inhibition and tolerance. The effect of variation in recruitment will differ according to the model of succession. Under facilitation, low recruitment of initial colonizers will slow succession. Tolerance is basically a recruitment-effect model, where the rate of succession depends simply on the rate of arrival of each new species. In contrast to these two models, under inhibition, recruitment rate of later stages will not affect the rate of succession.

Under the inhibition model, some form of disturbance is required for succession to continue. This can be in the form of biological or abiotic disturbance. For instance, on Californian rocky shores, succession from opportunistic green algae (mainly *Ulva*) to red algae is dependent on grapsid crabs grazing on early successional algae (Sousa, 1979a,b). Under these conditions recruitment of such "mediating species" (here, grapsid crabs) can affect the rate and direction of successional change. The mid-shore region of rocky shores in the UK is characterized by a cyclical mosaic of dense barnacles, fucoids, and bare rock with limpets (e.g. Lewis, 1964; Hawkins, 1981; Hawkins and Hartnoll, 1982, 1983; Harntoll and Hawkins, 1985). Patches cycle between these species at a periodicity of 6–7 years (Hartnoll and Hawkins, 1985). In a study of factors maintaining patchiness, Harntoll and Hawkins concluded that, in addition to species interactions, extrinsic stochastic factors affect the rate of change. They cited "abiotic factors which determine the levels of recruitment and mortality of important species, with recruitment being particularly significant" as key components in the maintenance of patchiness. Interactions between the three groups are complex (Fig. 3A). For instance, limpets bulldoze barnacles and consume algal sporelings; they also use algal canopy as a refuge. Fucoids inhibit barnacle settlement (by whiplash), and provide a refuge for limpets and thaiid gastropods (main predators on barnacles). Recruitment of all these species is variable (Southward, 1967, 1991; Barnes and Barnes, 1977; Bowman and Lewis, 1977; Bennell, 1981; Kendall et al., 1982; Hawkins and Hartnoll, 1982; Hartnoll and Hawkins, 1985). Thus heavy and or light recruitment of any one species can have major effects on the rate of change from barnacle dominance to fucoid dominance to limpet abundance; rate of change can be accelerated or decelerated, or the cycle may be reversed (Fig. 3B).

The concept of succession, implying predictability of species turnover into vacated habitat, appears inappropriate for reef fishes, perhaps owing to their unpredictable larval supply and greater mobility than intertidal invertebrates. However, the presence of territorial reef fishes may alter the successional pathways of turf algal communities (e.g. Hixon and Brostoff, 1983). Colonization studies reinforce the idea that unpredictable reef fish species composition of small reefs may be a result of recruitment. Walsh (1985) found that the location of small reefs relative to larger reef tracts affected fish species diversity, inferring that factors other than habitat, such as stochastic recruitment, determined assemblage structure (see also Molles 1978; Talbot et al., 1978).

5. Integrated Approaches to Community Structure

Menge and Sutherland (1987) have developed a model that incorporates recruitment variability into community regulation. This environmental stress model was designed to be general and to be tested in a variety of communities. The model suggests that variation in community structure is directly dependent on competition, predation and abiotic disturbance. Variation in recruitment intensity is considered to have indirect effects, by altering the importance of these factors. The model assumes that mobile species are more affected by environmental stress than are sessile species. Consequently, the interaction web is predicted to increase in complexity as environmental stress decreases. Recruitment is scored as high or low, and determined by a composite measure of recruitment levels of the dominant species in the community. In addition, recruitment is considered to be the same at all trophic levels. That is, if it is high for carnivores, it is also high for plants.

There are several predictions of this model: When recruitment is high (Fig. 4), and physical conditions are harsh, stress determines community structure, and the community will be a simple, stress-resistant one. As

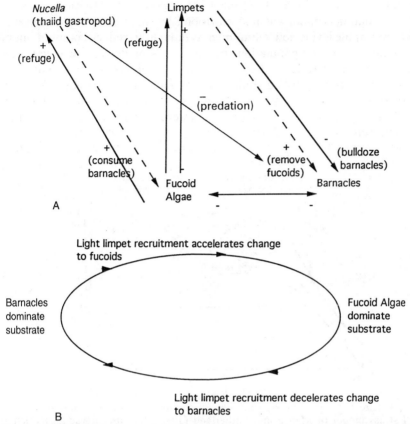

Fig. 3. A. Main interaction web between species on rocky shores in the UK. Solid lines are direct interactions (trophic and non trophic). Dashed lines represent indirect interactions. Patches cycle between barnacles, fucoid algae, and limpets on primary substrate. Variable recruitment in any of these species can affect the rate and direction of change. B. When the substrate is dominated by barnacles, light limpet recruitment accelerates the change to fucoids: When few limpets are present fucoids can establish, and whiplash prevents barnacle recruitment. Conversely when fucoids dominate the substrate, light limpet recruitment decelerates the change to barnacles. This is because whiplash continues to affect barnacles, and in low limpet densities fucoids escape limpet grazing. (Figure based on model by Hartnoll and Hawkins, 1985.)

conditions ameliorate, the abundance of sessile species increases and competition intensity increases. As environmental stress declines further, an influx of mobile consumers serves to reduce competition between sessile species. Finally in benign environments, predation becomes the dominant factor except at the top trophic level. With low recruitment, some of these predictions change (Fig. 4). Here, the relative importance of competition is

350   D.J. BOOTH AND D.M. BROSNAN

predicted to decrease at all trophic levels. Specifically, the model predicts that predation is important at lower trophic levels, in benign environments. At top trophic levels, low recruitment is thought to reduce competition, by slowing the rate of population increase. This model has received some support from intertidal studies. Menge and Farrell (1989) evaluated results from previously published studies, and found that, in the admittedly few data sets adequate for testing the model, some predictions were supported. In particular, competition and predation were less important at low recruitment levels.

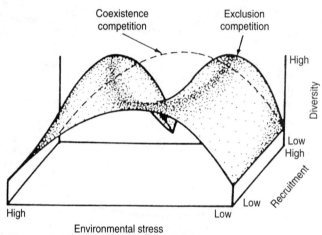

Fig. 4. Effect of high and low recruitment on community processes: the environmental stress model of Menge and Sutherland (1987). Species interactions, such as competition and predation, are more important at high recruitment. Differences in the importance of interactions cause diversity differences. For instance, at high recruitment, there is a bimodal diversity curve. This is because at intermediate levels of environmental stress, competition is high but predation is low, leading to competitive exclusion and low diversity. By contrast, when recruitment is low, competitive exclusion does not occur. Predation in benign environments leads to low diversity, and a unimodal diversity curve. (See text for full details.) Figure reproduced by kind permission of University of Chicago Press.

Menge and Farrell (1989) suggest that the structure of depauperate communities (or communities where space is not limiting) are more likely to be affected by variations on recruitment than that of speciose communities. We agree, but feel that this idea applies only under certain conditions. For instance, depending on the main interactions in a community, a speciose community may be less likely to develop when recruitment is low, but more likely to develop under high recruitment (see below). This prediction is based on the idea that an increase in recruitment of basal species can provide food for consumers, and high recruitment of habitat-providing species (e.g. mussels and corals) will also increase recruitment of dependent species. Some studies support this view (e.g. Roughgarden *et al.*, 1985; etc.; Sutherland and Ortega, 1986; Suchanek, 1978). However, others support the suggestions of Menge and Farrell (1989); on Panamanian rocky shores, recruitment is low, and species richness is higher than on New England shores (Menge and Farrell, 1989; Menge, 1991). Species richness was also high in the low recruitment communities studied by Keough and co-workers (see above). Sutherland (1990) suggested that recruitment will affect the number and importance of links in communities. In studies in Costa Rica, recruitment limitation was the most import-ant factor determining the abundance of the barnacle *Chthamalus fissus* (Sutherland and Ortega, 1986; Sutherland, 1987, 1990). *C. fissus* was rarely affected by competition, and only occasionally affected by predation (generally near crevices, and by the whelk *Acanthina brevidentata*). The interaction web between species in Costa Rica contains few links and many are weak (Fig. 5). By contrast, Sutherland (1990) noted that the Gulf of California has similar species composition to Costa Rica. However, recruitment is consistently high in this region (Dungan, 1985, 1986; Malusa, 1986; Lively and Raimondi, 1987; Raimondi, 1988). Here, many more ecological and evolutionary processes seem to be at work (Sutherland, 1990). Competition, predation, and in-direct effects are all important forces in this community (Fig. 5).

In addition, low recruitment of keystone species or dominant space-users may also lead to high diversity (see section C below).

C. The role of Recruitment Variability

Current models of community structure incorporating recruitment (Sale, 1974, 1975; Doherty, 1981; Menge and Sutherland, 1987) do not ad-equately address temporal and spatial variation in recruitment, or inter-specific differences in recruitment patterns within communities. Models stressing competition and other density dependent effects may be more appropriate when recruitment levels are high, while models highlighting the role of recruitment may apply at lower recruitment rates.

In the following sections we will consider possible effects on communi-ties when such variation in recruitment is considered. We will also discuss how the scale of recruitment processes may confound comparisons be-tween systems.

1. Effects of Overall Variability in Recruitment

Table 3 documents some examples of the extensive variability in recruit-ment at all scales. Causes of variability include factors that affect reproduc-

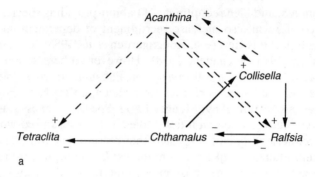

a

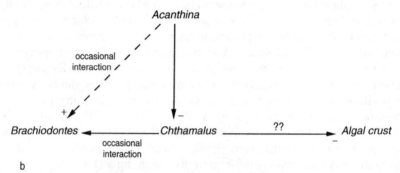

b

Fig. 5. A comparison of the main species interactions in two intertidal communities of similar species composition but with different recruitment levels. Solid lines denote direct interactions and dashed lines denote indirect effects. In a high recruitment area of the Gulf of California (Fig. 5a) the number and frequency of important interactions is higher than in Costa Rica where recruitment is often limiting (Fig. 5b).

a. Interaction web for an intertidal community in Gulf of California. Overall recruitment is high, and summer recruitment of barnacles (*Chthamalus*) consistently saturates space in the lower intertidal (Dungan, 1985; Raimondi, 1990). (Figure drawn from results by Dungan, 1985, 1986). The whelk *Acanthina* consumes *Chthamalus*, and thus indirectly benefits the competitively inferior barnacle *Tetraclita*, the limpet *Collisella* and algal crust *Ralfsia*. (Abundance of these species increases in the absence of *Chthamalus*, Dungan 1985, 1986.) In addition *Chthamalus* and *Ralfsia* negatively affect each other through pre-emptive competition for space; *Chthamalus* also pre-empts *Collisella*. *Collisella* indirectly benefits *Acanthina* (by consuming *Ralfsia*, and freeing space for *Chthamalus*), and *Ralfsia* indirectly harms *Acanthina* by pre-empting space from *Chthamalus*.

b. Species interaction web for the intertidal community in Costa Rica. Recruitment of barnacles is often limiting in this community (Sutherland, 1990) (Figure drawn from results of Sutherland and Ortega, 1986; Sutherland, 1987, 1990, and discussion in Sutherland, 1990.) *Acanthina* predation on *Chthamalus* is intense only near crevices. Competition is infrequent and occurs between *Chthamalus* and the mussel *Brachiodontes semilaevis*. *Chthamalus* may inhibit encrusting algae.

tion, survival and dispersal. Some of these factors are predictable and may apply broadly across the community, for instance seasonal changes in ocean currents. Others may be episodic and more specific. For instance, the barnacles *Semibalanus cariosus* is a large intertidal species covering extensive areas of the mid-intertidal in the San Juan Islands, USA. In 1982, most barnacles were 2–9 mm in diameter. However, a sizeable cohort of large barnacles 20–25 mm were found on southern and western facing slopes (Sebens and Lewis, 1985). This unusual and disjunct size pattern could be traced to freezing storms of 1969–1970 which killed most of the predatory whelks (*Nucella* spp). This led to enhanced recruitment of *S. cariosus*, and this strong year-class dominated the populations in these areas for at least a decade. Effects on dispersal and settlement may also be episodic. For instance, a single recruitment event of an abalone from California to Oregon has maintained a single cohort for over a decade (Frank, pers. comm.). Fisheries literature documents many cases where strong year-classes dominate adult populations for many years (see Cushing, 1975), and may at least temporarily rescue the fishery from collapse (e.g. referenced in Doherty and Williams, 1988). Coral reef fish populations similarly may be dominated by strong year-classes (see Doherty and Williams, 1988).

The effects of recruitment variability on communities will be to introduce substantial time lags, and perhaps to alter successional rates. No community models have evaluated this effect and it should be incorporated into future models. A potential mechanism for quantitative study of recruitment variation at the community level is to study the variance (Menge, pers. comm.)

2. Effects of Species-specific Recruitment Patterns

One consequence of recruitment variability is that different species may recruit at different rates and at different times (Brosnan, 1992). Studies have shown that high recruitment of one species does not necessarily imply high recruitment of all species (e.g. Keough and Downes, 1982; Watanabe and Harrold, 1991; Hurlbut, 1991a; Brosnan, in prep). Most models and studies of recruitment in communities do not account for individual species differences. These must be incorporated into future models. Brosnan (1992) explored how changes in the relative importance of predation and herbivory result, when fish recruitment varies independent of their intertidal prey species.

Variation in recruitment can destabilize communities and lead to large-scale fluctuations. For example, the urchin *Strongylocentrotus purpuratus*, shows high inter-annual recruitment variability in central California. A high recruitment pulse followed by subsequent recruitment failure led to large-scale fluctuations in algal and urchin abundance in a two-year period

(Watanabe and Harrold, 1991). There is no evidence to suggest that algal recruitment changed during the study period, because algal settlement was recorded throughout the study. At a similar site about 0·5 km away, urchin abundances did not fluctuate, and community structure remained largely unchanged. Thus differences can occur on a local scale.

3. Species-specific Differences in Population Openness

The relationship between a predator and its prey species depends critically on the population structure of both. Classical models of predator–prey interactions assume that all populations are closed (see reviews by May, 1981; Begon et al., 1990). Recruitment of both prey and predator then depends on the numbers of the other. If, however, either or both have an open population structure, the relationship between the two is effectively decoupled. For instance, the main prey of thaiid gastropods are barnacles and small mussels. Here the predator recruits locally, and has a closed population; the prey species recruit at long distances, and have open populations. By contrast, some algal species have spores which disperse distances of only a few metres; their consumers (herbivorous fish and invertebrates) are more mobile. In some cases, both prey and predator may recruit over long distances, for example, the crown-of-thorns seastar *Acanthaster planci* and its prey (primarily pocilloporid corals) (e.g. Dight et al., 1988; Glynn, 1989; Birkland and Lucas. 1991).

The effects of decoupling the relationship between local numbers of prey and predators are several. In closed systems, a predator that drives its prey extinct will then decline in numbers itself. Conversely, an increase in prey abundance can lead to increased reproduction and hence recruitment of the predator. These relationships do not necessarily hold in systems with one or more open populations. If predator numbers are determined by distant events, there may be relatively few or many predators per prey. We would therefore expect either local extinctions or a surge in the prey population. If the prey has an open population structure, the predator may be subject to a large variance in food availability. These effects will be most extreme when all species have open populations. We therefore predict that one consequence of open population structures will be increased variability in abundance, with frequent local extinctions caused by over-exploitation of prey. Massive irruptions of *Acanthaser planci* and destruction of coral fits this model.

Unpredictablity of food selects for polyphagy (Chew and Courtney, 1991), including switching. Unpredictability may also select for a bet-hedging lifestyle (Begon, 1985). Frank (1982) has discussed how variability in recruitment of intertidal invertebrates (limpets) may result in an unpredictable food supply for Black Oystercatchers *Haematopus bachmani*.

When limpets are scarce, oystercatchers may be forced to feed on less preferred food items or to emigrate.

Generalist foraging strategies and switching tend to buffer predator populations against variation in prey recruitment. Predator numbers may then be relatively constant, while individual prey species are more variable in abundance. In these circumstances, we may predict inverse density dependent mortality of prey. Hurlbut (1991a) has documented the effect of settlement density on predation rates in seven co-occurring invertebrates. Three species (the most abundant) showed positive density dependent mortality due to fish predation (Hurlbut 1991a). Four species (the rarer taxa) showed inverse density dependence (our Spearman rank correlations on published data, excluding samples without predation: *Phallusia nigra* r_s = ·964; *Didenium candidum* r_s = ·812; *Diplosoma listerianum* r_s = ·626; *Diplosoma sp.* r_s = ·701; $p < 0.01$ in all cases; all taxa are ascidians). We interpret these patterns as due firstly to aggregation of predators (fish) at high densities of common prey; and secondly to lack of aggregation in response to rare prey. The net outcome is that rare species suffer least mortality when settling close to conspecifics; the converse is true for common prey species.

The outcome of other interactions may also depend on the structure of the interacting populations. For instance, classical closed models of competition assume that the number of new individuals of a species is related to past success of that species. This may be the case for some marine species with restricted dispersal, such as the damselfish *Acanthochromis polyacanthus*. In open communities, the number of new recruits of a species is unrelated to current density. If a competitively inferior, open species recruits heavily, it may persist longer or even "win" an interaction with a competitively superior, closed species. Note also that classical competition models are sensitive to initial conditions for population densities. In open communities, initial conditions may vary considerably.

4. Keystone Species

It is widely recognized that some species have disproportionate effects on community structure. If such keystone species are recruitment-limited, their recruitment will have profound effects on the community. Brosnan (1992) noted that, since fish are dominant consumers in many tropical intertidal areas, biological interactions, such as competition among prey items, will be determined by recruitment of the major predator. (Studies show that some tropical fish can have a keystone effect, e.g. Menge *et al.*, 1986a.) Similarly, Hixon and Brostoff (1983) found that the territorial Hawaiian damselfish (*Stegastes fasciolatus*) could enhance algal diversity (and change the successional state of the algal assemblage) within its territories, and can thus be considered keystone species. *S. fasciolatus* appears to be recruitment-limited in Kaneohe Bay, Hawaii (Booth, pers. obs.) and so may be a good candidate to test the indirect role of recruitment of a keystone species on community structure. Similarly the episodic outbreaks of the crown-of-thorns starfish *Acanthaster planci* results in destruction of coral and cascading effects throughout the community (e.g. Dight *et al.*, 1990). Hay (1984) suggested that urchin numbers on Caribbean reefs may depend on the numbers of herbivorous competitor (acanthurids and scarids) and predatory (balistids and pomadasyid) fishes. On reefs overfished for these taxa, urchin abundances increased. Recruitment limitation of any of these taxa may have ramifications through the entire community. For instance, sea urchins are known to have major effects on kelp abundance. When urchins recruit successfully, the dynamics of subtidal kelp forests may be controlled by factors affecting urchin recruitment (Watanabe and Harrold, 1991). These authors found that, in central California, urchin recruitment was episodic. Between 1972 and 1981, recruitment of *S. purpuratus* occurred only once and *S. francisancus* did not recruit at all. Algal deforestations are rare here, and can be directly related to urchin recruitment events. By contrast, there is regular annual recruitment for both urchin species in southern California (Tegner and Dayton, 1981), and in the Aleutian Islands (Estes and VanBlaricom unpublished, cited in Watanabe and Harrold, 1991). In these two geographic areas deforestations are frequent. Watanabe and Harrold (1991) suggest that infrequent recruitment in central California may maintain urchin populations below a threshold effect level most of the time.

While recruitment variability in keystone species can affect all other species in a community, adults of keystone species may be unaffected by recruitment limitation of other species and so there may be an asymmetry of effects (e.g. Johnson and Mann, 1988).

5. Temporal and Spatial Scale

The effects of scale will depend on dispersal distances of propagules and mobility of adults (c.f. Section V). The effect of spatial variability may be swamped by highly mobile predators, that can aggregate to high recruitment areas. For example, on tropical shores in Panama, predation is intense and predators include highly mobile fish (e.g. Menge *et al.*, 1983, 1986a,b; Menge, 1991). Predation is spatially and temporally uniform (Menge, 1991). In this situation, we predict that local variation in recruitment intensity (e.g. between adjacent shores) would be swamped by the actions of mobile predators. By contrast, fish predation is not important on many temperate shores, and slow-moving invertebrates are the main consumers. Under these circumstances we predict higher between shore variance in prey abundance due to recruitment variability.

The lottery hypothesis has received little attention by intertidal researchers. One reason for this may be the spatial scale of study. A key prediction of the lottery model is that species composition is unpredictable at a small spatial scale but predictable at a larger spatial scale. In coral reef systems small scales are typically considered to be at the level of the individual reef; large scales encompass a number of reefs of geographic regions. In intertidal systems large and small scales may have very different dimensions. Many authors note that intertidal community structure and composition is predictable at the scale of individual shores as well as the larger geographic scale. For instance, exposed rocky shores in the northwest coast of America have characteristic communities (e.g. Stephenson and Stephenson, 1972; Kozloff, 1983; Ricketts et al., 1985). However, other research highlights unpredictability of rocky shores systems (e.g. Sutherland, 1974; Foster, 1975a,b; Underwood, 1981; Underwood and Jernakoff, 1984). It may be that this discrepancy is due to a confusion of scale. Intertidal communities may be less predictable on smaller scales (e.g. individual rock, small plot, or individual tidepool) than they are at the larger scales which are typically studied by ecologists. The difference between the rocky intertidal and coral reef systems is one of dimension: "small" in an intertidal system may be less than $1\,\mathrm{m}^2$, while small scale on a coral reef may be defined in the order of $10–100\,\mathrm{m}^2$. This difference may simply reflect the difference in relative mobility of space-occupiers. If this is so, then the lottery hypothesis may be more applicable to communities than is presently recognized, but it must be tested at the appropriate scale. Note that the model may be especially appropriate in tropical areas where communities are more speciose. See Strong (1992) for a counter-argument.

For coral reef fish, variability in species diversity may decrease with temporal and spatial scale within biogeographical boundaries, but variation is considerable at all scales examined (see review by Sale, 1991). Recruitment effects have been implicated in community studies of coral reef fishes at a number of scales. Williams and Hatcher (1983) found differences in fish assemblages (mainly pomacentrid species) in an 80 km transect across the central GBR. They suggested that differences could be attributed to patterns of larval dispersal. Thresher (1991) suggested that differences between Caribbean and Pacific fish community structure may be partly due to longer pelagic larval duration of Caribbean species, coupled with their general pattern of initial recruitment to seagrass. Possible differences in the trophic role of coral reef fishes in the Caribbean compared with the Indo-Pacific region (e.g. Harmelin-Vivien, 1989) may mean the effects of recruitment of fishes and other factors such as competition and predation on community structure may differ biogeographically.

Note also that extrapolating results from local studies to biogeographic and perhaps more ecologically meaningful scales, is questionable (see discussion in Doherty and Williams, 1988, and references therein).

6. Spatial Variability and Consumer Mobility

Consumer mobility can affect the relative importance of spatial variability in recruitment. Recruitment variability may be relatively unimportant where consumers are highly mobile and can move between areas of differing recruitment intensities (Brosnan, 1992). We expect that in areas with highly mobile predators, there will be little spatial variability in recruitment across all scales. Community structure will not vary as settlement and recruitment fluctuate. Conversely, areas with less mobile predators will show variability at all scales. For example, studies in Panama (Menge and Lubchenco, 1981; Menge et al., 1983; 1986a,b; Lubchenco et al., 1984) showed that recruitment limitation was severe; probably because a suite of predators including highly mobile fish were abundant in the community. In this situation, fluctuations in recruitment intensity, and local geographic variations in recruitment would probably be undetected, and unimportant to community structure. This is because predators can quickly move from areas of low to high recruitment. For instance, when, at the end of predator exclusion experiments, Menge and Lubchenco removed exclusion cages, predators and herbivores rapidly consumed most of the algae and invertebrates that had recruited to the plots. Similarly, when territorial damselfishes are removed, algal mats are quickly decimated by schooling surgeonfishes and parrotfishes (Low, 1971). In contrast, fish seem less common as consumers on temperate rocky shores (Menge, 1982, 1991) and particularly of algae on temperate shores (Hawkins and Hartnoll, 1983). Instead slow-moving invertebrates seem to predominate there (e.g. Menge, 1976, 1991; Lubchenco, 1978, 1980, 1982, 1983; Underwood 1980, 1984; Branch 1981, 1986) and these cannot migrate easily over large distances or between shores. In communities such as temperate shores, where consumers have limited mobility, we predict that local and large-scale variability in settlement followed by recruitment will be more easily detected. Consequently, recruitment variability will be more important to spatial and temporal community patterns. Some studies from temperate areas tend to support this prediction. However, few studies are available from tropical shores, making a full comparison difficult (Brosnan, 1992).

D. Summary of Models, Predictions and Tests

We conclude that recruitment can affect community structure in several general ways (see Table 4).

(i) In a community where interaction strengths are low, and especially where space is not limiting, each species could be considered independently (i.e. a community can be considered a collection of independent species). In this scenario, recruitment limitation of some or all species may be the primary regulatory mechanism (Doherty, 1981). By contrast, in communities characterized by strong interactions, and/ or intense competition for space, this model, or any recruitment-limitation model, will be inadequate. Models such as those of Menge and Sutherland (1987) above may be more appropriate here.

(ii) The scenario in (i) above may vary temporally (both seasonally and between years) within a particular system. In poor recruitment years, Doherty's recruitment limitation model may apply (e.g. Milicich, 1988). In good years, Doherty's model would not hold. The long-term effect of recruitment would depend on the relative frequency of "good" and "bad" years (see Jones (1990) for a coral reef fish example and Sebens and Lewis (1985) for an intertidal example) e.g. models 4, 8.

(iii) The order of arrival and relative recruitment levels of individual species, through space pre-emption, may determine the nature and strengths of interactions in the community, and the importance of other factors (e.g. disturbance models, Sutherland (1974); also lottery model of Sale, and Sutherland (1987, 1990).

(iv) Not all species have equal effects on community structure. Keystone species have a disproportionately large effect on species diversity. Consequently, if they are recruitment limited, a uni-modal relationship between their recruitment and overall species diversity may exist (extending Hixon and Brostoff 1983; model 9). Conversely, recruitment limitation of minor species will have relatively little effect on community structure (e.g. model 7).

(v) The classic view of species interactions will be altered when interacting species show different levels of open and local recruitment patterns. For instance, open recruitment by a prey species or subordinate competitors could determine the relative importance of outcome of species interactions (this paper).

(vi) Different processes will apply at different spatial scales (Brosnan, 1992 and Section VI.C.5, this paper).

Targeting communities containing species that fluctuate widely in recruitment and spatial patterns of species composition may allow testing of recruitment hypotheses of community regulation. However, as Sale, 1991 points out, most patterns may also be generated through differential predation, disturbance and competition. Conversely, communities composed of species with strong habitat requirements, or that are long-lived or show high post-settlement mortality are less likely to be regulated by recruitment (e.g. Hixon, 1991).

VII. MANAGEMENT IMPLICATIONS, CONSERVATION ISSUES AND FUTURE DIRECTIONS

A. Management and Conservation

The effects of recruitment on harvested species has received much attention in fisheries literature (see Cushing, 1975), see earlier. By contrast, little consideration has been given to the role of recruitment in maintaining threatened communities. This is probably because most emphasis has been given to protection and management of organisms that have more closed populations (such as large mammals), and because most models are for closed populations. Clearly, if management and species protection efforts are to be successful, then the effects of recruitment, and its interaction with human impact on ecological systems must be integrated into any management plan.

Any marine species that exhibits an open structure will need a protection strategy that is based on a metapopulation model. Recently, metapopulation dynamics have received considerable attention, particularly in relation to conservation biology, and here it has been mostly for terrestrial species (e.g. Gilpin, 1987; Harrison, 1991, and references in a review by Hanski, 1991). Metapopulation models have been developed for marine (intertidal) species by Roughgarden et al. (1984, 1985, 1988), Roughgarden (1986), Possingham and Roughgarden (1990) and for coral reef fish by Pulliam (1988). One potential application has been discussed above for the sea palm P. palmaeformis. This species exhibits temporal and spatial patterns of colonization and extinction. Other metapopulation models include the source-sink structure. An important consequence of this model for marine management is that source populations will need special protection. Setting up marine reserves in sink areas will not protect a species (Brosnan and Crumrine, 1992a, 1994). It is outside the scope of this paper to review in detail the various consequences of incorporating or ignoring metapopulation dynamics in marine environmental management, and species protection. The reader is directed to the above references for a more complete review of metapopulations.

In some marine environments, human disturbance can actually enhance the role of recruitment in regulating community structure. For example, Lehtinen et al. (1988) found that chlorates from coastal pollution enhanced the amount of kelp shelter available for fish, so fish populations became

Table 4

Summary of models incorporating recruitment into community structure

Model	Reference	Predictions/Conditions
1. Recruitment limitation	Doherty, 1981	Distribution and abundance of all species is limited by larval supply. Community structure the sum of recruitment and density independent mortality
2. Lottery	Sale, 1974	Larval supply unlimiting but stochastic. Species composition of community related to relative densities of each species in larval pool. Residents pre-empt space
3. Environmental stress model	Menge and Sutherland, 1987	At high recruitment, competition and predation are more important. The relationship between diversity and environmental stress is determined by recruitment levels
4. Composite successional model	Hughes, 1984	Initial colonizations do not involve important species interactions. Post colonization events are characterized by increasing intensity of species interactions. Competition leads to dominance, but usually disturbance prevents this and thus disturbance maintains diversity
5. Multiple stable points	Sutherland, 1974	Communities have multiple stable points. Order of recruitment can determine community composition
6. Recruitment and competition	Sherman, 1982	Recruitment limitation can alter species dominance, allowing coexistence
7. Keystone recruitment limitation	this study	If keystone species are recruitment limited, then their patterns of larval supply will strongly affect community structure and dynamics
8. Recruitment limitation and storage events	Warner and Chesson, 1985	Recruitment limitation interacts with age structure of populations to determine community structure
9. Maintenance of mosaic pattern	Hartnoll and Hawkins, 1985	Interaction between recruitment levels of key species, and composition of resident species affects the rate and direction of change

more recruitment-limited. Pauly (1979) suggested that overfishing releases some communities from predation, so other factors such as recruitment limitation may become important. Overfishing on "source reefs" for fish larvae (e.g. Pulliam, 1988) could lead to "recruitment overfishing" (Cushing, 1975), and so such source reefs should be identified and protected (see above).

If recruitment is important to maintaining communities, it complicates the delineation and establishment of "harvest refuges" for exploited stocks (e.g. Carr and Reed, 1993). Multi-species refuges, where each species differs in dispersal patterns and location of source areas, will be particularly hard to define. Widely dispersing marine organisms may cross national boundaries, complicating their effective management. If keystone species are recruitment limited, efforts to make more space available within the habitat will probably be unsuccessful in enhancing their density and thus role in the community.

Human impact has a variety of effects on marine intertidal systems which range from oil spills, and their associated clean-ups (e.g. Hawkins *et al.*, 1992) to harvesting (e.g. Ortega, 1987; Olivia and Castilla, 1986; Castilla and Bustamente, 1989; Duran and Castilla, 1989; Godoy and Moreno, 1989; Kingsford *et al.*, 1991) trampling, and general overuse of intertidal areas (e.g. Boalche *et al.*, 1974; Beauchamp and Gowing, 1982; Liddle, 1991; Brosnan, 1993; Brosnan and Crumrine, 1994). Recruitment can play a key role in how the community responds to, or recovers from, human effects. For instance, massive recruitment of limpets after the Torrey Canyon oil spill led to instability in limpet populations (Hawkins and Southward, 1992). Human impact can change the range and intensity of disturbance (Brosnan and Crumrine, 1992a, 1994; Brosnan, 1993). If affected populations are open and not recruitment limited, and if the spatial scale of the disturbance is less than the dispersal distance of propagules, then recovery can be rapid. By contrast, in recruitment limited species or in closed populations, human impact may have a long lasting effect. Recovery may only be possible if there is a source of recruits nearby. For example, Brosnan and Crumrine (1992b, 1994) and Brosnan (1993) found that algal populations could recover rapidly from trampling when adult plants were nearby. By contrast, recovery was much slower when few adult plants were present (Grubba and Brosnan, in prep.). In addition, if human impact adversely affects keystone or dominant species in the community, and if such species are recruitment limited then the entire community may be altered, or driven locally extinct. Brosnan and Crumrine (1992b, 1994) found that trampling removes mussels (*M. californianus*, a dominant species) from rocky shores, causing the loss of mussels and their associated species (over 300 species live associated with mussels, Suchanek (1979)). Mussel recruitment is sporadic (Peterson, 1984a,b) and does not occur if trampling continues (Brosnan and Crumrine, 1992a; Grubba and Brosnan, in prep.). Even two years after trampling had stopped, no mussel recruitment was recorded (Brosnan, pers. obs.).

In conclusion, marine ecosystems are coming under increasing threat from human use and disturbance. Management and protection plans need to incorporate recruitment patterns and processes (Brosnan, 1995).

B. Future Directions

The following are some suggestions for possible fruitful lines of study in the role of recruitment in the structuring of communities.

(i) Processes occurring during larval dispersal (that may affect recruitment) need to be determined. For example, larval behaviour (e.g. aggregation) may indirectly affect adult community structure. In such cases, studies of larval assemblages would be important (e.g. Breitburg, 1989, 1991; Leis, 1991).

(ii) Recruitment should not be studied in isolation, because it interacts with other factors such as predation (Fairweather, 1988), and disturbance (e.g. Dayton, 1971, 1975; Sale's (1975) lottery model; Connell, 1979). The nature and strength of such interactions may vary spatially and temporally (e.g. Fairweather, 1988; Hixon, 1991; Hixon and Beets, 1993; Menge and Sutherland, 1987).

(iii) Studies specifically designed to look at the effect of consumers and scale on recruitment effects should be conducted, by manipulating both recruitment and predator access and examining the scale of variability (e.g. Sale, 1985; Doherty, 1987b).

(iv) Studies which look at the effect of variability in recruitment of a keystone species would be useful. For instance, for starfish in intertidal systems (Paine, 1966) and territorial damselfish on coral reefs (e.g. Hixon and Brostoff, 1983) we might profitably ask: How large must recruitment variation be to have an effect on the community? Is there a time lag before the effects are noticeable in the community? Studies such as those of Fairweather (1988) and Jernakoff (1983) (both in marine intertidal communities) should be carried out on a larger scale.

(v) We suggest a large-scale interdisciplinary project to study metapopulation structure and dynamics. Such a study would include genetic analysis, for identification of isolated population segments and metapopulations, satellite imagery to follow water movements, experiments on shore or reefs to monitor recruitment variation and to test for its effects.

(vi) Additional studies such as Jones (1990), which determine threshold levels of recruitment above which recruitment limitation does not occur, may allow estimates of the temporal change in significance of recruitment to community structure to be assessed.

(vii) We suggest studying the effects of recruitment limitation on community structure in simple (but whole) communities (e.g. rocky intertidal), or at least localized interaction webs (e.g. damselfish territory). These studies should include more than one trophic level.

(viii) Population genetic studies will be vital components of any programme that considers effects of recruitment on community structure of benthic marine organisms, but their application and interpretation needs careful verification (see caveats in Section III.B. above). Recent advances in this area, such as microsatellite DNA studies, offer exciting alternatives to traditional measures of genetic diversity.

(ix) Studies on the role of recruitment in community structure and dynamics should be carried out at a range of spatial scales, and across different hydrographic and production regimes.

VIII. CONCLUDING REMARKS

Clearly, recruitment has come of age as a recognized factor affecting population and community structure of marine organisms. This recognition is somewhat belated, given the pioneering work of early researchers (e.g. Thorson, 1950, 1966; Crisp and Barnes, 1954; Barnes, 1956a,b; Southward and Crisp, 1956; Lewis, 1964; Crisp, 1974). Despite the recent advances in understanding recruitment processes, there have been few conceptual or empirical studies that consider the implications of variable levels of recruitment to community structure (but see e.g. Hartnoll and Hawkins, 1985). One possible explanation for this is that both recruitment research and community research are specialized and complex areas. In this regard, we suggest the following.

(i) Studies of community level effects of recruitment in coral reef communities need to consider a range of trophic levels. Few researchers specializing in coral reef fishes have considered the effects of fish recruitment on other, non-fish, members of the communities. Although the most tractable experiments will involve few species, the investigators should remain mindful of the extrapolation of their results to the wide community.

(ii) Marine intertidal ecologists might profitably consider the interaction between spatial scale and recruitment patterns and processes.

(iii) There is a general need to incorporate the effects of recruitment variability (including between species variability) and population structure into studies and models at all levels.

In conclusion, rocky shores and coral reef fish communities do share ecological features. More frequent exchange of ideas, results and tests of models developed in each other's systems will benefit researchers in both fields.

ACKNOWLEDGEMENTS

We thank Steven Courtney, Bruce Menge, Jane Lubchenco, Sylvia Behrens-Yamada, Sergio Navarette, Steve Hawkins, Mark Hixon, Peter Doherty, Gigi Beretta, Mark Carr and Phil Light, for valuable comments, suggestions and information. However, we accept full responsibility for any omissions and inaccuracies. D.M. Brosnan also thanks Eric Berlow and Carol Blanchette for early comments, Maurice James for information and copies of data, Steven Courtney for many late night brain-stormings, and acknowledges partial support from Oregon Sea Grant, and Sustainable Ecosystems Institute. D.J. Booth wishes to thank the Australian Institute of Marine Science and University of Technology, Sydney, and The University of the Virgin Islands, for support.

REFERENCES

Aldenhoven, J. (1986). Local variation in mortality rates and life expectancy estimates of the coral reef fish *Centropyge bicolor*. *Mar. Biol.* **92**, 237–244.

Amsler, C.D. and Searles, R.B. (1980). Vertical distribution of seaweed spores in a water column offshore of North Carolina. *J. Phycol.* **16**, 617–619.

Anderson, E.K. and North, W.J. (1966). *In situ* studies of spore production and dispersal in the giant kelp *Macrocystis*. *Proc. Int. Seaweed Symp.* **5**, 73–86.

Anderson, G.R.V., Ehrlich, A.H., Ehrlich, P.R. and Roughgarden, J.D. (1981). The community ecology of coral reef fishes. *Am. Nat.* **117**, 476–495.

Andrewartha, H. and Birch, L.C. (1954). *The distribution and abundance of animals*. Univ. of Chicago Press, Chicago.

Avise, J.C. and Shapiro, D.Y. (1986). Evaluating kinship of newly-settled juveniles within social groups of the coral reef fish *Anthias squamipinnis*. *Evolution* **40**, 1051–1059.

Barkai, A. and Branch, G.K. (1988). The influence of predation and substratal complexity on recruitment to settlement plates: a test of the theory of alternative states. *J. Exp. Mar. Biol. Ecol.* **124**, 215–237.

Barnes, H. (1955). Further observations on the rugophilic behavior in *Balanus balanoides* (L.). *Vidensk, Meddr. dansk. naturh. Foren.* **117**, 341–348.

Barnes, H. (1956a) *Balanus balanoides* (L.) in the Firth of Clyde: the development and annual variation of the larval population, and the causative factors. *J. Anim. Ecol.* **43**, 717–727.

Barnes, H. (1956b). *Balanus balanoides* (L.) in the Firth of Clyde: the development and annual variation of the larval population and causative factors. *J. Anim. Ecol.* **25**, 72–84.

Barnes, H. (1957). Processes of restoration and synchronization in marine ecology, The spring diatom outburst and the "spawning" of the common barnacle *Valanus balanoides* (L.). *Annee biol.* **33**, 67–85.

Barnes, H. and Powell, H.T. (1950). The development, general morphology and

366 D.J. BOOTH AND D.M. BROSNAN

subsequent elimination of barnacle populations, *Balanus crenatus* and *Balanus balanoides* after a heavy initial settlement. *J. Anim. Ecol.* **19**, 175–179.

Barnes, H. and Barnes, M. (1977). The importance of being a "littoral" nauplius. In: *Biology of benthic organisms. Proc. 11th Eur. Mar. Biol. Symp.* (Ed. by B.F. Keegan, P. O'Ceidigh and P.J.S. Boaden). Oxford University Press, Oxford.

Barnett, B.E. and Crisp, D.J. (1979). Laboratory studies of gregarious settlement in *Balanus balanoides* and *Elminius modestus* in relation to competition between these species. *J. Mar. Biol. Ass. U.K.* **59**, 581–590.

Bayne, B.L. (1964). Primary and secondary settlement in *Mytilus edulis* L. (Mollusca). *J. Anim. Ecol.* **33**, 513–523.

Bayne, B.L. (1965). Growth and the delay of metamorphosis of the larvae of *Mytilus edulis* (L.). *Ophelia* **2**, 1–47.

Bayne, B.L. (1976). The biology of mussel larvae. In: *Marine mussels and their ecology*. (Ed. by B.L. Bayne) pp. 81–120. Cambridge University Press, Cambridge.

Beauchamp, K.A. and Gowing, M.M. (1982). A quantitative assessment of human trampling effects on a rocky intertidal community. *Mar. Envir. Res.*, **7**, 279–283.

Begon, M. (1985). A general theory of life-history variation. In: *Behavioral Ecology* (Ed. by R.M. Sibley and R.H. Smith) pp. 91–97. Blackwell Scientific Publications, Oxford.

Begon, M., Harper, J.L. and Townsend, C.R. (1990). *Ecology: individuals, populations and communities*, second edition. Blackwell Scientific Publications, Oxford.

Behrens-Yamada, S. (1989) Are direct developers more locally adapted than planktonic developers? *Mar. Biol.* **103**, 403–411.

Bennell, S.J. (1981). Some observations on the littoral barnacle populations of north Wales. *Mar. Environ. Res.* **5**, 227–240.

Bertness, M.D., Gaines, S.D., Stephens, E.G. and Yund, P.O. (1992). Components of recruitment in populations of the acorn barnacles *Semibalanus balanoides* (Linnaeus). *J. Exp. Mar. Biol. Ecol.* **156**, 199–215.

Beverton, R.J.H. and Holt, S.J. (1957) On the dynamics of exploited fish populations. *Fishery Investigations, London* **19**, 1–533.

Birch, L.C. (1957). The meanings of competition. *Am. Nat.* 91, 5–18.

Birkland, C. and Lucas, J.S. (1991). *Acanthaster planci: Management Problem of Coral Reefs*. CRC Press Boca Raton, USA.

Black, K.P. and Gay, S.L. (1987). Hydrodynamic control of dispersal of crown-of-thorns starfish larvae: I. Small scale hydrodynamics on and around schematized and actual reefs. *Victorian Inst. Mar. Sci. Tech. Rep.* **8**. 1–67.

Black, K., Gay, S. and Andrews, J. (1990). Residence times of neutrally buoyant matter such as larvae, sewage or nutrients on coral reefs. *Coral Reefs* **9**, 105–114.

Boalche, G.T., Holme, N.A., Jephson, N. and Sidwell, J.M. (1974). A resurvey of Colemans intertidal traverse at Wembury, South Devon. *Jour. Mar. Biol. Assoc. U.K.* **54**, 551–553.

Bohnsack, J.A. and Talbot, F.H. (1980). Species-packing by reef fishes on Australian and Caribbean reefs: an experimental approach. *Bull. Mar. Sci.* **30**, 710–723.

Booth, D.J. (1991a). The effects of sampling frequency on estimates of recruitment of the domino damselfish (*Dascyllus albisella* Gill). *J. exp. Mar. Bol. Ecol.* **145**, 149–159.

Booth, D.J. (1991b). Larval settlement and juvenile group dynamics in the domino damselfish (*Dascyllus albisella*). PhD thesis, Oregon State University. 183 pp.

Booth, D.J. and Beretta, G.A. (1994) Seasonal recruitment and habitat associations in a guild of pomacentrid reef fishes at two sites around St. Thomas, USVI. *Coral Reefs* **13**, 81–89.

Booth, D.J. (1992). Larval settlement and preferences by domino damselfish *Dascyllus albisella* Gill. *J. exp. Mar. Biol. Ecol.* **155**, 85–104.

Booth, D.J. (1995). Juvenile groups in a coral-reef damselfish: density-dependent effects on individual fitness and population demography. *Ecology* **76**, 91–106.

Bowman, R.A. (1986). The biology of the limpet *Patella vulgata* (L). in the British Isles: spawning time as a factor determining recruitment success. In: *The ecology of rocky coasts* (Ed. by P.G. Moore and R. Seed) pp. 178–193. Hodder and Stoughton, Sevenoaks, UK.

Bowman, R.A. and Lewis, J.R. (1977). Annual fluctuations in recruitment of *P. vulgata*. *J. Mar. Biol. Ass. U.K.* **57**, 799–815.

Bowman, R.A. and Lewis, J.R. (1986). Geographical variation in the breeding cycles and recruitment of *Patella* species. *Hydrobiol.* **142**, 41–56.

Branch, G.M. (1981). The biology of limpets: physical factors, energy flow and ecological interactions. *Oceanogr. Mar. Biol. Ann. Rev.* **19**, 235–379.

Branch, G.M. (1985). The impact of predation by kelp gulls (*Larus dominicanus*) on the sub-Atlantic limpet *Nacelta delesseris*. *Polar Biol.* **4**, 171–177.

Branch, G.M. (1986). Limpets: their role in littoral and sublittoral community dynamics. In: *The ecology of rocky coasts* (Ed. by P.G. Moore and R. Seed) pp. 97–116. Columbia University Press, New York.

Breitburg, D.L. (1989). Demersal schooling prior to settlement by larvae of the naked goby. *Environ. Biol. Fishes* **26**, 97–103.

Breitburg, D.L. (1991). Settlement patterns and presettlement behavior of the naked goby, *Gobisoma bosci*, a temperate oyster reef fish. *Mar. Biol.* **106**, 213–222.

Brosnan, D.M. (1990). Fairweather friends: relationships between plants and animals can change as environmental conditions alter. In: *Plant–Animal interactions in the marine benthos* (Ed. by D.M. John, S.J. Hawkins and J.H. Price) Abstract: Systematics Symposium, Liverpool 1990.

Brosnan, D.M. (1992). Ecology of tropical rocky shores: plant–animal interactions in tropical and temperate latitudes. In: *Plant–Animal interactions in the marine benthos* (Ed. by D.M. John, S.J. Hawkins and J.H. Price) pp. 101–131 Clarendon Press, Oxford.

Brosnan, D.M. (1993). The effect of human trampling on biodiversity of rocky shores: monitoring and management strategies. *Rec. Adv. Mari. Sci. Tech.* pp. 333–341.

Brosnan, D.M. and Crumrine, L.L. (1992a). Human Impact and a management plan for the shore at Yaquina Head Outstanding Natural Area. *Report to the Bureau of Land Management, U.S. Dept. of the Interior USA* 94 pp.

Brosnan, D.M. and Crumrine, L.L. (1992b). A study of human impact on four shores on Oregon coast: implications for management and development. *Report to the Department of Land Conservation and Development, State of Oregon* 36 pp.

Brosnan, D.M. and Crumrine, L.L. (1994). Effects of human trampling on marine rocky shore communities. *J. Exp. Mar. Biol. Ecol.* **177**, 79–97.

Brosnan, D.M. (1995). Bridging gaps among ecology, law and policy. *Wildlife Soc. Bull.* in press.

Brothers, E.B. and Thresher, R.E. (1985). Pelagic-duration, dispersal and the distribution of Indo-Pacific coral-reef fishes. pp. 53–69. In: *The ecology of coral reefs* (Ed. by M.L. Reaka) NOAA Symp. Ser. Undersea Res. 3(1). Rockville, Md.

Burrows, E.M. and Lodge, S.M. (1950). A note on the interrelationships of *Patella*, *Balanus* and *Fucus* on a semi-exposed coast. *Ann. Rept. for 1949, Mar. Biol. Stat. Port Erin* **62**, 30–34.

Busek, D. (1988). Settlement as a major determinant of intertidal oyster and barnacle distributions along a horizontal gradient. *J. Exp. Mar. Biol. Ecol.* **122**, 1–18.

Buss, L.W. and Jackson, J.B.C. (1979). Competitive networks: non-transitive competitive relationships in cryptic coral reef environments. *Am. Nat.* **113**, 223–234.

Caffey, H.M. (1982). No effect of naturally-occurring rock types on settlement or survival in the intertidal barnacle *Tesseropora rosea* (Krauss). *J. Exp. Mar. Biol. Ecol.* **63**, 119–132.

Caffey, H.M. (1985). Spatial and temporal variation in settlement and recruitment of intertidal barnacles. *Ecol. Monogr.* **55**, 313–332.

Caley, M.J. (1993). Predation, recruitment and the dynamics of communities of coral-reef fishes. *Marine Biology* **117**, 33–43.

Carr, M. (1991). Habitat selection and recruitment of an assemblage of temperate zone reef fishes. *J. Exp. Mar. Biol. Ecol.* **146**, 113–137.

Carr, M. and Reed, D. (1993). Conceptual issues relevant to marine harvest refuges: examples from temperate reef fishes. *Can. J. Fish. Aquatic Sci.* **49**, 423–443.

Castilla, J.C. and Bustamente, R.H. (1989). Human exclusion from rocky intertidal of Las Cruces, Central Chile: effects on *Durvillea antarctica* (*Phaeophyta, Durvillealii*). *Mar. Ecol. Prog. Ser.* **50**, 203–214.

Chabot, R. and Bouget, E. (1988). Influence of substratum heterogeneity and settled barnacle density on the settlement of cypris larvae. *Mar. Biol.* **97**, 45–56.

Chapman, A.R.O. (1974). The ecology of macroscopic marine algae. *Ann. Rev. Ecol. Syst.* **5**, 65–80.

Chapman, A.R.O. (1986). Population and community ecology of seaweeds. *Adv. Mar. Biol.* **23**, 1–161.

Chesson, P. (1981). Models of spatially distributed populations: the effects of within-patch variability. *Theoret. Pop. Biol.* **19**, 288–325.

Chew, F.S. and Courtney, S.P. (1991). Plant Apparency and evolutionary escape from insect herbivory. *Am. Nat.* **138**, 729–750.

Cody, M.L. (1968). On the methods of resource division in grassland bird communities. *Am. Nat.* **102**, 107–147.

Cohen, E.B., Sissenwine, M.P. and Laurence, G.C. (1988). The "recruitment problem" for marine fish populations with emphasis on Georges Bank. pp. 373–393. In: *Toward a theory on Biological-physical interactions in the world oceans* (Ed. by B.J. Rothschild). NATO ASI Ser. C. Phys. Sc. vol. 239. Kluwer Acad. Publ., Dordecht, Netherlands.

Connell, J.H. (1961). The influence of interspecific competition and other factors on the distribution of the barnacle *Chthamalus stellatus*. *Ecology* **42**, 710–723.

Connell, J.H. (1972). Community interactions on marine rocky intertidal shores. *Ann. Rev. Ecol. Syst.* **3**, 169–192.

Connell. J.H. (1978). Diversity in tropical rain forests and coral reefs. *Science* **199**, 1302–1310.

Connell, J.H. (1979). Tropical rain forests and coral reefs as open nonequilbrium systems. In: *Population dynamics* (Ed. by R.M. Anderson, B.D. Turner and L.R. Taylor) pp. 141–163. Blackwell Scientific Publications, Oxford.

Connell, J.H. (1985). The consequences of variation in initial settlement vs. post-settlement mortality in rocky intertidal communities. *J. Exp. Mar. Biol. Ecol.* **93**, 11–45.

Connell, J.H. and Keough, M.J. (1985). Disturbance and patch dynamics of subtidal marine animals on hard substrates. In: *The ecology of natural disturbance and patch dynamics* (Ed. by S.T.A. Pickett and P.S. White) pp. 125–151. Academic Press, New York.

Connell, J.H. and Slatyer, R.O. (1977). Mechanisms of succession in natural communities and their role in community stability and organization. *Am. Nat.* **111**, 1119–1143.

Cooper, K. (1982). A model to explain the induction of settlement and metamorphosis of planktonic eyed pedivelegers of the blue mussel *Mytilus edulis* L. by chemical and tactile cues. *J. Shellfish Res.* **2**, 117.

Cooper, K. (1983). Potential for the application of the chemical DOPA to commercial bivalve setting systems. *J. Shellfish Res.* **3**, 110–111.

Cowen, R. (1985). Large scale patterns of recruitment by the labrid, *Semicossyphys pulcher*: causes and implications. *J. Mar. Res.* **43**, 719–742.

Crisp, D.J. (1955). The behavior of barnacle cyprids in relation to water movement over a surface. *J. Exp. Biol.* **32**, 569–590.

Crisp, D.J. (1958). The spread of *Elminius modestus* Darwin in north-west Europe. *J. Mar. Biol. Ass. U.K.* **37**, 483–520.

Crisp, D.J. (1961). Territorial behavior in barnacles settlement. *J. exp. Biol*, **38**, 429–446.

Crisp, D.J. (1974). Factors influencing the settlement of marine invertebrate larvae. In: *Chemoreception in marine organisms*. (Ed. by P.T. Grant and A.M. Makie) pp. 177–265. Academic Press, New York.

Crisp, D.J. (1976a). Settlement responses in marine organisms. In: *Adaptations to environment: Essays on the Physiology of Marine Animals*. (Ed. by R.C. Newell) pp. 93–124. Butterworths, London.

Crisp, D.J. (1976b). The role of pelagic larvae. In: *Perspectives in experimental zoology* (Ed. by T. Spencer-Davies) pp. 145–155. Pergamon Press, New York.

Crisp, D.J. and Barnes, H. (1954). The orientation and distribution of barnacles at settlement with particular reference to surface contour. *J. Anim. Ecol.* **23**, 142–162.

Crisp, D.J. and Meadows, P.S. (1962). The chemical basis of gregariousness in cirripedes. *Proc. Roy. Soc. (B).* **158**, 364–387.

Crisp, D.J. and Spencer, C.P. (1958). The control of the hatching process in barnacles. *Proc. Roy. Soc. (B).* **148**, 278–299.

Crisp, D.J. and Williams, G.B. (1960). Effect of extracts from fucoids in promoting settlement of epiphytic Polyzoa. *Nature London.* **188**, 1206–1207.

Cushing, P. (1975). *Marine Ecology and Fisheries*. Cambridge University Press, Cambridge 278 pp.

Davis, M. and Olla, B. (1987). Aggression and variation in growth of chum salmon (*Oncorynchus keta*) juveniles: effects of limited rations. *Can. J. Fis. Aquatic Sci.* **44**, 192–197.

Day, R. and McEdward, L. (1984). Aspects of the physiology and ecology of pelagic larvae of marine benthic invertebrates. In: *Marine Plankton Life Cycle Strategies* (Ed. by K.A. Steidinger and L.M. Walker) pp. 93–120. CRC Press, Boca Raton, Florida.

Dayton, P.K. (1971). Competition, disturbance and community organization: the provision and subsequent utilization of space in a rocky intertidal community. *Ecol. Monogr.* **41**, 351–389.

Dayton, P.K. (1973). Dispersion, dispersal and persistence of the annual intertidal alga, *Postelsia palmaeformis* Rurprecht. *Ecology* **54**, 433–438.

Dayton, P.K. (1975). Experimental evaluation of ecological dominance in a rocky intertidal algal community. *Ecol. Monogr.* **45**, 137–159.

Deevey, E. (1947). Life tables for natural populations of animals. *Quart. Rev. Biol.* **22**, 283–314.

Denely, E.J. and Underwood, A.J. (1979). Experiments on factors influencing settlement, survival and growth of two species of barnacles in New South Wales. *J. Exp. Mar. Biol. Ecol.* **36**, 269–293.

Dethier, M.N. and Duggins, D.O. (1988). Variation in strong interactions in the intertidal zone along a geographical gradient: a Washington–Alaska comparison. *Mar. Ecol. Prog. Ser.* **50**, 97–105.

Diamond, J. (1978). Niche shifts and the rediscovery of interspecific competition. *Am. Sci.* **66**, 322–331.

Dight, I., James, M. and Bode, L. (1988). Models of larval dispersal within the central Great Barrier Reef: patterns of connectivity and their implications for species distributions. *Proc. Sixth Int. coral Reef Symp., Australia* **3**, 217–224.

Dight, I.J., James, M.K. and Bode, L. (1990). Modelling the larval dispersal of *Acanthaster planci*. II Patterns of reef connectivity. *Coral Reefs* **9**, 125–134.

Dittman, D. and Robles, C. (1991). Effect of algal epiphytes on the mussel *Mytilus californianus*. *Ecology* **72**, 286–296.

Doherty, P.J. (1981). Coral reef fishes: Recruitment-limited assemblages? *Proc. Intl. Coral Reef Symp. 4th*, **2**, 465–470.

Doherty, P.J. (1983a). Tropical territorial damselfishes: is density limited by aggression or recruitment? *Ecology* **64**, 176–190.

Doherty, P.J. (1983b). Diet, lunar and seasonal rhythms in the reproduction of two tropical damselfishes, *Pomacentrus flavicauda* and *P. wardi*. *Mar. Biol.* **75**, 215–224.

Doherty, P.J. (1987a). The replenishment of populations of coral reef fishes, recruitment surveys and the problems of variability manifest on multiple scales. *Bull. Mar. Sci.* **44**, 411–422.

Doherty, P.J. (1987b). Light traps—selective but useful devices for quantifying the distributions and abundances of larval fishes. *Bull. Mar. Sci.* **41**, 423–443.

Doherty, P.J. (1991). Spatial and temporal patterns in recruitment. In: *The Ecology of Fishes on Coral Reefs* (Ed. by P.F. Sale) pp. 261–293. Academic Press, London.

Doherty, P. and Fowler, A. (1994). An empirical test of recruitment-limitation in a coral reef fish. *Science* **263**, 935–939.

Doherty, P.J. and Sale, P.F. (1986). Predation on juvenile coral reef fishes: an exclusion experiment. *Coral Reefs* **4**, 225–234.

Doherty, P.J. and Williams, D.M. (1988). The replenishment of coral reef fish populations. *Ocean. and Mar. Biol. Ann. Rev.* **26**, 487–551.

Druehl, L.D. (1981). Geographic distribution. In: *The biology of seaweeds*. (Ed. by C.S. Lobban and M.J. Wynne) pp. 306–325. University of California Press, Berkeley and Los Angeles.

Dumas, J.V.D. and Witman, J.D. (1993). Predation by herring gulls (*Larus argentatus* Coues) on two rocky intertidal crab species (*Carcinus maenas* (L.) and *Cancer irroratus* Say). *J. Exp. Mar. Biol. Ecol.* **169**, 89–101.

Dungan, M.L. (1985). Competition and the morphology, ecology, and evolution of acorn barnacles: an experimental test. *Paleobiology* **11**, 165–173.

Dungan, M.L. (1986). Three-way interactions: barnacles, limpets, and algae in a Sonoran Desert rocky intertidal zone. *Am. Nat.* **127**, 292–316.

Duran, L.R. and Castilla, J.C. (1989). Variation and persistence of the middle rocky intertidal community of central Chile, with and without human harvesting. *Mar. Biol.* **53**, 555–562.

Durante, K. (1991). Larval behavior, settlement preference and induction of metamorphosis in the temperate solitary ascidian *Molgula citrina*. *J. exp. Mar. Biol. Ecol.* **145**, 175–187.

Ebert, T.A. and Russell, M.P. (1988). Latitudinal variation in size structure of the west coast purple sea urchin: A correlation with headlands. *Limnol. Oceanogr.* **33**, 286–294.

Eckert, G.J. (1984). Annual and spatial variation in recruitment of labroid fishes among seven reefs in the Capricorn/Bunker group, Great Barrier reef. *Mar Biol.* **78**(2), 129–138.

Eckert, G.J. (1987). Estimates of adult and juvenile mortality for labrid fishes at One Tree reef. *Mar. Biol.* **95**, 167–171.

Edwards, D.C., Conover, D.O. and Sutter, F. III. (1982). Mobile predators and the structure of marine intertidal communities. *Ecology* **63**, 1175–1180.

Ehrlich, P. (1975). The population biology of coral reef fishes. *Annu. Rev. Ecol. Syst.* **6**, 211–247.

Elton, C.S. (1958). *The ecology of invasions by animals and plants*. Methuen, London.

Eyster, L.S. and Pechenik, J.A. (1987). Attachment of *Mytilus edulis* L. larvae on algal and byssla filaments is enhanced by water agitation. *J. Exp. Mar. Biol. Ecol.* **114**, 99–110.

Fairweather, P.G. (1988). Consequences of supply-side ecology: manipulating the recruitment of intertidal barnacles affects the intensity of predation upon them. *Biol. Bull.* **175**, 349–354.

Farrell, T.M. (1989). Succession in a rocky intertidal community: the importance of disturbance size and position within a disturbed patch. *J. Exp. Mar. Biol. Ecol.* **128**, 57–73.

Farrell, T.M. (1991). Models and mechanisms of succession: an example from a rocky intertidal community. *Ecol. Monogr.* **61**, 95–113.

Farrell, T.M., Bracher, D. and Roughgarden, J. (1991). Cross-shelf transport causes recruitment to intertidal populations in central California. *Limnol. Oceanogr.* **36**, 279–288.

Foster, M.S. (1975a). Algal succession in a *Macrocystis pyrifera* forest. *Mar. Biol.* **32**, 313–329.

Foster, M.S. (1975b). Regulation of algal community development in a *Macrocystis pyrifera* forest. *Mar. Biol.* **32**, 331–342.

Foster, M.S. (1982). Factors controlling the intertidal zonation of *Iridaea flaccida* (Rhodophyta). *J. Phycol.* **18**, 285–294.

Frank, P.W. (1982). Effects of winter feeding on limpets by black oystercatchers *Haematopus bachmani*. *Ecology* **63**, 1352–1362.

Gaines, S.D. and Bertness, M.D. (1992). Dispersal of juveniles and variable recruitment in sessile marine species. *Nature, London* **360**, 579–580.

Gaines, S.D. and Roughgarden, J. (1985). Larval settlement rate: a leading determinant of structure in an ecological community of the marine intertidal zone. *Proc. Natl. Acad. Sci.* **82**, 3707–3711.

Gaines, S.D. and Roughgarden, J. (1987). Fish predation in offshore kelp forests affect recruitment to intertidal barnacle populations. *Science* **235**, 479–481.

Galzin, R., Planes, S., Dufour, V. and Salvat, B. (1994). Variation in diversity of coral reef fish between French Polynesian atolls. *Coral Reefs* **13**, 175–180.

Glynn, P.W. (1989). Coral mortality and disturbances to coral reefs in the tropical eastern pacific. In: *Global ecological consequences of the 1982–3 El Nino-southern oscillation* (Ed. by P.W. Glynn). Elsevier Oceanography Series.

Gilpin, M.E. (1987). Spatial structure and population vulnerability. In: *Viable populations for conservation* (Ed. by M.E. Soule) pp. 125–140. Cambridge University Press, Cambridge.

Godoy, C. and Moreno, C.A. (1989). Indirect effects of human exclusion from the rocky intertidal in southern Chile: a case of cross-linkage between herbivores. *Oikos* **54**, 101–106.

Gosling, E.M. (1992). Genetics of *Mytilus*. In: *The mussel Mytilus: ecology, physiology, genetics and culture* (Ed. by E. Gosling) pp. 309–382. Elsevier.

Grosberg, R.K. and Quinn, J.F. (1986). The genetic control and consequences of kin recognition by the larvae of a colonial marine invertebrate. *Nature* **322**, 457–459.

Hamner, W.M., Jones, M.S., Carleton, J.H., Hauri, I.R. and Williams, D.M. (1988). Zooplankton, planktivorous fish, and water currents on a windward reef face: Great Barrier Reef, Australia. *Bull. Mar. Sci.* **42**, 459–479.

Hancock, D.A. (1973). The relationship between stock and recruitment in exploited invertebrates. *J. Cons. Int. Explor. Mer.* **164**, 113–131.

Hanski, I. (1991). Single species metapopulation dynamics: concepts models and observations. *Biol. Jour. Linn. Soc.* **42**, 17–38.

Harger, J.R.E. (1970). The effect of wave impact on some aspects of the biology of mussels. *Veliger* **12**, 401–414.

Harmelin-Vivien, M. (1989). Reef fish community structure: An Indo-Pacific comparison. *Ecol. Stud.* **69**, 21–60.

Harrison, S. (1991). Local extinction in a metapopulation context: an empirical evaluation. *Biol. Jour. Linn. Soc.* **42**, 73–88.

Hartnoll, R.G. and Hawkins, S.J. (1985). Patchiness and fluctuations on a moderately exposed rocky shore. *Ophelia* **24**, 53–63.

Hatton, H. (1938). Essai de bionomie explicative sur quelques especes intercotidales d'algues et d'animaux. *Annals Inst. oceanogr. Monaco* **27**, 242–348.

Havenhand, J.N., Thorpe, J.P., Todal, C.D. (1986). Estimates of Biochemical genetic diversity within and between the nudibranch molluscs *Adalaria proxima* and *Onchideris muricata*. *J. Exp. Mar. Biol. Ecol.* **95**, 105–111.

Hawkins, S.J. (1981). The influence of season and barnacles on the algal colonization of *Patella vulgata* exclusion areas. *J. Mar. Biol. Ass. UK* **61**, 1–15.

Hawkins, S.J. (1983). Interaction of *Patella* and macroalgae with settling *Semibalanus balanoides* (L.) in the Isle of Man (1977–1981). *J. Exp. Mar. Biol. Ecol.* **62**, 271–283.

Hawkins, S.J. and Hartnoll, K. (1983). Some anomalies on Lundy in the distribution of common eulittoral prosobranchs with planktonic larvae. *J. Moll. Stud.* **49**, 86–88.

Hawkins, S.J. and Hartnoll, R.G. (1982). Settlement patterns of *Semibalanus balanoides* (L.) in the Isle of Man (1977–1981). *J. Exp. Mar. Biol. Ecol.* **62**, 271–283.

Hawkins, S.J. and Hartnoll, R.G. (1983). Grazing of intertidal algae by marine invertebrates. *Oceanogr. Mar. Biol. Ann. Rev.* **21**, 131–181.

Hawkins, S.J. and Southward, A.J. (1992). Lessons from the Torrey Canyon oil spill: recovery and stability of rocky shore communities. In: *Symposium on marine habitat restoration*. USA National Oceanographic and Atmospheric Administration.

Hawkins, S.J., Hartnoll, R.G., Kain, J.M. and Norton, T.A. (1992). Plant-animal interactions on hard substrata in the north-east Atlantic. In: *Plant–Animal interactions in the marine benthos*. (Ed. by D.M. John, S.J. Hawkins and J.H. Price) pp. 1–32. Clarendon Press, Oxford.

Hawkins, S.J., Watson, D.C., Hill, A.S., Harding, S.P., Kyriakides, M.A., Hutchinson, S. and Norton, T.A. (1989). A comparison of feeding mechanisms in microphagus herbivorous intertidal prosobranchs in relation to resource partitioning. *J. Moll. Stud.* **55**, 151–161.

Hay, M.E. (1984). Patterns of fish and urchin grazing on Caribbean coral reefs: are previous results typical? *Ecology* **65**, 446–454.

Hilbish, T.J. and Koehn, R.K. (1985). The physiological basis for selection at LAP locus. *Evol.* **39**, 1302–1317.

Hixon, M.A. (1991). Predation as a process structuring coral-reef fish communities. In: *The Ecology of coral reef fishes*. (Ed. by P.F. Sale) pp. 475–508. Academic Press, San Diego.

Hixon, M.A. and Beets, J. (1993). Predation, prey refuges and the structure of coral reef assemblages. *Ecol. Monog.* **63**, 77–101.

Hixon, M.A. and Brostoff, W.N. (1983). Damselfish as keystone species in reverse: intermediate disturbance and diversity of reef algae. *Science* **220**, 511–513.

Hjort, J. (1914). Fluctuations in the great fisheries of northern Europe. *Rapp. P.-V. Reun., Cons. Int. Explor. Mer.* **20**, 1–13.

Hockey, P.A.R. and Branch, G.M. (1984). Oystercatchers and limpets: impact and implications. A preliminary assessment. *Ardea* **72**, 199–206.

Hoffmann, A.J. (1987). The arrival of seaweed propagules at the shore: a review. *Bot. Mar.* **30**, 151–165.

Hoffmann, A.J. and Santelices, B. (1991). Banks of algal macroscopic forms: hypotheses on their functioning and comparisons with seed banks. *Mar. Ecol. Prog. Ser.* **79**, 185–194.

Holm, E. (1990). Effects of density-dependent mortality on the relationship between recruitment and larval settlement. *Mar. Ecol. Prog. Ser.* **60**, 141–146.

Hughes, R.G. (1984). A model of the structure and dynamics of benthic marine invertebrate communities. *Mar. Ecol. (Prog. Ser.)* **15**, 1–11.

Hurlbut, C.J. (1991a). Community recruitment: settlement and juvenile survival of seven co-occurring species of sessile marine invertebrates. *Mar. Biol.* **109**, 507–515.

Hurlbut, C.J. (1991b). Larval substrate selection and post settlement mortality as determinants of the distribution of two bryozoans. *J. Exp. Mar. Biol. Ecol.* **147**, 103–119.

Jernakoff, P. (1983). Factors affecting the recruitment of algae in a midshore region dominated by barnacles. *J. Exp. Mar. Biol. Ecol.* **67**, 17–31.

Jernakoff, P. (1985a). An experimental evaluation of the influence of barnacles, crevices and seasonal patterns of grazing on algal diversity and cover in an intertidal barnacle zone. *J. Exp. Mar. Biol. Ecol.* **88**, 287–302.

Jernakoff, P. (1985b). The effect of overgrowth by algae on the survival of the intertidal barnacle *Tesseropora rosea* Krauss. *J. Exp. Mar. Biol. Ecol.* **94**, 89–97.

Johnson, C.R. and Mann, K.H. (1988). Diversity, patterns of adaptation and stability of Nova Scotian kelp beds. *Ecol. Monogr.* **58**, 129–154.

Johnson, L. and Strathmann, R.R. (1989). Settling barnacle larvae avoid substrata previously occupied by a mobile predator. *J. Exp. Mar. Biol. Ecol.* **128**, 87–103.

Jones, G.P. (1984). Population ecology of the temperate reef fish *Pseudolabrus celidotus* Block & Schneider (Pisces: Labridae). I. Factors influencing recruitment. *J. Exp. Mar. Biol. Ecol.* **75**, 257–276.

Jones, G.P. (1987). Some interactions between residents and recruits in two coral reef fishes. *J. Exp. Mar. Biol. and Ecol.* **114**, 169–182.

Jones, G.P. (1988). Experimental evaluation of the effects of habitat structure and competitive interactions on juveniles of two coral reef fishes. *J. Exp. Mar. Biol. Ecol.* **123**, 115–126.

Jones, G.P. (1990). The importance of recruitment to the dynamics of a coral reef fish population. *Ecology* **71**, 1691–1698.

Jones, G.P. (1991). Postrecruitment processes in the ecology of coral reef fish populations: a multifactorial perspective. In: *The ecology of fishes on coral reefs* (Ed. by P.F. Sale) pp. 294–328. Academic Press Inc., San Diego.

Jorgenson, C.B. (1981). Mortality, growth, and grazing impact on a cohort of bivalve larvae *Mytilus edulis* L. *Ophelia*, **20**, 185–192.

Judge, M.L., Quinn, J.F. and Wolin, C.L. (1988). Variability in recruitment of *Balanus glandula* (Darwin, 1854) along the central Californian coast. *J. Exp. Mar. Biol. Ecol.* **119**, 235–251.

Kami, H.T. and Ikehara, I.I. (1976). Notes on the annual juvenile siganid harvest in Guam. *Micronesica* **12**, 323–325.

Kay, A.M. and Keough, M.J. (1981). Occupation of patches in the epifaunal communities on pier pilings and the bivalve *Pinna bicolor* at Edinburgh, South Australia. *Oecologia* **48**, 123–130.

Kay, A.M. and Liddle, M.J. (1989). Impact of human trampling in different zones of a coral reef. *Environ. Manage.* **13**, 509–520.

Kendall, M.A., Bowman, R.S., Williamson, P. and Lewis, J. (1982). Settlement patterns, density and stability in the barnacle *Balanus balanoides*. *Neth. J. Sea Res.* **16**, 119–126.

Keough, M.J. (1983). Patterns of recruitment of sessile invertebrates in two sub-tidal habitats. *J. Exp. Mar. Biol. Ecol.* **66**, 213–245.

Keough, M.J. (1984a). Effects of patch size on the abundance of sessile marine invertebrates. *Ecology* **65**, 423–437.

Keough, M.J. (1984b). The dynamics of the epifauna of *Pinna bicolor:* interactions between recruitment, predation, and competition. *Ecology* **65**, 677–688.

Keough, M.J. and Butler, A.J. (1983). Temporal changes in species number in an assemblage of sessile marine invertebrates. *J. Biogeog.* **10**, 317–330.

Keough, M.J. and Downes, B.J. (1982). Recruitment of marine invertebrates: the role of active larval choice and early mortality. *Oecologia* **54**, 348–352.

King, P.A., McGrath, D. and Gosling, E.M. (1989). Reproduction and settlement of *Mytilus edulis* on an exposed rocky shore in Galway Bay, west coast of Ireland. *J. mar. biol. Ass. UK* **55**, 477–495.

Kingsford, M.J. (1988). The early life history of fishes in coastal waters of New Zealand: a review. *N.Z. J. Mar. FW Res.* **22**, 463–479.

Kingsford, M.J. (1990). Linear oceanographic features: A focus for research on recruitment processes. *Aust. J. Ecol.* **15**, 391–401.

Kingsford, M.J. and Choat, J.H. (1986). Influence of surface slicks on the distribution and onshore movement of small fish. *Mar. Biol.* **91**, 161–171.

Kingsford, M.J. and Suthers, I.M. (In press). The distribution of larval fish at estuarine plume fronts off eastern Australia. *Estuarine Cont. Shelf Res.*

Kingsford, M.J., Underwood, A.J. and Kennelly, S.J. (1991). Humans as predators on rocky reefs in New South Wales, Australia. *Mar. Ecol. Prog. Ser.* **72**, 1–14.

Kingsford, M.J., Wolanski, E. and Choat, J.H. (1991). Influence of tidally-induced fronts and Langmuir circulation on distribution and movements of presettlement fishes around a coral reef. *Mar. Biol.* **109**, 167–180.

Knight-Jones, E.W. (1951). Gregariousness and some other aspects of the setting behavior of Spirorbis. *J. mar. biol. Ass. UK* **30**, 210–222.

Knight-Jones, E.W. (1953). Laboratory experiments on gregariousness during setting in *Balanus balanoides* and other barnacles. *J. exp. Biol.* **30**, 584–598.

Knight-Jones, E.W. (1955). The gregarious setting reaction of barnacles as a measure of systematic affinity. *Nature, London* **174**, 226.

Knight-Jones, E.W. and Crisp, D.J. (1953). Gregariousness in barnacles in relation to fouling of ships and to anti-fouling research. *Nature, London* **171**, 1109.

Knight-Jones, E.W. and Stevenson, J.P. (1950). Gregariousness during settlement of the barnacle *Elminius modestus* Darwin. *J. mar. biol. Ass. UK* **29**, 281–297.

Kobayashi, D.R. (1989). Fine-scale distribution of larval fishes: patterns and processes adjacent to coral reefs in Kaneohe Bay, Hawaii. *Mar. Biol.* **100**, 285–293.

Koehn, R.K. (1975). Migration and population structure in the pelagically dispersing marine invertebrate *Mytilus edulis*. In: *Proceedings of the third International Congress on Isozymes* (Ed. by C. Market) pp. 945–959. Academic Press, New York.

Koehn, R.K. (1991). The genetics and taxonomy of species in the genus *Mytilus*. *Aquaculture* **94**, 123–145.

Koehn, R.K., Milkman, R. and Mitton, J.B. (1976). Population genetics of marine pelecypods. IV Selection, migration and differentiation in the blue mussel *Mytilus edulis*. *Evol.* **30**, 2–32.

Kozloff, E. (1983). *Seashore life of the northern Pacific coast* (second edition). University of Washington Press, Seattle, London. 370 pp.

Lasker, R. (1985). What limits clupeoid production? *Can. J. Fish. Aquat. Sci.* **42**, 31–38.

Lassig, B. (1982). The minor role of large transient fishes in structuring full scale coral patch reef assemblages. PhD dissertation, Macquarie University, Sydney, Australia.

Lee, H. and Ambrose, W.G. (1989). Life after competitive exclusion: an alternative strategy for a competitive inferior. *Oikos* **56**, 424–430.

Lehtinen, K.J., Notini, M., Mattsson, J. and Landner, L. (1988). Disappearance of bladder-wrack (*Fucus vesiculosus*) in the Baltic Sea: relation to pulp mill chlorate. *Ambio* **17**, 387–393.

Leis, J. (1986). Vertical and horizontal distribution of fish larvae near coral reefs at Lizard Island, Great Barrier Reef. *Mar. Biol.* **90**, 505–516.

Leis, J. (1991). The pelagic stages of reef fishes: the larval biology of coral reef fishes. In: *The ecology of fishes on coral reefs* (Ed. by P.F. Sale), pp. 183–230. Academic Press Inc., San Diego.

Leis, J. and Goldman, C. (1987). Composition and distribution of larval fish assemblages in the Great Barrier Reef lagoon at Lizard Island, Australia. *Aust. J. Mar. FW Res.* **38**, 211–223.

Le Tourneux, F. and Bourget, E. (1988). Importance of physical and biological settlement cues used at different spatial scales by the larvae of *Semibalanus balanoides*. *Mar. Biol.* **97**, 57–66.

Levin, P.S. (1993) Habitat structure, conspecific presence and spatial variation in the recuitment of a temperate reef fish, *Oecologia*, **94**, 176–185.

Levinton, J.S. and Suchanek, T.H. (1978). Geographic variation, niche breadth and geographic differentiation at different geographic scales in the mussels *Mytilus californianus* and *M. edulis*. *Mar. Biol.* **49**, 363–375.

Lewis, J.R. (1964). *The ecology of rocky shores*. English University Press, London, 323 pp.

Lewis, J.R. and Bowman, R.S. (1975). Local habitat induced variations in the population dynamics of *P. vulagata. J. Exp. Mar. Biol. Ecol.* **17**, 165–203.

Liddle, M.J. (1991). Recreation Ecology: Effects of trampling on plants and corals. *Trends in Ecology and Evolution* **6**, 13–16.

Lively, C.M. and Raimondi, P.T. (1987). Desiccation, predation, and mussel-barnacle interactions in the northern Gulf of California. *Oecologia* **74**, 304–309.

Lobel, P. (1988). Larval retention by currents off the Hawaiian Islands. *Env. Biol. Fish.* **15**, 122–135.

Low, R.M. (1971). Interspecific territoriality in a pomacentrid reef fish, *Pomacentrus flavicauda. Ecology* **52**, 648–654.

Lubchenco, J. (1978). Plant species diversity in a marine intertidal community: importance of herbivore food preference and algal competitive abilities. *Am. Nat.* **112**, 23–39.

Lubchenco, J. (1980). Algal zonation in the New England rocky intertidal community: an experimental analysis. *Ecology* **61**, 333–344.

Lubchenco, J. (1982). Effects of grazers and algal competitors on fucoid colonization in tidepools. *J. Phycol.* **18**, 544–550.

Lubchenco, J. (1983). *Littorina* and *Fucus*: effects of herbivores, substratum heterogeneity, and plant escapes during succession. *Ecology* **64**, 1116–1123.

Lubchenco, J. and Menge, B.A. (1978). Community development and persistence in a low rocky intertidal zone. *Ecol. Monogr.* **48**, 67–94.

Lubchenco, J., Menge, B.A., Garrity, S.D., Askenas, L.R., Gaines, S.D., Emlet, R., Lucas, J. and Strauss, S. (1984). Structure, persistence and role of consumers in a tropical rocky intertidal community (Taboguilla Island, Bay of Panama). *Jour. Exp. Mar. Biol. Ecol.* **78**, 23–73.

Luckhurst, B. and Luckhurst, K. (1978). Analysis of the influence of the substrate variables on coral reef fish communities. *Mar. Biol.* **49**, 317–323.

Lutz, R.A. and Kennish, M.J. (1992). Ecology and morphology of larval and early postlarval mussels. In: *The mussel Mytilus: ecology, physiology, genetics and culture* (Ed. by E. Gosling) pp. 53–86. Elsevier.

MacArthur, R.H. and MacArthur, J.W. (1961). On bird species diversity. *Ecology* **42**, 594–598.

MacArthur, R.H., Recher, H. and Cody, M. (1966). On the relation between habitat selection and species diversity. *Am. Nat.* **100**, 319–332.

Malusa, J.R. (1986). Life history and environment of two species of intertidal barnacles. *Biol. Bull.* **170**, 409–428.

Mapstone, B. and Fowler, A. (1988). Recruitment and the study of assemblages of fishes on coral reefs. *Trends in Evol. and Ecol.* **3**, 72–77.

Marsh, C.P. (1986a). Impact of avian predators on high intertidal limpet populations. *J. Exp. Mar. Biol. Ecol.* **104**, 185–201.

Marsh, C.P. (1986b). Rocky intertidal community organization: the impact of avian predators on mussel recruitment. *Ecology* **67**, 771–786.

May, R.M. (1981). Models for two interacting populations. In: *Theoretical ecology: principles and applications* (Ed. by R.M. May) pp. 78–104. Sinauer, Sunderland, Mass.

Mayr, E. (1963). *Animal species and evolution.* Belknap Press of Harvard University Press, Cambridge, Massachusetts.

McFarland, W., Brothers, E., Ogden, J., Shulman, M., Bermingham, E. and Kotchian-Prentiss, N. (1985). Recruitment patterns in young French grunts *Haemulon flavolineatum* (Family Haemulidae) at St. Croix, U.S.V.I. *Fish. Bull.* **83**, 413–426.

McFarland, W.N. and Ogden, J.C. (1985). Recruitment of young coral reef fishes from the plankton. In: *The ecology of coral reefs* (Ed. by M.L. Reaka) pp. 37–51. NOAA Symp. Ser. Undersea Res. 3(1). Rockville, Md.

McGrath, D., King, P.A. and Gosling, E.M. (1988). Evidence for direct settlement of *Mytilus edulis* L., larvae on adult mussel beds. *Mar. Ecol. Prog. Ser.* **47**, 103–106.

Meadows, P.S. and Campbell, J.I. (1972). Habitat selection by aquatic invertebrates. *Adv. in Mar. Biol.* **10**, 271–382.

Menge, B.A. (1976). Organization of the New England rocky intertidal community: role of predation, competition, and environmental heterogeneity. *Ecol. Monogr.* **46**, 355–393.

Menge, B.A. (1978). Predation intensity in a rocky intertidal community; effect of an algal canopy, wave action and desiccation on predator feeding rates. *Oecologia* **34**, 17–35.

Menge, B.A. (1982). Reply to a comment by Edwards, Conover, and Sutter. *Ecology* **63**, 1180–1184.

Menge, B.A. (1983). Components of predation intensity in the low zone of the New England rocky intertidal region. *Oecologia* **58**, 141–155.

Menge, B.A. (1991). Relative importance of recruitment and other causes of variation in rocky intertidal community structure. *J. Exp. Mar. Biol. Ecol.* **146**, 69–100.

Menge, B.A., Ashkenas, L.R. and Matson, A. (1983). Use of artificial holes in studying community development in cryptic marine habitats in a tropical rocky intertidal region. *Mar. Biol.* **77**, 129–142.

Menge, B.A. and Farrell, T.M. (1989). Community structure and interaction webs in shallow marine hard bottom communities: tests of an environmental stress model. *Adv. Ecol. Res.* **19**, 189–262.

Menge, B.A., Farrell, T.M., Olsen, A.M., van Tamelen, P. and Turner, T. (1993). Algal recruitment and the maintenance at a plant mosaic in the low intertidal region on the Oregon Coast. *J. Exp. Biol. Ecol.* **170**, 91–116.

Menge, B.A. and Lubchenco, J. (1981). Community organization in temperate and tropical rocky intertidal habitats: prey refuges in relation to consumer pressure gradients. *Ecol. monogr.* **51**, 429–450.

Menge, B.A., Lubchenco, J., Ashkenas, L.R., and Ramsey, F. (1986a). Experimental separation of effects of consumers on sessile prey in the low zone of a rocky shore in the Bay of Panama: direct and indirect consequences of food web complexity. *J. Exp. Mar. Biol. Ecol.* **100**, 225–269.

Menge, B.A., Lubchenco, J., Gaines, S.D. and Askenas, L.R. (1986b). A test of the Menge-Sutherland model of community organization in a tropical rocky intertidal food web. *Oecologia* **71**, 75–89.

Menge, B.A. and Sutherland, J.P. (1976). Species diversity gradients: synthesis of the roles of predation, competition, and temporal heterogeneity. *Am. Nat.* **110**, 351–369.

Menge, B.A. and Sutherland, J.P. (1987). Community regulation: variation in disturbance, competition, and predation in relation to environmental stress and recruitment. *Am. Nat.* **130**, 730–757.

Milicich, M.J. (1988). The distribution and abundance of pre-settlement fish in the nearshore waters of Lizard Island. *Proc. 6th Int. Coral Reef Symp.* **2**, 785–790.

Minchinton, T.E. and Scheibling, R.E. (1993). Variations in sampling procedure and frequency affect estimates of recruitment of barnacles. *Mar. Ecol. Prog. Ser.* **99**, 83–88.

378   D.J. BOOTH AND D.M. BROSNAN

Molles, M. (1978). Fish species diversity on model and natural reef patches: experimental insular biogeography. *Ecol. Monogr.* **48**, 289–305.

Moran, M.J., Fairweather, P.G. and Underwood, A.J. (1984). Growth and mortality of the predatory intertidal whelk *Morula marginalba* Blainville (Muricidae): The effects of different species of prey. *Jour. Exp. Mar. Biol. Ecol.* **75**, 1–17.

Moreno, C.A., Sutherland, J.P. and Jara, F.J. (1984). Man as a predator in the intertidal zone of central Chile. *Oikos* **42**, 155–160.

Morse, A.N.C. (1992). Role of algae in recruitment of marine invertebrate larvae. In: *Plant-Animal interactions in the marine benthos*. (Ed. by D.M. John, S.J. Hawkins and J.H. Price) pp. 385–406. Clarendon Press, Oxford.

Munro, J. (1983). Caribbean coral reef fishery resources. *ICLARM Stud. Rev.* **7**, 1–276.

Navarrete, S.A. and Castilla, J.C. (1990). Barnacle walls as mediators of intertidal mussel recruitment: effects of patch size on the utilization of space. *Mar. Ecol. Prog. Ser.* **68**, 113–119.

Nelson, W.G. (1981). Inhibition of barnacle settlement by Ekofisk crude oil. *Mar. Ecol. Prog. Ser.* **5**, 41–43.

Ochi, H. (1986). Growth of the anemonefish *Amphiprion clarkii* in temperate waters, with special reference to the influence of settling time on the growth of 0-year-olds. *Mar. Biol.* **92**, 223–229.

Olivia, D. and Castilla, J.C. (1986). The effects of human exclusion on the population structure of key-hole limpets *Fisurella crassa* and *F. limbata* on the coast of central Chile. *PSZNI Mar. Ecol.* **7**, 201–217.

Olson, R. (1985). The consequences of short distance larval dispersal in sessile marine invertebrates. *Ecology* **66**(1), 30–39.

Ortega, S. (1986). Fish predation on gastropods on the Pacific coast of Costa Rica. *J. Exp. Mar. Biol. Ecol.* **97**, 181–191.

Ortega, S. (1987). The effect of human predation on the size distribution of *Siphonaria gigas* (Mollusca, Pulmonata) on the Pacific coast of Costa Rica. *The Veliger* **29**, 251–255.

Paine, R.T. (1966). Food web complexity and species diversity. *Am. Nat.* **100**, 65–75.

Paine, R.T. (1969). A note on trophic complexity and community stability. *Am. Nat.* **103**, 91–93.

Paine, R.T. (1974). Intertidal community structure: experimental studies on the relationship between a dominant competitor and its principal predator. *Oecologia* **15**, 93–120.

Paine, R.T. (1976). Size-limited predation: an observational and experimental approach with the *Mytilus–Pisaster* interaction. *Ecology* **57**, 858–873.

Paine, R.T. (1979). Disaster, catastrophe and local persistence of the sea palm *Postelsia palmaeformis*. *Science* **205**, 685–687.

Paine, R.T. (1980). Food webs: linkage, interaction strength and community infrastructure. *J. Anim. Ecol.* **49**, 667–685.

Paine, R.T. and Levin, S.A. (1981). Intertidal landscapes: disturbance and the dynamics of pattern. *Ecol. Monogr.* **51**, 145–178.

Paine, R.T. and Vadas, R.L. (1969). The effects of grazing by sea urchins, *Strongylocentrotus* spp., on benthic algal populations. *Limnol. Oceanogr.* **14**, 710–719.

Pauly, D. (1979). Theory and management of multispecies stocks: a review with emphasis on South East Asian demersal fisheries. *ICLARM Stud. Rev.* **1**, Manilla, Phillipines, 35 pp.

Pawlick, J.R. (1992). Chemical ecology of the settlement of benthic marine inverte-brates. *Oceanogr. Mar. Biol. Annu. Rev.* **30**, 273–335.

Pechenek, J.A. (1984). The relationship between temperature, growth rate, and duration of planktonic life for larvae of the gastropod *Crepidula fornicata* (L.) *J. Exp. Mar. Biol. Ecol.* **82**, 147–159.

Peterson, J.H. (1984a). Larval settlement behavior in competing species: *Mytilus californianus* Conrad and *M. edulis* L. *J. Exp. Mar. Biol. Ecol.* **82**, 147–159.

Peterson, J.H. (1984b). Establishment of mussel beds; attachment behavior and distribution of recently settled mussels (*Mytilus californianus*) *Veliger* **27**, 7–13.

Pickett, S.T.A. and White, P.S. (Eds.) (1985). *The ecology of natural disturbance and patch dynamics*. Academic Press. Orlando, USA.

Pineda, J. (1991). Predictable upwelling and the shoreward transport of planktonic larvae by internal tidal bores. *Science* **253**, 548–551.

Pitcher, C.R. (1988a). Spatial variation in the temporal pattern of recruitment of a coral reef damselfish. *Proc. 6th Int. Coral Reef Symp., Australia* **2**, 811–816.

Pitcher, C.R. (1988b). Validation of a technique for reconstructing daily patterns in the recruitment of a coral reef damselfish. *Coral Reefs* **7**, 105–111.

Possingham, H.P. and Roughgarden, J. (1990). Spatial population dynamics of a marine organism with a complex life cycle. *Ecology* **71**, 973–985.

Povey, A. and Keough, M.J. (1991). Effects of trampling on plant and animal populations on rocky shores. *Oikos* **61**, 355–368.

Pulliam, H. (1988). Sources, sinks and population regulation. *Am. Nat.* **132**, 652–661.

Queller, D.C., Strassmann, J.E. and Hughes, C.R. (1992). Microsatellite and kinship. *Trends Ecol. Evol.* **8**, 285–289.

Quinn, G.P. (1988). Effects of conspecific adults, macroalgae, and height on the shore on recruitment of an intertidal limpet. *Mar. Ecol. Prog. Ser.* **48**, 305–308.

Raimondi, P.T. (1988). Rock type affects settlement, and zonation of the barnacle *Chthamalus anisopoma* Pilsbury. *J. Exp. Mar. Biol. Ecol.* **123**, 253–267.

Raimondi, P.T. (1990). Patterns, mechanisms, consequences of variability in settle-ment and recruitment of an intertidal barnacle. *Ecol. Monogr.* **60**, 283–309.

Reed, D.C. (1990). The effects of variable settlement and early competition on patterns of kelp recruitment. *Ecology* 776–787.

Reed, D.C., Laur, D.R. and Ebling, A.W. (1988). Variation in algal dispersal and recruitment: the importance of episodic events. *Ecol. Monogr.* **58**, 321–335.

Richards, W. and Lindeman, K. (1987). Recruitment dynamics of reef fishes: planktonic processes, settlement and dispersal ecologies and fishery analysis. *Bull. Mar. Sci.* **41**, 397–410.

Ricker, W.E. (1954). Stock and recruitment. *J. Fish. Res. Bd. Can.* **11**, 559–623.

Ricketts, B.P.T., Calvin, J. and Hedgepeth, J.W. revised by Phillips, D.W. (1985). *Between Pacific Tides* (fifth edition) Stanford University Press, California, USA 652 pp.

Roberts, C. and Ormond, R. (1987). Habitat complexity and coral reef fish diversity and abundance on Red Sea fringing reefs. *Mar. Ecol. Prog. Ser.* **41**, 1–8.

Robertson, D.R. (1973). Field observations on the reproductive behavior of a pomacentrid fish, *Acanthochromis polyacanthus*. *Z. Tierpsychol.* **37**, 319–324.

Robertson, D.R. (1988). Abundances of surgeonfishes in Caribbean Panama: due to settlement or post-settlement processes? *Mar. Biol.* **97**, 495–501.

Robertson, D.R. (1992). Patterns of lunar settlement and early recruitment in Caribbean reef fishes at Panama. *Mar. Biol.* **114**, 527–537.

Robertson, D.R. and Lassig, B. (1980). Spatial distribution patterns and coexistence of a group of territorial damselfishes from the Great Barrier Reef. *Bull. Mar. Sci.* **30**, 187–203.

Robertson, D.R., Green, D. and Victor, B.C. (1988). Temporal coupling of the production and recruitment of larvae of a Caribbean reef fish. *Ecology* **69**, 370–381.

Root, R. (1967). The niche exploitation pattern of the blue-grey gnatcatcher. *Ecol. Monog.* **37**, 317–350.

Roughgarden, J. (1986). A comparison of food-limited and space-limited animal competition communities. In: *Community ecology* (Ed. by J. Diamond and T.J. Case) pp. 492–516. Harper and Row, New York.

Roughgarden, J., Gaines, S.D. and Iwasa, Y. (1984). Dynamics and evolution of marine populations with pelagic larval dispersal. In: *Exploitation of Marine Communities* (Ed. by R.M. May). Dahlem Konferenzen, Springer-Verlag, Berlin.

Roughgarden, J., Iwasa, Y. and Baxter, C. (1985). Demographic theory for an open marine population with space-limited recruitment. *Ecology* **66**, 54–67.

Roughgarden, J., Gaines, S.D. and Possingham, H. (1988). Recruitment dynamics in complex life cycles. *Science* **241**, 1460–1466.

Rumrill, S.S. (1987). Differential predation upon embryos and larvae of temperate Pacific echinoderms. Ph.D. thesis University of Alberta, Edmonton.

Rumrill, S.S. (1988). Temporal and spatial variability in the intensity of recruitment of a sea star: frequent recruitment and demise. *Am. Zool.* **28**, 123A.

Russ, G.R. (1991). Coral reef fisheries: effects and yields. In: *The ecology of fishes on coral reefs* (Ed. by P.F. Sale) pp. 601–636. Academic Press Inc., San Diego.

Ryland, J.S. (1959). Experiments on the selection of algal substrates by Polyzoan larvae. *J. exp. Biol.* **36**, 613–631.

Sale, P.F. (1974) Mechanisms of coexistence in a guild of territorial fishes at Heron Island, *Proc. 2nd Int. Coral reef symp*, **1**, 193–206.

Sale, P.F. (1975). Patterns of use of space in a guild of territorial reef fishes. *Mar. Biol.* **29**, 89–97.

Sale, P.F. (1978). Coexistence of coral reef fish—a lottery for living space. *Env. Biol. Fish.* **3**, 85–102.

Sale, P.F. (1980). The ecology of fishes on coral reefs. *Oceanogr. Mar. Biol. Ann. Rev.* **18**, 367–421.

Sale, P.F. (1985). Patterns of recruitment in coral reef fishes. *Proc. Int. Coral Reef Symp., 5th* **5**, 391–396.

Sale, P.F. (1990). Recruitment of marine species: is the bandwagon rolling in the right direction? *Trends Ecol. Evol.* **5**, 25–27.

Sale, P.F. (1991). Reef fish communities: open nonequilibrial systems. In: *The ecology of fishes on coral reefs* (Ed. by P.F. Sale) pp. 564–600. Academic Press Inc., San Diego.

Sale, P.F. and Douglas, W.A. (1984). Temporal variability in the community structure of fish on coral patch reefs and the relation of community structure to reef structure. *Ecology* **65**, 409–422.

Sale, P. and Steele, W. (1986). Random placement and the structure of reef fish communities. *Mar. ecol. Prog. Ser.* **28**, 165–274.

Sale, P.F., Doherty, P.J., Eckert, G., Douglas, W. and Ferrell, D. (1984). Large scale spatial and temporal variation in recruitment to fish populations on coral reefs. *Oecologia* **64**, 191–198.

Sammarco, P. and Andrews, J. (1989). The helix experiment: differential localized dispersal and recruitment patterns in Great Barrier Reef corals. *Limnol. Oceanogr.* **34**, 896–912.

Santelices, B. (1990). Patterns of reproduction, dispersal and recruitment in seaweeds. *Oceanogr. Mar. Bio. Ann. Rev.* **28**, 177–276.

Santelices, B. and Martinez, E. (1988). Effects of filter-feeders and grazers on algal settlement and growth in mussel beds. *J. Exp. Mar. Biol. Ecol.* **188**, 281–306.

Sebens, K.P. (1983). The larval and juvenile ecology of the temperate octocoral *Alcyonium siderium* Verrill. I. Substratum selection by benthic larvae. *J. Exp. Mar. Biol. Ecol.* **71**, 73–89.

Sebens, K.P. and Lewis, J.R. (1985). Rare events and population structure of the barnacle *Semibalanus cariosus* (Pallas, 1788). *J. Exp. Mar. Biol. Ecol.* **87**, 55–65.

Seed, R. and Suchanek, T.H. (1992). Population and community ecology of Mytilus. In: *The mussel Mytilus: ecology, physiology, genetics and culture.* (Ed. by E.M. Gosling) pp. 87–169. Elsevier Press, Amsterdam.

Sewell, M.A. and Watson, J.C. (1993). A "source" for asteroid larvae?: recruitment of *Pisaster ochraceus. Pycnopodia helianthoides* and *Dermasterias imbricata* in Nootka Sound British Columbia. *Mar. Biol.* **117**, 387–398.

Shaklee, J. (1984). Genetic variation and population structure in the damselfish, *Stegastes fasciolatus*, throughout the Hawaiian archipelago. *Copeia* 1984, 629–640.

Shanks, A.L. (1983). Surface slicks associated with tidally forced internal waves may transport pelagic larvae of benthic invertebrates shoreward. *Mar. Ecol. Prog. Ser.* **13**, 311–315.

Shanks, A.L. (1986). Tidal periodicity in the daily settlement of intertidal barnacle larvae and an hypothesized mechanism for the cross-shelf transport of cyprids. *Biol. Bull.* **170**, 429–440.

Shanks, A.L. and Wright, W.G. (1987). Internal-wave-mediated shoreward transport of cyprids, megalopae, and gammarids and correlated longshore differences in the settling rate of intertidal barnacles. *J. Exp. Mar. Biol. Ecol.* **114**, 1–13.

Shulman, M.J. (1984). Resource limitation and recruitment patterns in coral reef fish assemblages. *J. Exp. Mar. Biol. Ecol.* **74**, 85–109.

Shulman, M.J. (1985). Recruitment of coral reef fishes: Effects of distribution of predators and shelter. *Ecology* **66**, 1056–1066.

Shulman, M.J. and Ogden, J.C. (1987). What controls tropical reef fish populations: recruitment or benthic mortality? An example in the Caribbean reef fish *Haemulon flavolineatum. Mar. Ecol. Prog. Ser.* **39**, 233–242.

Shulman, M.J., Ogden, J.C., Ebersole, J.P., McFarland, W.N., Miller, S.L. and Wolf, N.F. (1983). Priority effects on the recruitment of coral reef fish. *Ecology* **64**, 1508–1513.

Smith, C. and Tyler, J. (1975). Succession and stability in fish communities of the dome-shaped patch reefs in the West Indies. *Am. Must. Nov.* **2572**, 1–18.

Sousa, W.P. (1979a). Experimental investigations of disturbance and ecological succession in a rocky intertidal algal community. *Ecol. Monogr.* **49**, 227–254.

Sousa, W.P. (1979b). Disturbance in marine intertidal boulder fields: the non-equilibrium maintenance of species diversity. *Ecology* **60**, 1225–1239.

Sousa, W.P. (1980). The responses of a community to disturbance: the importance of successional age and species' life histories. *Oecologia* **45**, 72–81.

Sousa, W.P. (1984a). The role of disturbance in natural communities. *Ann. Rev. Ecol. Syst.* **15**, 353–391.

Sousa, W.P. (1984b). Intertidal mosaics: patch size, propagule availability and spatially variable patterns of succession. *Ecology* **65**, 1918–1935.

Sousa, W.P. (1985). Disturbance and patch dynamics on rocky intertidal shores. In: *The ecology of natural disturbance and patch dynamics*. (Ed. by S.T.A. Pickett and P.S. White) pp. 101–124. Academic Press, New York, USA.

Southward, A.J. (1967). Recent changes in abundance of intertidal barnacles in south-west England: a possible effect of climatic deterioration. *J. Mar. Biol. Ass. UK* **47**, 81–95.

Southward, A.J. (1991). Forty years of changes in species composition and population density of barnacles on a rocky shore near Plymouth. *J. mar. biol. Ass. UK* **71**, 495–513.

Southward, A.J. and Crisp, D.J. (1954). Recent changes in the distribution of the intertidal barnacles *Chthamalus stellatus* (Poli) and *Balanus balanoides* L. in the British Isles. *J. Anim. Ecol.* **23**, 163–177.

Southward, A.J. and Crisp, D.J. (1956). Fluctuations in the distribution and abundance of intertidal barnacles. *J. Mar. Biol. Ass. UK* **35**, 211–229.

Sprung, M. (1984). Physiological energetics of mussel larvae I. Shell growth and biomass. *Mar. Ecol. Prog. Ser.* **17**, 295–305.

Stephens, E.G. and Bertness, M.D. (1991). Mussel facilitation of barnacle survival in a sheltered bay habitat. *J. Exp. Mar. Biol. Ecol.* **145**, 33–48.

Stephenson, T.A. and Stephenson, A. (1972). *Life between tidemarks on rocky shores*. W.H. Freeman, San Francisco, USA.

Stimson, J., Blum, S. and Brock, R. (1983). An experimental study of the influence of muraenid eels on reef fish sizes and abundance. *Univ. Hawaii Sea Grant Quarterly* **4**, 1–6.

Stoner, D.S. (1990). Recruitment of a tropical colonial ascidian: relative importance of pre-settlement vs post-settlement processes. *Ecology* **71**, 1682–1690.

Strathmann, M.F. (1987). *Reproduction and development of marine invertebrates of the northern Pacific coast*. University of Washington Press, Seattle. 670 pp.

Strathmann, R.R. and Branscombe, E.S. (1979). Adequacy of cues to favorable sites used by settling larvae of two intertidal barnacles. In: *Reproductive ecology of marine invertebrates* (Ed. by S.E. Stancyk) pp. 77–89. University of South Carolina Press, Columbia.

Strathmann, R.R., Branscombe, E.S. and Vedder, K. (1981). Fatal errors in set as a cost of dispersal and the influence of intertidal flora on set of barnacles. *Oecologia* **48**, 13–18.

Strong, D.J. Jr. (1992). Are trophic cascades all wet? Differentiation and donor control in speciose ecosystems. *Ecology* **73**(3), 747–754.

Suchanek, T.H. (1978). The ecology of *Mytilus edulis* L. in exposed rocky intertidal communities. *J. Exp. Mar. Biol. Ecol.* **31**, 105–120.

Suchanek, T.H. (1979). The *Mytilus californianus* community: studies on the composition, structure, organization, and dynamics of a mussel bed. Ph.D. dissertation, Department of Zoology, University of Washington, Seattle.

Suchanek, T.H. (1981). The role of disturbance in the evolution of life history strategies in the intertidal mussels *Mytilus edulis* and *Mytilus californianus*. *Oecologia* **50**, 143–152.

Suchanek, T.H. (1985). Mussels and their role in structuring rocky shore communities. In: *The ecology of rocky coasts* (Ed. by P.G. Moore and R. Seed). pp. 70–96. Hodder and Stoughton, Sevenoaks, UK.

Sundene, O. (1962). The implications of transplant and culture experiments on the growth and distribution of *Alaria esculenta*. *Nytt. Mag. Bot.* **9**, 155–174.

Sutherland, J.P. (1974). Multiple stable points in natural communities. *Am. Nat.* **18**, 859–873.

Sutherland, J.P. (1987). Recruitment limitation in a tropical intertidal barnacle: *Tetraclita panamensis* (Pilsbry) on the Pacific coast of Costa Rica. *J. Exp. Mar. Biol. Ecol.* **113**, 267–282.
Sutherland, J.P. (1990). Recruitment regulates demographic variation in a tropical intertidal barnacle. *Ecology* **71**, 955–972.
Sutherland, J.P. and Karlson, R.H. (1977). Development and stability of the fouling community at Beaufort, North Carolina. *Ecol. Monogr.* **47**, 425–446.
Sutherland, J.P. and Ortega, S. (1986). Competition conditional on recruitment and temporary escape from predators on a tropical rocky shore. *J. Exp. Mar. Biol. Ecol.* **95**, 155–166.
Suthers, I.M. (1989). Dispersal and growth of post-larval cod (*Gadus morhua*) off southwestern Nova Scotia. Ph.D. thesis, Dalhousie University. 173 pp.
Sweatman, H.P.A. (1983). Influence of conspecifics on choice of settlement sites by larvae of two pomacentrid fishes (*Dascyllus aruanus* and *D. reticulatus*) on coral reefs. *Mar. Biol.* **75**, 225–230.
Sweatman, H.P.A. (1985a). The influence of adults of some coral reef fishes on larval recruitment. *Ecol. Monog.* **55**, 469–485.
Sweatman, H.P.A. (1985b). The timing of settlement of larval *Dascyllus aruanus*: some consequences for larval selection. *Proc. Int. Coral Reef Congr., 5th* **5**, 367–371.
Talbot, F.H., Russell, B.C. and Anderson, G.R.V. (1978). Coral reef fish communities: unstable, high-diversity systems? *Ecol. Monogr.* **48**, 425–440.
Tegner, M.J. and Dayton, P.K. (1981). Population structure, recruitment and mortality of two sea urchins *Strongylocentrotus franciscanus* and *S. purpuratus* in a kelp forest. *Mar. Ecol. Prog. Ser.* **5**, 255–268.
Thorrold, S.D. and Milicich, M.J. (1990). Comparison of larval duration and pre- and post-settlement growth in two species of damselfish, *Chromis atripectoralis* and *Pomacentrus coelestis* from the Great Barrier Reef. *Mar. Biol.* **105**, 375–385.
Thorson, G. (1946). Reproduction and larval development of Danish marine invertebrates. *Medd. Danske Vidensk Selsk. Skrifter, Naturv. og Math. Afd.* **7**, 1–59.
Thorson, G. (1950). Reproductive and larval ecology of marine bottom invertebrates. *Biol. Rev.* **25**, 1–45.
Thorson, G. (1966). Some factors influencing the recruitment and establishment of marine benthic communities. *Neth. J. Sea Res.* **3**, 267–293.
Thresher, R. (1985). Distribution, abundance and reproductive success in the coral reef fish *Acanthochromis polyacanthus*. *Ecology* **66**, 1139–1150.
Thresher, R. (1991). Geographic variability in the ecology of coral reef fishes: Evidence, evolution and possible implications. In: *The ecology of fishes on coral reefs* (Ed. by P.F. Sale) pp. 401–436. Academic Press Inc., San Diego.
Thresher, R., Harris, G., Gunn, J. and Clementson, L. (1989). Phytoplankton production pulses and episodic settlement of a temperate marine fish. *Nature. London.* **341**, 641–643.
Todd, C.D., Havenhand, J.N. and Thorpe, J.P. (1988). Genetic differentiation, pelagic larval transport and gene flow between local populations of the intertidal marine mollusc, *Adalaria proxima* (Alder and Hancock). *Funct. Ecol.* **2**, 441–451.
Tonn, W., Paszkowski, C. and Holopainen, I. (1992). Piscivory and recruitment: mechanisms structuring prey populations in small lakes. *Ecology* **73**, 951–958.
Underwood, A.J. (1980). The effects of grazing by gastropods and physical factors on the upper limits of distribution of intertidal macroalgae. *Oecologia* **46**, 201–213.

384 D.J. BOOTH AND D.M. BROSNAN

Underwood, A.J. (1981). Structure of a rocky intertidal community in New South Wales: patterns of vertical distribution and seasonal changes. *J. Exp. Mar. Biol. Ecol.* **51**, 57–85.

Underwood, A.J. (1984). Vertical and seasonal patterns in competition for microalgae between intertidal gastropods. *Oecologia* **64**, 211–222.

Underwood, A.J. and Denley, E.J. (1984). Paradigms, explanations, and generalizations in models for the structure of intertidal communities on rocky shores. In: *Ecological communities: conceptual issues and the evidence.* (Ed. by D.R. Strong, Jr., D. Simberloff, L.G. Abele and A.B. Thistle) pp. 151–180. Princeton University Press, Princeton New Jersey.

Underwood, A.J. and Fairweather, P.G. (1989). Supply-side ecology and benthic marine assemblages. *Trends Ecol. Evol.* **4**, 16–19.

Underwood, A.J. and Jernakoff, P. (1981). Effects of interactions between algae and grazing gastropods on the structure of a low-shore intertidal algal community. *Oecologia* **48**, 221–233.

Underwood, A.J. and Jernakoff, P. (1984). The effect of tidal height, wave exposure, seasonality and rockpools on grazing and distribution of intertidal algae in New South Wales. *J. Exp. Mar. Biol. Ecol.* **75**, 71–96.

Underwood, A.J., Denley, E.J. and Moran, M.J. (1983). Experimental analyses of the structure and dynamics of mid-shore rocky intertidal communities in New South Wales. *Oecologia* **56**, 202–219.

Victor, B.C. (1982). Daily otolith increments and recruitment in two coral reef wrasses, *Thalassoma bifasciatum* and *Halichoeres bivittatus. Mar. Biol.* **71**, 203–208.

Victor, B.C. (1983). Recruitment and population dynamics of a coral reef fish. *Science* **219**, 419–420.

Victor, B.C. (1984). Coral reef fish larvae: patch size estimation and mixing in the plankton. *Limnol. and Oceanog.* **29**, 1116–1119.

Victor, B.C. (1986). Larval settlement and juvenile mortality in a recruitment-limited coral reef fish population. *Ecol. Monogr.* **56**, 145–160.

Victor, B.C. (1991). Settlement strategies and biogeography of reef fishes. In: *The ecology of fishes on coral reefs* (Ed. by P.F. Sale) pp. 231–293. Academic Press Inc., San Diego.

Walsh, W.J. (1985). Reef fish community dynamics on small artificial reefs: the influence of isolation, habitat structure, and biogeography. *Bull. Mar. Sci.* **36**, 357–376.

Walsh, W. (1987). Patterns of recruitment and spawning in Hawaiian reef fishes. *Env. Biol. Fish.* **18**, 257–276.

Warner, R. and Chesson, P. (1985). Coexistence mediated by recruitment fluctuations: a field guide to the storage effect. *Am. Nat.* **125**, 769–787.

Warner, R. and Hughes, T. (1988). The population dynamics of reef fishes. *Proc. 6th Int. coral Reef Symp. (Australia)* **1**, 149–155.

Watanabe, J.M. (1984). The influence of recruitment, competition, and benthic predation on spatial distributions of three species of kelp forest gastropods (Trochidae: *Tegula*). *Ecology* **65**, 920–936.

Watanabe, J.M. and Harrold, C. (1991). Destructive grazing by sea urchins *Strongylocentrotus* spp. in a central Californian kelp forest: potential roles of recruitment, depth, and predation. *Mar. Ecol. Prog. Ser.* **71**, 125–141.

Wellington, G.B. and Victor, B.C. (1985). El Nino mass coral mortality: a test of resource limitation in a coral reef damselfish population. *Oecologia* **68**, 15–19.

Wellington, G.B. and Victor, B.C. (1989). Planktonic larval duration of 100 species of Pacific and Atlantic damselfishes (Pomacentridae). *Mar. Biol.* **101**, 557–567.

Wethey, D.S. (1984). Spatial pattern in barnacle settlement: day to day changes during the settlement season. *J. mar. biol. Ass. UK* **64**, 687–696.

Wethey, D.S. (1986). Local and regional variation in settlement and survival in the littoral barnacle *Semibalanus balanoides* (L.): patterns and consequences. In: *The ecology of rocky coasts.* (Ed. by P.G. Moore and R. Seed) pp. 194–202. Hodder and Stoughton, Sevenoaks, UK.

Williams, D.M. (1982). Patterns in the distribution of fish communities across the central Great Barrier Reef. *Coral Reefs* **1**, 35–43.

Williams, D.M. (1983). Daily, monthly and yearly variation in recruitment of a guild of coral reef fishes. *Mar. Ecol. Prog. Ser.* **10**, 231–237.

Williams, D.M. (1991). Patterns and processes in the distribution of coral reef fishes. In: *The ecology of fishes on coral reefs* (Ed. by P.F. Sale) pp. 437–474. Academic Press Inc., San Diego.

Williams, D.M. and Hatcher, A. (1983). Structure of fish communities on outer slopes of inshore, mid-shelf and outer shelf reefs of the Great Barrier Reef. *Mar. Ecol. Prog. Ser.* **10**, 239–250.

Williams, D.M. and Sale, P. (1981). Spatial and temporal patterns of recruitment of juvenile coral reef fishes to coral habitats within "One Tree Lagoon", Great Barrier Reef. *Mar. Biol.* **65**, 245–253.

Williams, D.M., Wolanski, E. and Andrews, J. (1984). Transport mechanisms and the potential movement of planktonic larvae in the central region of the Great Barrier Reef. *Coral Reefs* **3**, 229–236.

Wilson, M. and Botkin, B. (1990). Models of simple microcosms: emergent properties and the effects of complexity on stability. *Am. Nat.* **135**, 414–434.

Witman, J.D. and Suchanek, T.H. (1984). Mussels in flow: drag and dislodgement by epizoans. *Mar. Ecol. Prog. Ser.* **16**, 259–268.

Yoshioka, P.M. (1982). Role of planktonic and benthic factors in the population dynamics of the bryozoan. *Membranipora membranacea. Ecology* **63**, 457–468.

Young, C.M. and Chia, F.S. (1981). Laboratory evidence for delay of larval settlement in response to a dominant competitor. *Intl. Jour. Invert. Repro.* **3**, 221–226.

Young, C.M. and Chia, F.S. (1984). Microhabitat-associated variability in survival and growth of subtidal solitary ascidians during the first twenty-one days after settlement. *Mar. Biol.* **81**, 61–68.

Index

Note: Figures and Tables are indicated in index by *italic page numbers*

Acanthaster planci 354
Acanthina spp *352*
  A. brevidentata 351
Acanthochromis polyacanthus 316, 320, 355
Adalaria proxima 325
aerodynamic method [of measuring CO_2 flux] 19–20
African Great Lakes
  cichlid species
    evolutionary responses to ecological pressures 208–11
    feeding habits 223
    intralacustrine speciation 200
  faunal differences 210
  habitat similarities 208–10
  predation pressure differences 210–11
  see also Malawi; Tanganyika; Victoria
agricultural disturbance
  effects on soil biota 109–36
  and food web theory 150–6
  see also tillage-induced disturbance
aircraft-based measurements, CO_2 flux 40, 43
Alaria escuelanta 323
albedos, plant canopy 16
algae, marine, dispersal 320, 323
algal eating Tanganyikan cichlid fishes 227
allopatric evolution, Tanganyikan cichlid fishes *194–5*, 200–5
alternative agricultural systems 147–8
Altolamprologus spp *191*, 200, 203
  A. compressiceps 198, 212, *225*, 226, *232*, 238
Anas platyrhynchos 283
Anthias squamipinnis 319
arbuscular mycorrhizal fungi
  consumption by soil fauna 119–20, 122
  effect of herbicides 141–2

effect of soil tillage 115
  symbiosis involving *72*, 81–2, 87, 88, 114
Archilochus alexandri 266
Asprotilapia spp *191*, 203
  A. leptura 207
Astatoreochromis spp *191*
Astatotilapia spp *191*
atoll lagoons 329
Aulonocranus spp *191*, 203
  A. dewindti 198, *207*, 216–17, *225*, 237
Azorhizobium spp *72*, 77

bacterial-feeding nematodes
  effect of mulching *145*
  effect of soil tillage *112*, 115–16, 117, *131*, 132
bacteroids 92
  see also nitrogen-fixing bacteria
Baileychromis spp *191*
Balanus eburneus 333
Bathybates spp *191*, 196, 203, 206
  B. fasciatus 195
  B. ferox 195
Bayesian models, patch assessment models 281–2
Beer–Lambert law 21
benthivores, Tanganyikan cichlid species 227
Benthochromis spp *191*, 224
biodiversity *see* species diversity
biodynamic farming 147–8
birds, predation of soil macrofauna by 124, 158
bottom-up process models 5
Boulengerochromis spp *192*, 210, 212, 213, 214
  B. microlepis 196, *199*, 235
boundary value analysis, PPFD–CO_2 flux 24
Bowen ratio 19

387

Brachiodontes semilaevis 352
Brachydanio rerio 293
Bradyrhizobium spp *72*, 77, 90
breeding habits, Tanganyikan cichlid
fishes 211–19
broadleaf forest, canopy
photosynthesis compared with
needleleaf forest 14–15
buccal incubation 215
see also mouthbrooding cichlid fishes
Buchnera spp, transmission modes 78,
80
Bufo bufo 293

Callochromis spp *191*, 203
C. macrops 197, 201, *207*, *225*
C. pleurospilus 200
canopy flux, carbon dioxide 8
canopy net photosynthetic rate 8
factors affecting 14
carabid beetles
effect of herbicides 142
effect of soil tillage *112*, 126, 127,
128, *131*, *134*
Carasius auratus 283–4
carbon budget
equivalence with CO$_2$ flux
measurements *9*
of stand 6–9
symbols used 3
carbon dioxide flux
and carbon budget *7*
effect of CO$_2$ enrichment 42–3
response to PPFD
micrometeorological and enclosure
techniques compared 27–9, 43
nutrient-availability effects 36, 44
respiration effects 29–31
seasonality effects 35–6, 44
statistical analysis 37–40
temperature effects 37
vegetation-class effects 32–4, 44
water-availability effects 36–7, 44
carbon dioxide flux measurements
applications
bottom-up process models 5
canopy and stand physiology 5
net carbon balance of ecosystems
4–5
top-down parametric models 5

data sets
background 21–2
heterogeneity of data 44–5
sources and description listed 52–
62
statistical analysis
grouped data sets 23, 26
instantaneous data sets 22–3
vegetation classification 22
early studies 4
equivalence with terms of carbon
budget *9*
measurement techniques
aerodynamic method 19–20
Bowen ratio method 18–19
eddy correlation/covariance
method 20
enclosures method 4, 20
energy balance method 18–19
models used 21
velocity profile method 19–20
scaling up
in space 40–1
in time 41–2
symbols used 3
carbon dioxide storage flux 9–12
Cardiopharynx spp *191*
C. schoutedeni 207
Caribbean coral reef fishes *335*, *336*, 339
Chaetodon miliaris 345
Chalinochromis spp *191*, 203
C. brichardi 198
chaperonin 94
chemoautotrophic bacteria
evolutionary origins 76
symbiosis involving *72*, 76
Chlorella spp., symbiosis involving *72*,
74, 90
Chromis spp 330
Chthamalus spp 342, *352*
C. anisopoma 334, 341
C. fissus 325, 326, 331–2, *333*, 340,
341, 351
Cichlasoma nigrofasciatum 261
cichlid fishes
effect of breeding mode on
speciation 219–20
in Lake Tanganyika 190–2
breeding habits 211–20
coexistence within rocky shore
communities 220–34

community composition and
    species diversity 234–40
evolution 201–11
feeding habits 223–9
habitat 194–200
mouthbrooders
    colour morphs 203
    diversity within group 215
    intermediate behaviour of
        Perissodini 215–16
    lekking behaviour 216–18
    midwater spawning 216
    parental care 211, 218–19
    shell brooders 212–13
substrate spawners
    colour morphs 203–4
    diversity within group 212–14
    helpers 213–15
    shell brooders 212–13
    spawning synchroneity 209, 213
clupeid fishes 189, 196, 211
Cobisoma bosci 342
Collembola
    effect of soil tillage 112, 120, 128,
        132
    see also springtails
Collisella spp 352
colour morphs, Tanganyikan cichlid
    fishes 203–5
Colpula lativentris 291
Columba livia 291
commensalism, cichlid fishes 218, 227
community structure
    definition 337
    integrated approaches 349–51
    recruitment effects 342–53
    future study suggested 363–4
    Tanganyikan cichlid fishes 237–40
competition
    endosymbiotic micro-organisms 84–7
    and recruitment 342–5
competitive distributions 255
    difficulties in studying 296–7
    see also ideal free distribution
        models
connectedness food webs 150, 152–4
    effect of tillage 154
Connell's intermediate disturbance
    hypothesis 136, 313, 342
continuous input IFD models 257, 286
    resource wastage incorporated 260

and standing crops 273–4
unequal-competitor model 262–3
convective boundary layer (CBL) of
    plant canopy, CO$_2$
    concentration 10
Convoluta roscoffensis, symbiosis
    involving 85, 86
coral reefs
    bleaching due to expulsion of algae
        71, 91
    compared with African Great Lakes
        236
    compared with rocky shores 311
    effect of human activities 71
    fish recruitment
        effect on population structure 338
        historical background 312–13
        seasonality effects 330
    symbiosis in 70
corn (maize) crop, CO$_2$ flux
    measurements 35, 61–2
Crassostrea virginica 333
crops, CO$_2$ flux measurements 32–3,
    34, 44, 59–62, 66–8
Ctenochromis spp 191
    C. horei 198
Culex quinquefasciatus 274
Cunningtonia spp 191, 201
    C. longiventralis 198, 207
cyanobacteria 72
Cyathopharynx spp 191, 201, 203
    C. furcifer 198, 207, 208, 209, 217,
        225, 239–40
Cyphotilapia spp 192, 201, 203
    C. frontosa 202, 215
Cyprichromis spp 191, 201, 203, 216,
    218, 222, 224, 226
    C. leptosoma 198, 204

Daphnia pulex 290
dark respiration, in CO$_2$ flux 13, 30, 44
Dascyllus albisella 335, 346
decomposer-based food web see
    detritus food webs
despotism, in population distribution
    theory 260–2
detritus food webs
    response to tillage-based disturbance
        127–36
    three-dimensional structure 154–5

detritus food webs—*contd*
  weeds affecting 148–50
  within-habitat compartments in 155–6, *157*
dichromatism, piscivorous cichlid fishes 228
Didenium candidum 355
diel partitioning, Tanganyikan cichlid fishes 221–2
Diplosoma listerianum 355
Doherty's recruitment limitation model 338, 344, *361*
Douglas fir *see Pseudotsuga menziesii*
dynamic programming 275

earthworms
  effect of herbicides *138*, *139*, 140
  effect of soil tillage *112*, 124, *128*, *131*, 133, *134*
  soil fauna/flora affected by 125
earthworms–microfauna/microflora interactions, effect of soil tillage 124–5
ecofarming 147
ectodine cichlid fishes *191*
  spatial partitioning of mating sites 230
  see also Asprotilapia; *Aulonocranus*; *Callochromis*; *Cardiopharynx*; *Cunningtonia*; *Cyathopharynx*; *Ectodus*; *Enantiopus*; *Lestradea*; *Microdontochromis*; *Opthalmotilapia*; *Xenotilapia*
Ectodus spp *191*
  E. descampsi 207
ectomycorrhizal fungi *72*
  effect of pollution 71
  host range 81, *82*, 83, 84
eddy correlation/covariance method [of measuring CO_2 flux] 20, 40, 43
Enantiopus spp *191*
  E. melanogenys 198, *207*, 219
enchytraeids, effect of soil tillage *112*, 123, 130, *131*, 132
enclosure methods [for CO_2 flux measurement] 4, 20
  compared with micrometeorological methods 27, *28*

endophytic fungi
  evolutionary origins 77
  symbiosis involving *72*, 73
endosymbiotic micro-organisms
  characteristics 71, 73
  competition among endosymbionts 84–7
  in environment 88–93
  examples *72*
  habitat in host 93–4
  host range 79–84
  host-related determinants of distribution 78–87
  modes of transmission 78–9
  non-symbiotic factors affecting distribution 87–8
  phyletic diversity 74–7
  rarity in environment 88–90
Enteromorphha spp 323
environmental stress model 349–50, *351*, *361*
Eretmodus spp *191*, 200, 203
  E. cyanostictus 198
ericoid mycorrhizal fungi *72*
  host range 84
eucalyptus forest, CO_2 flux measurement 27–8, 54, 55
Euglenes fulgens 266
evolution, Tanganyikan cichlid fishes 201–11

feeding habits, Tanganyikan cichlid fishes 223–9
filamentous brown algae, recruitment variability 324, *334*
fisheries management 311–12, 360
fishes
  parental care by 211, 215, 218–19
  see also cichlid fishes
food chain length, in soil food webs 150–1
food webs
  decomposer/detritus-based system 107–9
    response to tillage-based disturbance 127–36
  effect of disturbance 106–7, 150–6
  see also connectedness food webs
forests
  CO_2 flux measurements 32–3, 34, 44, 54–5, 64–5

effect of pollution 71
Frankia spp, symbiosis involving 72, 76
Fulmarus glacialis 269
fungal-feeding nematodes
  effect of mulching *145*
  effect of soil tillage *112*, 115–16, *131*
fungi
  symbiosis involving 70, 72
  see also soil fungi
Fusarium spp, protection by symbiosis
  87

Gadus morhua 293
Garrulus glandarius 261, *290*
Gasterosteus aculeatus 283
genetic survey, rocky shore
  invertebrates 325–6
Gigaspora gigantea 88
global changes, in carbon dioxide,
  effects on quantum yield 43
glyphosphate [herbicide], soil fungi
  affected by 137
Gnathochromis spp *191*
  G. permaxillaris 195
  G. pfefferi 199, 227
Grammatotria lemairei 197, 200, *207*
grasses, symbiosis involving 72, 73, 87
grasslands, CO$_2$ flux measurements 32–
  3, 44, 56–8, 65–6
Great Barrier Reef, fish recruitment
  320, *335*, *336*, 338, 339, 343
green mulches 146–7
Greenwoodochromis spp *191*
gross photosynthetic rate 8, 13
  instantaneous vs integrated
  measurements 41–2
gross primary production (GPP), of
  carbon 8
group foraging 283–4
gyres, marine organism recruitment
  affected by 320, *321*

habitat, Lake Tanganyika, and faunal
  diversity 194–200
habitat selection models 255–7, 266
  see also ideal free distribution (IFD)
  models
Haematopus spp
  H. cachmani 355

H. ostralegus 258, *290*, *292*
Hairston–Smith–Slobodkin (HSS)
  theory 106, 119
Haplotaxodon spp *192*
H. microlepis 196, 215, 218
Hemibates spp *191*, 196
herbicides
  response of detritus food web
  organisms 137–43
  field studies compared with
  laboratory studies 137, *138*,
  139–40, *139*
Holmgren's kleptoparasitic IFD model
  270–3, *288*
homoserine 93
hummingbirds, habitat selection, with
  interspecific competition 266–
  8
Huxley's density-limiting hypothesis
  261
Hydrocynus spp 210

ideal despotic distribution (IDD) 260–2
ideal free distribution (IFD) models
  assessing success 289–300
  assumptions made *285*
  continuous input models 257, *286*
  resource wastage incorporated 260
  and standing crops 273–4
  unequal-competitor model 262–3
  despotism incorporated 260–2
  and form of competition 268–9, *287*
  interference models 257–9, *286*
  unequal-competitor model 263–5
  Kennedy/Gray review 300
  Milinski's critique 300
  kleptoparasitism incorporated 269–
  73, *287*
  patch assessment incorporated 278–
  84
  perceptual constraints incorporated
  277–8, *287*
  reasons for limited success 297–8
  resource dynamics incorporated 273–
  5, *286*
  simplest models 256–7
  state-dependent models 275–7, *287*
  summary 284–8
  survival incorporated as measure of
  fitness 275–7

IFD models—*contd*
  unequal-competitor models 262–9
    competitor distribution
      determined by form of
      competition 268–9, *287*
    continuous input model 262–3
    distributions resulting from models
      264–5
    interference model 263–4
    isoleg theory applied 265–8
    truncated distribution models 264–
      5
  ideal free distribution theory 255–6
IFD models *see* ideal free distribution
    models
integrated farming practices 147
interference IFD models 257–9, *286*
  unequal-competitor model 263–5
intermediate disturbance hypothesis
    136, 313, 342
intertidal communities
  metapopulation model 329
  population structure influenced by
      recruitment 339–40
  recruitment studies, historical
      background 311, 313
  recruitment variability 331, *333–4*
  survivorship curves *317*
  see also rocky shores
isoleg theory 265–8

Jaccard's similarity coefficient,
    Tanganyikan cichlid species
    200
Julidochromis spp *191*, 200, 203
  J. "dark" *197*
  J. marlieri 213
  J. ornatus *197*, 213
  J. regani *197*, 213

kelps
  dispersal patterns 323–4
  recruitment variability *334*
keystone recruitment limitation model
    356, *361*
keystone species, community structure
    affected by 356–7
kleptoparasitism 269
  incorporation into ideal free

distribution models 269–73
Korona's competition-form model 268–
    9

Labroides dimidiatus 336
lake-level changes
  effect on evolution of cichlid fishes
      201, 205
  Lake Tanganyika 201, 202
Lampornis clemenciae 266
lamprologine cichlid fishes *191*, *198–9*
  breeding habits 212, *212*, 213, 214,
      230–1
  piscivores 201, 226, 228
  plankter feeders 224, 226
  spatial partitioning of breeding sites
      230–1
  zoobenthos feeders 226
  see also Altolamprologus;
        Chalinochromis;
        Julidochromis; *Lamprologus*;
        Lepidiolamprologus;
        Neolamprologus;
        Telmatochromis
Lamprologus spp *191*, 200, 203
  L. brevis 238–9
  L. callipterus *197*, 213, 214, 218–19,
      225, 226, 227, *232*, 239
  L. kungweensis 232
  L. lemairii *197*, 213, 222
  L. modestus 226
  L. ocellatus *197*, *232*
  L. ornatipinnis *197*, *232*
  L. signatus 232
Larus argentatus *291*, *292*
larval stages *317*
  coral reef fishes 312
  delayed metamorphosis 327
  intertidal species 313, 324–5
Lates spp 195, 210
leaf area index (LAI) 16
  photon flux density affected by 35
  respiration rates affected by 35
legume nodules, nitrogen-fixing
      bacteria in 75, 77, 81, 92, 93
lekking behaviour, Tanganyikan cichlid
      fishes 216–18, 239–40
Lepidiolamprologus spp *191*, 203
  L. attenuatus *197*, *232*, 238
  L. cunningtoni 197

L. elongatus 197, 212, 218, 226, 238
L. nkambae 197
L. profundicola 197, 226, 228, 229
Lepomis macrochirus 274
Lestradea spp 191
  L. perspicax 201
  L. stappersi 201
lichens
  effect of pollution 71
  phyletic diversity in 74
  symbiosis in 70, 72, 73
Limnochromis spp 191, 196
Limnothrissa spp 189, 196, 209, 223
Limnotilapia spp 192
  L. dardennii 199
litter burial [by tillage], soil organisms
    affected by 106, 132, 156
littoral cichlids
  in Lake Tanganyika 197–201
    breeding habits 211–20
  see also Tanganyika, Lake, cichlid
    fishes
live mulches 146–7
Lobochilotes spp 192, 203
  L. labiatus 199, 202, 227
Lollium perenne 56
lottery hypothesis [for spatial
    competition] 343, 361
  assumptions 343, 344
  effect of spatial scale 357, 358
luminescent bacteria
  symbiosis involving 72, 88, 89, 91–2
  see also Photobacterium; Vibrio
lunar spawning species, ciclid fishes 213

macroallopatric evolution,
    Tanganyikan cichlid fishes
    194–5, 201–2
Macrocystis pyrifera 323
Malawi, Lake 188
  cichlid species 190, 208, 211, 215
marine organisms
  openness of systems considered 316–
    29
  recruitment
    definitions 315–16, 318
    factors affecting 321–2
    settlement 315
Menge–Sutherland environmental
    stress model 349–50, 351, 361

metapopulation models [for marine
    organisms] 327–9
  in coral reefs 327
  diffuse-source model 328, 329
  example in marine intertidal
    community 329
  source–sink model 328, 328
methanogenic bacteria, symbiosis
    involving 72
microallopatric evolution, Tanganyikan
    cichlid fishes 194–5, 202–5
microarthropods–microfauna–
    microflora interactions, effect
    of soil tillage 121–3
Microdontochromis spp 191, 224
  M. tenuidentatus 215, 218
microfauna–microflora interactions,
    effect of soil tillage 117–20
micrometeorological methods [for CO_2
    flux measurement] 4, 18–20
  compared with other methods 27, 28,
    40, 43
  convention used for flux 6–7
Micropterus salmoides 274
Microtus agrestis 291
midwater spawning, cichlid fishes 216
mites
  effect of mulching 145
  effect of soil tillage 112, 120–1, 128,
    130, 131, 134
Monteith model 5, 6
Monterey Bay, barnacle recruitment
    studies 325
Morula marginalba 345
mouthbrooding cichlid fishes 211, 215–18
mulches 143–7
  dead 144–6
  live 146–7
mussel beds, as settlement substrates
    34, 346
mycetocyte symbiotic micro-organisms
    72, 90
mycorrhizal fungi 72
  effect of pollution 71
  factors affecting distribution 88
  host range 81–2, 82, 83, 84
  productivity 70
  see also arbuscular . . .;
      ectomycorrhizal . . .;
      ericoid . . .; orchid mycorrhizal
      fungi

Mytilus spp
  M. californianus 325, 363
  M. edulis 317, 324, 326

nematodes
  effect of herbicides *138*, *139*, 140
  effect of mulching *145*, 146
  effect of soil tillage *112*, 115–17, *128*, *131*, 132, *134*
Neolamprologus spp *191*, 201, 203
  N. brevis *198*, 213, 218, *225*, *232*
  N. brichardi *198*, 201, 214, 215, 222, 226
  N. buescheri *197*
  N. caudopunctatus *197*, 218, 224, *232*
  N. cylindricus *197*, *225*
  N. fasciatus *197*, 218, *225*, 226, *232*
  N. furcifer *198*, *212*, *225*, 226
  N. leleupi 226
  N. meeli *198*, *225*, *232*
  N. modestus *198*
  N. mondabu *198*, *225*
  N. moorii 194, *198*, 202, *212*
  N. multifasciatus 213, *225*, *232*, 235
  N. mustax *198*
  N. niger *198*
  N. ocellatus *225*
  N. ornatipinnis *225*
  N. prochilus *199*, *225*
  N. pulcher *199*, 214, *225*
  N. savoryi *199*, 214, *225*, 226
  N. sexfasciatus *199*, 201–2, *225*
  N. tetracanthus *199*
  N. toae *199*, 202, *212*, 222, 226
  N. tretocephalus *199*, 201, *212*, 226
  N. wauthoni *232*
Neothauma spp 213
net ecosystem exchange (NEE) 9
net ecosystem production (NEP), of carbon 8
net primary productivity (NPP) 9
  estimation using remote sensing 17
night-time CO_2 respiration rate 10
nitrogen-fixing bacteria *72*
  abundance in environment due to symbiosis 92–3
  evolutionary origins 76–7
  host range 81
  phyletic diversity 74
  productivity 70
    see also Azorhizobium;
        Bradyrhizobium; *Frankia*;
        rhizobia
non-trophic interactions, and recruitment 346–7
no-tillage systems
  alternative agricultural systems 147–8
  compared with conventional tillage systems 109–36
  herbicides 137–43
  mulches 143–7
  see also tillage . . .
Notonecta hoffmanni 274
Nucella spp *349*, 353
nutrient availability, CO_2 flux affected by 36, 44

omnivorous nematodes, effect of soil tillage *112*, 116, *131*
omnivory, in detritus food web 151–2
Ondatra zibethicus *291*
open populations
  coral reef fishes 319–20
  marine species 316–29
  rocky shore marine organisms 323–6
Opthalmotilapia spp *191*, 201, 203, 206
  O. nasutus 207
  O. ventralis *197*, *207*, *225*
orchid mycorrhizal fungi *72*, 73, 89
Oreochromis spp *192*
  O. tanganicae 199
"organic" agricultural practices 147
Orthochromis spp *191*
overfishing
  predation affected by *209*, 240, 362
  prediction of consequences 311–12

Paracyprichromis spp *191*, 216, 222, 226
  P. brienni 216
Pararge aegaria 260–1
parasitism, cichlid fishes 218–19
parental care patterns, cichlid fishes 211, 215, 218–19
Parus major 256
patch assessment
  experimental investigation 282–4
  models 278–82
    Bayesian models 281–2

patchy environment, species diversity of cichlid fishes affected by 221
Patella vulgata 341
pelagic cichlids 195
Pemphigus betae 296
perceptual constraints 277
    incorporation into IFD models 277–8
perissodine scale-eating fishes
    breeding habits 215–16
    feeding/hunting behaviour 228, 229
Perissodus spp *192*, 206
    P. microlepsis 198, 216, 218, 228, 229
Peromyscus leucopus 291
Petrochromis spp *192*, 200, 203, 227
    P. famula 198
    P. fasciolatus 198
    P. macrognathus 198
    P. orthognathus 198
    P. polyodon 199, 227
    P. trewavasae 198
Phallusia nigra 355
Photobacterium spp, symbiosis involving *72*, 88, 89
photosynthetic algae, symbiosis involving *72*, 73
photosynthetically active radiation (PAR), ratio to biomass production 5–6
photosynthetic capacity 23, 44
    effect of vegetation class 32, 33, 44
photosynthetic efficiency 17, 18
photosynthetic photon flux density (PPFD) 5
    photosynthetic rates affected by 14–15
relationships
    with canopy photosynthetic rate 21
    with CO$_2$ flux 24, *25–6*
    factors affecting 27–37
Pianka's overlap index *225*
Picea spp
    P. abies 261
    P. sitchensis 14, *15*, *55*
Pinna spp 343
Pisaster ochraceus 326, 340
piscivorous cichlid fishes 226, 228
plankter-feeding fishes 224, 226
plant exudates, utilization by rhizobia 92–3
Plecodus spp *192*

P. paradoxus 198
P. straeleni 198, 228
ploughing
    effect on soil organisms 132–3
    see also tillage-induced disturbance
Pluvialis apricaria 292
Poecilia reticulata 292
polymorphism
    coral reef fishes 320
    Tanganyikan cichlid fishes 203–5
Pomacentrus spp.
    P. amboinensis 338
    P. moluccensis 339
    P. nagasakiensis 339
    P. wardi 335, 336
population, definition 337
population distributions 255–6
    see also ideal free distribution
population openness
    coral reef fishes 319–20
    rocky shore marine organisms 323–6
    species-specific differences 354–6
population structure
    recruitment effects
        coral reef fishes 338–9
        marine intertidal species 339–40
Postelsia palmaeformis 323, 329, 362
Prasinocladia spp, symbiosis involving *86*
predation
    decision-making prey 274–5
    Holmgren's competitive distribution model 270–3
    and recruitment *322*, 345–6
    spatial variability 357
predatory fishes
    cichlid breeding habits and social structure affected by *209*, 239–40
    evolution of cichlids affected by 210–11
predatory mesoarthropods, effect of soil tillage *112*, 126–7
predatory microarthropods 122, 123
predatory nematodes
    effect of mulching *145*
    effect of soil tillage *112*, 116, *131*
    feeding range 152–3
protozoa
    effect of herbicides *138*
    effect of soil tillage 115, *131*

Pseudolabrus celidotus 331
Pseudosimochromis spp *192*
P. curvifrons 198
Pseudotsuga menziesii 55, 83, 84

quantum flux density *see*
    photosynthetic photon flux
    density (PPFD)

radiation
    relationships between various
        approaches 15–17
    see also net radiation;
        photosynthetically active
        radiation; photosynthetic
        photon flux density; solar
        radiation
radiation use efficiency 17–18
Ralfsia spp *352*
recruitment
    definitions 311, 312, 315, *318*, 339
    effects on community structure 342–
        53
        future study suggested 363–4
    effects on population structure 311–
        14
recruitment dynamics
    and management/conservation plans
        360
    marine organisms, historical
        perspective 310–14
recruitment limitation model 338, 344,
    361
recruitment-limited populations, coral
    reef fishes 312, 338–9, 344
recruitment variability
    causes 353
    effects 353–4
        species-specific recruitment
            patterns 354
    examples
        coral reef fishes 330, 331, *334–6*
        intertidal species 330, 331, *333–4*
    in marine systems 330–6
    seasonality effects 330, 331
    temperate/tropical differences 331–2
refuges
    effect of recruitment 362
    Tanganyikan cichlid fishes 274, 275

Reganochromis spp *191*
regional scale processes, CO_2
    concentration affected by 10–
        11
relative payoff sum (RPS) rule 278–9
    limitations 279
remote sensing
    barnacle recruitment studies 324–5
    reflectance indices 17
respiration flux 12–13
rhizobia *72*
    abundance in environment 92
    competition among symbionts 85, 87,
        90
    evolutionary origins 77
    host range variation 81
rhizopines, synthesis in legume nodules
    93
rhizosphere effects, soil microflora
    affected by 114
risk-sensitive foraging 275
rocky shores
    compared with coral reefs 311
    Lake Tanganyika 196
    recruitment
        historical background of studies 313
        and population structure 339–40
        unpredictability 330

Salmo salar 262
salt marsh vegetation, CO_2 flux
    measurements 35, 57
saprophagous mesoarthropods, effect
    of tillage 126
satellite imagery, barnacle recruitment
    studies 324–5
savanna, CO_2 flux measurements 57–8
sawdust mulch, effect on soil organisms
    144, *145*, 146
scale-eating cichlid fishes
    breeding habits 215–16
    feeding/hunting behaviour 228, 229
    see also Perissodus; Plecodus
Scathophaga stercoraria 256
schooling [whilst feeding], cichlid fishes
    224, 227, 235
Scutellospora spp 88
seasonality
    marine organism recruitment
        affected by 330, 331

plant CO_2 flux affected by 35–6, 44
Sebastes spp 331
Semibalanus spp
  S. balanoides 334, 341
  S. cariosus 353
  S. glandula 333, 334
settlement, marine organisms 315, 316
Shannon diversity index, Tanganyikan
    cichlid species *200*
Shannon–Wiener diversity index 133
  soil biota
    effect of herbicides 143
    effect of mulches *145*
    effect of tillage *134*
    tillage-effect index based on 133,
      135–6
shell-brooding cichlid fishes 212–14
  interpopulation differences 238–9
  spatial partitioning 231, *232–3*
Simochromis spp *192*
  S. diagramma 198
  S. loocki 198
  S. marginatus 198
Sitka spruce
  canopy photosynthesis 14, *15*
  CO_2 flux measurements 55
  see also Picea sitchensis
size-structuring, Tanganyikan cichlid
    species 235
soil bacteria
  effect of herbicides 137, *138, 139*,
    140
  effect of mulching *145*
  effect of tillage 111, *112, 128, 131*
  symbiosis involving 70, 72, 92–3
  see also rhizobia
soil flux, carbon dioxide 8
soil fungi
  effect of herbicides 137, *138, 139*,
    140
  effect of mulching *145*
  effect of tillage 111, *112, 128, 131*,
    134
  symbiosis involving 70, 72
  see also mycorrhizal fungi
soil macrofauna
  effect of herbicides *138, 139*, 140–1
  effect of tillage *112, 128, 131*, 132,
    134, 135
    unpredictability of responses 129,
      136

see also carabid beetles; earthworms;
    spiders
soil mesoarthropods
  effect of tillage *112*, 126–7
  effect of weed-mulching 149
soil mesofauna
  effect of mulching *145*
  effect of tillage *112*, 120–1, 123, *128*,
    130, *131, 134, 135*
  see also collembola; enchytraeids;
    mites; springtails
soil microarthropods
  effect of tillage 120–1
  feeding on plant litter 153
soil microfauna
  effect of mulching *145*
  effect of tillage *112*, 115–17, *128*,
    131, 134, 135
  see also nematodes; protozoa
soil microflora
  effect of earthworms 125
  effect of herbicides 137, *138, 139*
  effect of mulching *145*
  effect of tillage 111, *112*, 113–15,
    128, 131, 134, 135
  effect of weed-mulching 148
soil organic matter
  effect of herbicides 140, 141
  effect of mulching *145*
  effect of tillage 110–11, *112, 128*
soil respiration, in CO_2 flux 13, 30
solar radiation, symbols used 3
Sorex araneus 261
sorting, symbionts 79
Spathodus spp *191*, 200, 203
spatial variability, and consumer
    mobility 358–9
species diversity
  cichlid fishes 190, *199*
    possible mechanisms 220–3
    size effects 234–5
  coral reef fishes 358
  in Lake Tanganyika 189
  soil biota
    effect of herbicides 142–3
    effect of mulching *145*, 146
    effect of tillage 133–6
    effect of weeds 149
spiders, effect of soil tillage *112*, 126–7,
    128, 131, 134
Spinachia spinachia 283, *293*

springtails
    effect of herbicides *138*, *139*, 142
    effect of mulching *145*
    effect of soil tillage 120, *131*, *134*
    see also Collembola
standing crops, and continuous input
        IFD models 273–4
state-dependent IFD models 275–7,
        287
statistical analysis
    CO$_2$ flux measurement data set 22–
        6, 37–40
    mulching study *145*
    symbols used 3
    tillage studies *128*, 129
Stegastes spp 330, 344
    S. fasciolatus 319, 356
    S. francisancus 356
    S. planifrons 330
    S. purpuratus 356
stochastic recruitment 236, 311
Stolothrissa spp 189, 196, *209*
storage flux, carbon dioxide 9–12
stress/adversity tolerators 94
stress protein 94
Strongylocentrotus purpuratus 354
Styela spp 343
substrate-spawning cichlid fishes 211,
        212–15
succession, and recruitment 347–8
survivorship curves, marine organisms
        317
Sutherland–Parker unequal competitor
        IFD models 262–4, 265
Symbiodinium spp *72*, 74
    expulsion from host 91
    host range 83
    phyletic diversity 75
    transmission modes 78–9, *80*
symbiosis
    examples 70
    impact on abundance of symbionts in
        environment 91–3
    impact of human activities 70–1
    micro-organisms released from 90–1
symbiotic micro-organisms 69–95
    methods of studying 70
    reason for study 69–71
    see also endosymbiotic micro-
        organisms
Synodontis multipunctatus 219

Tadorna tadorna 290, *292*
tall-grass prairie, CO$_2$ flux
        measurements 36, 40, 57, 58
Tangachromis spp *191*
Tanganicodus spp *191*, 200, 203
    T. irsacae 215
Tanganyika, Lake
    cichlid fishes 190–2
        breeding behaviour/strategies/
            tactics 211
        breeding habits 211–20
        coexistence within rocky shore
            communities
            effect of feeding habits 223–9
            mechanisms considered 220–3
            partitioning within communities
                231, 233–4
            spatial partitioning 230–1
        community composition
            biotic influences 239–40
            persistence 237
            physical influences 238–9
            predictability 237–40
            stability 237
        evolution
            mechanisms 201–5
            responses to ecological pressures
                208–11
        feeding habits 223–9
        habitat 195–201
        mouthbrooders 211, 215–19
        parental care 211, 218–19
        phylogenetic relationships 205–8
        species diversity
            and coexistence within rocky
                shore communities 220–34
            effect of size difference 234–5
            effect of species packing 236–7
        substrate spawners 211, 212–15
    deforestation affecting 242–3
    early studies [of fauna] 189–90
    geological history 192–3, *202*
    habitat and faunal diversity 194–201
        deepwater zone 195
        littoral zone 195–201
        pelagic zone 195
    hydrology 193
    limnology 193, 194
    species diversity 189
    water level variations *201*, 202
Tarucus theophrastus 260, *293*

taxonomy
  endosymbionts 75–7
  Tanganyikan cichlid fishes *191–2*,
      205–8
Telmatochromis spp *191*, 203
  T. bifrenatus 199
  T. dhonti 199, 218, 219, *232*
  T. temporalis 199, *232*
  T. vittatus 199, 218, 219
Telotrematocara spp *192*
temperature
  CO_2 flux affected by 37
  distribution of endosymbionts
      affected by 88
  inversion overnight under plant
      canopy 10
  nematodes affected by soil tillage 117
Tetraclita spp *352*
Tetrao tetrix 292
Tetraselmis spp, symbiosis involving 85,
      86
Thalassoma spp
  T. bifasciatum 335, *336*, 339
  T. lunare 336
Tilapia spp *192*
tillage
  alternatives 136–48
    alternative agricultural systems
        147–8
    herbicides 137–43
    mulches 143–7
  disadvantages 108
  purposes 108
  see also no-tillage systems
tillage-induced disturbance
  biodiversity affected by 106, 133–6
  responses of different trophic levels
      109–27, *128*
    earthworms *112*, 124, *128*
    earthworms–microfauna/
        microflora interactions 124–5
    enchytraeids *112*, 123
    microarthropod–microfauna–
        microflora interactions 121–3
    microfauna–microflora
        interactions 117–20
    predatory mesoarthropods *112*,
        126–7
    resource base 110–11, *112*, *128*
    saprophagous mesoarthropods 126
    soil macrofauna *112*, 124, *128*
    soil mesofauna *112*, 123, *128*
    soil microarthropods 120–1
    soil microfauna *112*, 115–17, *128*
    soil microflora 111, *112*, 113–15,
        128
    responses of overall food web 127–
        32
    responses of species assemblages
        132–6
top-down parametric models 5–6, 17
Trebouxia spp., symbiosis involving *72*,
      74
Trematocara spp *192*, 196, 222
Tribolium confusum 291
Trifolium subterraneum, symbiosis with
      rhizobia 93
Triglachromis spp *191*
  T. otostigma 197
Tropheus spp *192*, 199, 203, 227
  T. duboisi 215
  T. moorii 199, 203, 215, 223, 227
tropical species
  ecological specialization among 222
  see also cichlid fishes
Tsuga heterophylla 83
tundra, CO_2 flux measurements 30, 56,
      57
Tylochromis spp *192*
  T. polylepis 200

U index [for diversity effects of soil
      tillage] 133, 135
  values quoted for various trophic
      levels *135*
ULva spp 348
unequal competitor continuous input
      model 262–3
unequal competitor interference model
      263–4

Vanellus vanellus 292
vegetation classification, in CO_2 flux/
      solar radiation studies 22, 32–
      4, 44
vegetation indices 17
Vibrio spp, symbiosis involving *72*, 88,
      89, 91–2
Victoria, Lake 188
  cichlid species 190, 208, 211, 215

V index [for mass effects of soil tillage]
109–10
values quoted for various trophic
levels *112*, *128*, *131*
variance with organism size *128*, 129

water availability
CO_2 flux affected by 36–7, 44
distribution of endosymbionts
affected by 87–8
weed control/management
by herbicides 137–43
by mulches 143–7
by tillage 108
with low levels of weeds 149–50
weeds, detritus food webs affected by
148–50, 157

wheat crop
CO_2 flux measurements 35, 59–61
photosynthetic relationships 41, 42

Xenochromis spp *192*
X. hecqui 195
Xenotilapia spp *191*, 195, 203
X. "blackfin" *207*
X. boulengeri 197
X. "Chituta Bay yellow" 215
X. flavipinnis 197, 215
X. longispinis 197
X. ochrogenys 197, 205
X. omatipinnis 207
X. sima 197, 200, *207*
X. spilopterus 197, *207*, 215, 218,
224, 225

Advances in Ecological Research
Volumes 1–25

Cumulative List of Titles

Aerial heavy metal pollution and terrestrial ecosystems, **11**, 218

Analysis of processes involved in the natural control of insects, **2**, 1

Ant–plant–homopteran interactions, **16**, 53

Biological strategies of nutrient cycling in soil systems, **13**, 1

Bray-Curtis ordination: an effective strategy for analysis of multivariate ecological data, **14**, 1

Can a general hypothesis explain population cycles of forest lepidoptera? **18**, 179

Carbon allocation in trees: a review of concepts for modelling, **25**, 60

A century of evolution in *Spartina anglica*, **21**, 1

The climatic response to greenhouse gases, **22**, 1

Communities of parasitoids associated with leafhoppers and planthoppers in Europe, **17**, 282

Community structure and interaction webs in shallow marine hard-bottom communities: tests of an environmental stress model, **19**, 189

The decomposition of emergent macrophytes in fresh water, **14**, 115

Dendroecology: a tool for evaluating variations in past and present forest environments, **19**, 111

The development of regional climate scenarios and the ecological impact of greenhouse gas warming, **22**, 33

Developments in ecophysiological research on soil invertebrates, **16**, 175

The direct effects of increase in the global atmospheric CO_2 concentration on natural and commercial temperate trees and forests, **19**, 2

The distribution and abundance of lake-dwelling Triclads—towards a hypothesis, **3**, 1

The dynamics of aquatic ecosystems, **6**, 1

The dynamics of field population of the pine looper, *Bupalis piniarius* L. (Lep., Geom.), **3**, 207

Earthworm biotechnology and global biogeochemistry, **15**, 379

Ecological aspects of fishery research, **7**, 114

Ecological conditions affecting the production of wild herbivorous mammals on grasslands, **6**, 137

Ecological implications of dividing plants into groups with distinct photosynthetic production capabilities, **7**, 87

Ecological implications of specificity between plants and rhizosphere microorganisms, **21**, 122

Ecological studies at Lough Ine, **4**, 198
Ecological studies at Lough Hyne, **17**, 115
Ecology of mushroom-feeding Drosophilidae, **20**, 225
The ecology of the Cinnabar moth, **12**, 1
Ecology of coarse woody debris in temperature ecosystems, **15**, 133
Ecology, evolution and energetics: a study in metabolic adaptation, **10**, 1
Ecology of fire in grasslands, **5**, 209
The ecology of pierid butterflies: dynamics and interactions, **15**, 51
The ecology of serpentine soils, **9**, 225
Ecology, systematics and evolution of Australian frogs, **5**, 37
Effects of climatic change on the population dynamics of crop pests, **22**, 117
The effects of modern agriculture, nest predation and game management on the
    population ecology of partridges (*Perdix perdix* and *Alectoris rufa*), **11**, 2
El Niño effects on Southern California kelp forest communities, **17**, 243
Energetics, terrestrial field studies and animal productivity, **3**, 73
Energy in animal ecology, **1**, 69
Estimating forest growth and efficiency in relation to canopy leaf area, **13**, 327
Evolutionary and ecophysiological responses of mountain plants to the growing
    season environment, **20**, 60
The evolutionary consequences of interspecific competition, **12**, 127
The exchange of ammonia between the atmosphere and plant communities, **26**, 302
Fire frequency models, methods and interpretations, **25**, 239
Food webs: theory and reality, **26**, 187
Forty years of genecology, **2**, 159
Foraging in plants: the role of morphological plasticity in resource acquisition, **25**,
    160
Fossil pollen analysis and the reconstruction of plant invasions, **26**, 67
The general biology and thermal balance of penguins, **4**, 131
General ecological principles which are illustrated by population studies of Uropo-
    did mites, **19**, 304
Genetic and phenotypic aspects of life-history evolution in animals, **21**, 63
Geochemical monitoring of atmospheric heavy metal pollution: theory and appli-
    cations, **18**, 65
Heavy metal tolerance in plants, **7**, 2
Herbivores and plant tannins, **19**, 263
Human ecology as an interdisciplinary concept: a critical inquiry, **8**, 2
Industrial melanism and the urban environment, **11**, 373
Inherent variation in growth rate between higher plants: a search for physiological
    causes and ecological consequences, **23**, 188
Insect herbivory below ground, **20**, 1
Integration, identity and stability in the plant association, **6**, 84
Isopods and their terrestrial environment, **17**, 188
Landscape ecology as an emerging branch of human ecosystem science, **12**, 189
Litter production in forests of the world, **2**, 101
Mathematical model building with an application to determine the distribution of
    Dursban® insecticide added to a simulated ecosystem, **9**, 133
Mechanisms of microarthropod–microbial interactions in soil, **23**, 1
Mechanisms of primary succession: insights resulting from the eruption of Mount St
    Helens, **26**, 1
The method of successive approximation in descriptive ecology, **1**, 35
Modelling the potential response of vegetation to global climate change, **22**, 93

Module and metamer dynamics and virtual plants, **25**, 105

Mutualistic interactions in freshwater modular systems with molluscan components, **20**, 126

Mycorrhizal links between plants: their functioning and ecological significances, **18**, 243

Mycorrhizas in natural ecosystems, **21**, 171

Nutrient cycles and H^+ budgets of forest ecosystems, **16**, 1

On the evolutionary pathways resulting in C_4 photosynthesis and Crassulacean acid metabolism (CAM), **19**, 58

Oxygen availability as an ecological limit to plant distribution, **23**, 93

The past as a key to the future: the use of palaeoenvironmental understanding to predict the effects of man on the biosphere, **22**, 257

Pattern and process of competition, **4**, 11

Phytophages of xylem and phloem: a comparison of animal and plant sap-feeders, **13**, 135

The population biology and turbellaria with special reference to the freshwater triclads of the British Isles, **13**, 235

Population cycles in small mammals, **8**, 268

Population regulation in animals with complex life-histories: formulation and analysis of damselfly model, **17**, 1

Positive-feedback switches in plant communities, **23**, 264

The potential effect of climatic changes on agriculture and land use, **22**, 63

Predation and population stability, **9**, 1

Predicting the responses of the coastal zone to global change, **22**, 212

The pressure chamber as an instrument for ecological research, **9**, 165

Principles of predator-prey interaction in theoretical experimental and natural population systems, **16**, 249

The production of marine plankton, **3**, 117

Production, turnover, and nutrient dynamics of above- and belowground detritus of world forests, **15**, 303

Quantitative ecology and the woodland ecosystem concept, **1**, 103

Realistic models in population ecology, **8**, 200

Relative risks of microbial rot for fleshy fruits: significance with respect to dispersal and selection for secondary defence, **23**, 35

Renewable energy from plants: bypassing fossilization, **14**, 57

Responses of soils to climate change, **22**, 163

Rodent long distance orientation ("homing"), **10**, 63

Secondary production in inland waters, **10**, 91

The self-thinning rule, **14**, 167

A simulation model of animal movement patterns, **6**, 185

Soil arthropod sampling, **1**, 1

Soil diversity in the Tropics, **21**, 316

Soil fertility and nature conservation in Europe: theoretical considerations and practical management solutions, **26**, 242

Species abundance patterns and community structure, **26**, 112

Stomatal control of transpiration: Scaling up from leaf to regions, **15**, 1

Structure and function of microphytic soil crusts in wildland ecosystems of arid to semi-arid regions, **20**, 180

Studies on the cereal ecosystems, **8**, 108

Studies on grassland leafhoppers (Auchenorrhyncha, Homoptera) and their natural enemies, **11**, 82

Studies on the insect fauna on Scotch Broom *Sarothamnus scoparius* (L.) Wimmer, **5**, 88

Sunflecks and their importance to forest understorey plants, **18**, 1

A synopsis of the pesticide problem, **4**, 75

Temperature and organism size—a biological law for ecotherms?, **25**, 1

Theories dealing with the ecology of landbirds on islands, **11**, 329

A theory of gradient analysis, **18**, 271

Throughfall and stemflow in the forest nutrient cycle, **13**, 57

Towards understanding ecosystems, **5**, 1

The use of statistics in phytosociology, **2**, 59

Vegetation, fire and herbivore interactions in heathland, **16**, 87

Vegetational distribution, tree growth and crop success in relation to recent climate change, **7**, 177

The zonation of plants in freshwater lakes, **12**, 37